Advanced Web Metrics with Google Analytics™

Advanced Web Metrics with Google Analytics™

Brian Clifton

Wiley Publishing, Inc.

Acquisitions Editor: Willem Knibbe
Development Editor: Dick Margulis
Technical Editor: Alex Ortiz-Rosado
Production Editor: Rachel McConlogue
Copy Editor: Luann Rouff
Production Manager: Tim Tate
Vice President and Executive Group Publisher: Richard Swadley
Vice President and Executive Publisher: Joseph B. Wikert
Vice President and Publisher: Neil Edde
Book Designer: Franz Baumhackl
Compositor: Simmy Cover, Happenstance Type-O-Rama
Proofreader: Jen Larsen, Word One
Indexer: Ted Laux
Cover Designer: Ryan Sneed
Cover Image: iStockPhoto

Published simultaneously in Canada

ISBN: 978-0-470-25312-0

For general information on our other products and services or to obtain technical support, please contact our Customer Care Department within the U.S. at (800) 762-2974, outside the U.S. at (317) 572-3993 or fax (317) 572-4002.

Wiley also publishes its books in a variety of electronic formats. Some content that appears in print may not be available in electronic books.

Library of Congress Cataloging-in-Publication Data

Clifton, Brian, 1969-
Advanced Web metrics with Google Analytics / Brian Clifton.
p. cm.
ISBN 978-0-470-25312-0 (pbk.)
1. Google Analytics. 2. Web usage mining. 3. Internet users--Statistics--Data processing. I. Title.
TK5105.885.G66C55 2008
006.3--dc22
2008001405

10 9 8 7 6 5 4 3 2 1

Dear Reader,

Thank you for choosing *Advanced Web Metrics with Google Analytics.* This book is part of a family of premium-quality Sybex books, all written by outstanding authors who combine practical experience with a gift for teaching.

Sybex was founded in 1976. More than thirty years later, we're still committed to producing consistently exceptional books. With each of our titles we're working hard to set a new standard for the industry. From the paper we print on, to the authors we work with, our goal is to bring you the best books available.

I hope you see all that reflected in these pages. I'd be very interested to hear your comments and get your feedback on how we're doing. Feel free to let me know what you think about this or any other Sybex book by sending me an e-mail at nedde@wiley.com, or if you think you've found a technical error in this book, please visit http://sybex.custhelp.com. Customer feedback is critical to our efforts at Sybex.

Best regards,

NEIL EDDE
Vice President & Publisher
Sybex, an Imprint of Wiley

Eight out of 10 implementations of web analytics solutions are incorrectly set up.

—Bill Hunt, CEO, Global Strategies International

Acknowledgments

Writing this, my first book, has been both very rewarding and very hard work; but above all it has been enjoyable. Many people have contributed either directly or indirectly to its content. Some inspired me, some sanity checked my coding, some proofread my English, some contributed ideas, and some simply encouraged me to dig deep and work through those many late nights. I hope I have remembered everyone. If I have missed any, my apologies; I will add your name to the book blog website (www.advanced-web-metrics.com) and any future print editions.

First, many thanks to the Wiley publishing team: Willem Knibbe, who I first discussed the content with many moons ago and who subsequently convinced the Wiley group that such a book was both worthy of producing and needed by the army of online marketers that use Google Analytics; Dick Margulis, who did such sterling work at taking my initial stab at producing a book and restructured it into something much better; Pete Gaughan for managing the whole process; Rachel McConlogue and Luann Rouff and many other people at Wiley who work tirelessly in the background to help create and polish what I hope you will consider an easy, yet enlightening, read.

Significant feedback, help, and brainstorming were also freely provided by Nina Privetera Hoyt, a former colleague and great friend now working for Médecins Sans Frontières; Dave Mumford, Andrew Miles, and Nikki Rae from Omega Digital Media Ltd, who helped with providing data and screenshots from their Google Analytics accounts as well commenting on the first draft and helping with the book blog; Sara Andersson for her generous advice and strategic thinking regarding integrating offline and online marketing, and for sharing her ideas on search input; Daniel Silander of Neo@Ogilvy for his thoughtful feedback from an agency's perspective; Dennis R. Mortensen for his hawk eyes at proofreading and his honest opinion from a competitor's point of view; Chris Sherman for reviewing this book and for honoring me by writing the foreword; and Timo Aden, Jean-Baptiste Creusat "Jee Bee," Alan Boydell, Rene Nijhuis, Estela Oliva, and Philip Walford of the Google Analytics Team (EMEA), for their stimulating discussions, experiences, and thoughts about implementing Google Analytics for their clients.

Last but not least, special thanks go to Tomas Remotigue and Alex Ortiz-Rosado, both of Google, who have significantly contributed to my knowledge and understanding of the internal workings of Google Analytics over the years. Both worked late and on their own time to sanity check and expand upon the technical aspects of this book, with Alex becoming my much appreciated technical editor.

About the Author

Brian Clifton is an established search engine marketing and web analytics expert who has worked in these fields since 1997. Specializing in search engine optimization (SEO) and web analytics, his business was the first U.K. partner for Urchin Software Inc., the company that later became Google Analytics. Brian joined Google in 2005 to define, develop, and lead the web analytics team for Europe, Middle East, and Africa.

Brian received a BSc in chemistry from the University of Bristol in 1991 and a Ph.D. in physical and theoretical chemistry in 1996. Further work as a postdoctoral researcher culminated in publishing several scientific papers in journals, including *Molecular Physics*, *Colloids and Surfaces*, and *Langmuir*. During that time he was also an international weightlifter, representing Great Britain at world and European championships.

Studying science at university during the early nineties meant witnessing the incredible beginnings of the Web. In 1991, Tim Berners-Lee, a scientist working at the CERN laboratory in Switzerland, launched the first web browser and web server to the academic community, therefore sowing the first seeds of the World Wide Web.

Although the communication potential of the Web was immediately clear to Brian, it took a little while for ideas to formulate around business opportunities. In 1997 he left academia to found Omega Digital Media, Ltd., a U.K. company specializing in the provision of professional services to organizations wishing to utilize the new digital medium.

Since leaving the field of chemical research (and weightlifting), Brian has continued to write. Whitepapers include "How Search Engine Optimization (SEO) Works," "Web Analytics Data Sources," and "Web Analytics: Increasing Accuracy for Business Growth." As with most of his Mosaic–Netscape peers, Brian is also an avid writer on his own blog: www.advanced-web-metrics.com; this is his first book.

Brian holds the title of Associate Instructor at the University of British Columbia for his contribution to teaching modules in support of the "Award of Achievement in Web Analytics." You can also hear him speak at numerous conferences around the word—particularly in Europe, where he presents on search marketing, web analytics, website optimization through testing, and how these can all interlink to create a successful online business strategy. Brian currently lives in West Sussex, United Kingdom.

Foreword

You know a book is going to be a winner when it begins with two simple, but incredibly powerful words: "measuring success." Measuring success is what distinguishes winners from losers on the web, and all savvy website owners will tell you that understanding and making use of web analytics data is absolutely key to their success.

Why? Because analytics data offers a wealth of information about what visitors are doing on your website: what they're reading, how they navigate, what they're buying, and what they're ignoring. By capturing, analyzing, and taking action on this gold mine of information, you can tune your website for maximum performance.

Advanced Web Metrics with Google Analytics is a compelling guide to this process. It's a behind-the-scenes look at some of the most powerful tools available to anyone who runs a website. Author Brian Clifton knows his stuff, and while he offers a thorough look at the technical aspects of Google Analytics, his straightforward style makes the book accessible to anyone who's interested in improving the performance of a website.

Importantly, Clifton also understands that while mastering technical details is crucial, the primary reason you would want to dive in deep with web analytics is to achieve goals. At its heart, *Advanced Web Metrics with Google Analytics* is as much about developing effective online business practices as it is about mastering a set of tools.

Some might wonder why they should go to all the trouble to understand user behavior and take on the task of optimizing a website when it's easy and relatively cheap to simply purchase search advertising. The answer is simple: Searchers still overwhelmingly prefer natural search results to search ads, by a factor of 3:1, according to Jupiter Research. Taking the time to optimize your website based on the data collected with analytics tools almost always pays off with increased traffic, sales, and profitability—effects that can last for years.

Search advertising is getting more expensive and competitive. By contrast, most website owners still don't do much with site optimization, so the playing field is relatively level for everyone, regardless of whether you have a small website or one with millions of pages.

If you're looking for a way to improve your website and enhance the results of your online efforts, *Advanced Web Metrics with Google Analytics* is an excellent guide—one that you can put to good use right away to help you achieve, and even surpass, your boldest goals.

Chris Sherman
Executive Editor
SearchEngineLand.com

Contents

Introduction

Although the birth of Web took place in August 1991, it did not become commercial until around 1995. In those early days it was kind of fun to have a spinning logo, a few pictures, and your contact details as the basis of your online presence. My first website was just that—no more than my curriculum vitae online at the University of Bristol. Then companies decided to copy (or worse, scan) their paper catalogs and brochures and simply dump these on their websites. This was a step forward in providing more content, but the user experience was poor to say the least, and no one was really measuring conversions. The most anyone kept track of was visits, and these were often confused with hits.

Around the year 2000, fueled by the dot-com boom, people suddenly seemed to realize the potential of the Web as a useful medium to find information; the number of visitors using it grew rapidly. Organizations started to think about fundamental questions, such as "What is the purpose of having a website?" and considered how to build relevant content for their online presence. With that, user experience improved. Then, when widespread broadband adoption began to happen, those organizations wanted to attract the huge audience that was now online. Hence the reason for the rapid growth in search engine marketing that followed.

Now, with businesses accepting the growing importance of their online presence, comes the need to measure the effects—and success or not—of a website on the rest of the business. Put simply, this is what web analytics tools, such as Google Analytics, attempt to do. By measuring the ability of your online and offline marketing to attract visitors, the resulting user experience, conversion rate, and ROI enables you to continually benchmark yourself and improve your online strategy.

But what can be measured, how accurate is this, and how can a business be benchmarked? In other words, how do you measure success? Using best practice principles I gained as a professional practitioner, this book uses real-world examples that clearly demonstrate how to manage Google Analytics. This includes not only installation and configuration guides, but also how to turn data into information that enables you to understand your website visitor's experience. With this understanding, you can then build business action items to drive improvements in visitor acquisition (both online and offline), conversion rates, repeat visit rates, customer retention, and ultimately your bottom line.

Who Should Read This Book?

If you have ever wondered whether your checkout system is off-putting to potential new customers, how differently your website might be perceived by a new versus a returning visitor, if paid advertising yields better conversions than free organic search listings, whether you can better qualify leads by fine-tuning your search marketing strategy, or simply how to measure the performance of your website, then this book is for you. The most important prerequisite for reading this book is an inquisitive mind with the drive and desire to improve the user experience—that is, engagements and conversions on your website.

I have attempted to make this book's subject matter accessible to a broad spectrum of readers, including marketers, webmasters, CEOs, and anyone with a business interest in making their website work. After all, the concept of measuring success is a universal desire. Although the content is not aimed at the complete web novice, don't worry; nor is it aimed at engineers. I am not one myself; and installing, configuring, or using Google Analytics does not require an engineer! Rather, I hope that *Advanced Web Metrics with Google Analytics* will appeal to existing web analysts as well as readers new to the field of web measurement.

This book describes the best practice techniques you can use to set up and configure Google Analytics. The purpose is simple: to give you the information you need to maximize your website's potential. With a better understanding of your website visitors, you will be able to tailor page content and marketing budgets with laserlike precision for a better return on investment. I also discuss advanced configurations (Chapter 9, "Google Analytics Hacks"), which are not documented elsewhere. These provide you with an even greater understanding of your website visitors so that you can dive into the metrics that make sense for your organization. In as many areas as possible, I include real-world practical examples that are currently in use by advanced users.

The book's content is primarily aimed at an organization's marketer and webmaster, who would work in conjunction with each other. Many chapters focus on integrating your analytical skills with your marketing and webmaster skills, and require no coding ability. There are also sections and exercises in this book that require you to modify your web page content; after all, web analytics is all about instigating change using reliable metrics as your guide. Therefore, knowledge of HTML (the ability to read browser source code) and experience with online marketing methods (for example, pay-per-click, e-mail marketing, organic search, etc.) is required. Some advanced techniques also require an understanding of JavaScript.

How This Book Is Organized

There are four parts to this book: "Measuring Success," "Using Google Analytics Reports," "Implementing Google Analytics," and "Using Visitor Data to Drive Website Improvement."

Each part begins with the fundamentals that need to be considered for that topic. Then we build in the detail, followed by real-world examples that demonstrate how to apply what has been presented in that chapter. As a former implementer, analyst, and consultant myself, I cram in as many useful tips, workarounds, and practical suggestions as I possibly can.

Beginning with Chapter 4, you will be viewing reports in detail. As each subsequent chapter extends your skills, the examples become more involved and sophisticated, so try not to skip chapters!

By the final chapter you will have a thorough understanding of best-practice Google Analytics techniques and be well on your way to measuring the success (or otherwise) of your own website through a clear understanding of the processes involved. I have tried to present the material so that readers may explore the possibilities of Google Analytics further and perhaps even add their own contributions to this book via the book blog: www.advanced-web-metrics.com/blog.

Note: For help with terminology throughout this book, you may find the following link useful: www.google.com/support/analytics/bin/static.py?page=glossary.html

All scripts presented in this book or on the website www.advanced-web-metrics.com have been tested and validated by the author and are believed to be correct as of the date of publication or posting. The Google Analytics software on which they depend is subject to change, however; therefore, no warranty is expressed or implied guaranteeing that they will work as described in the future. Always check the most current Google Analytics documentation.

The views expressed in this book are my own and do not represent those of Google. The names of actual companies and products mentioned herein may be trademarks of their respective owners.

Measuring Success

I

Lord Kelvin is often quoted as the reason why metrics are so important: "If you cannot measure it, you cannot improve it." That statement is ultimately the purpose of web analytics. By enabling you to identify what works and what doesn't from a visitor's point of view, web analytics is the foundation for running a successful website. Even if you get those decisions wrong, web analytics provides the feedback mechanism that enables you to identify mistakes quickly.

In Part I, you will learn the following:

Why Understanding Your Web Traffic Is Important to Your Business

Web analytics is a thermometer for your website—constantly checking and monitoring your online health. As a methodology, it is the study of online experience in order to improve it, and without it you are flying blind. How else would you determine whether your search engine marketing is effective at capturing your maximum potential audience or whether negative blog comments are hindering conversions? Is the user experience a good one, encouraging engagement and return visits, or are visitors bouncing off your website after viewing only a single page?

1

In Chapter 1, you will learn:

The kinds of information you can obtain from analyzing traffic on your site
The kinds of decisions that web analytics can help you make
The ROI of web analytics
How web analytics helps you understand your web traffic

Information Web Analytics Can Provide

In order to do business effectively on the Web, you need to continually refine and optimize your online marketing strategy, site navigation, and page content. A low-performing website will starve your return on investment (ROI) and can damage your brand. But you need to understand what is performing poorly—the targeting of your marketing campaigns or your website's ability to convert? Web analytics provides the tools for gathering this information about what happens on your website, and enables you to benchmark the effects.

Note that I have deliberately used the word *tools* in its plural form. This is because the term web analytics covers many areas that require different methodologies or data collection techniques. For example, *offsite tools* are used to measure the size of your potential audience (opportunity), your share of voice (visibility), and the buzz (comments) that is happening on the Internet as a whole. These are relevant metrics regardless of your website's existence. Conversely, *onsite tools* measure the visitor's journey, its drivers, and your website's performance. These are directly related to your website's existence.

Google Analytics is an onsite visitor reporting tool. From here on, when I use the general term web analytics, I am referring to onsite measurement tools.

If you have already experienced looking at metrics from pay-per-click advertising campaigns, Google Analytics is simply the widening of that report view to see all referrals and behavior of visitors. If you are new to any kind of web metrics reporting, then the amount of information available can at first feel overwhelming. However, bear with me—this book is intended to guide you through the important aspects of what you need to know in order to be up and running with Google Analytics quickly and efficiently.

Keep in mind that web analytics are tools—not ends in themselves. They cannot tell you why visitors behave the way they do or which improvements you should make. For that you need to invest in report analysis; and that means hiring expertise, training existing staff, using the services of an external consultant, or a combination of all of these.

Consider Figure 1.1, a typical model that most websites fit. It illustrates that the vast majority of websites have single-figure conversion rates. Why is that, and can it be improved? I can say with certainty that in my 15 years of writing and viewing web pages, there has always been room for improvement from a user experience point of view—including on my own websites. Ultimately, it is the user experience of your visitors that will determine the success of your website; and web analytics tools provide the means to investigate this.

Note: The average conversion rate reported by the e-tailing group corresponds closely with Forrester Research, July 2007, and the Fireclick Index (http://index.fireclick.com/fireindex.php?segment=0).

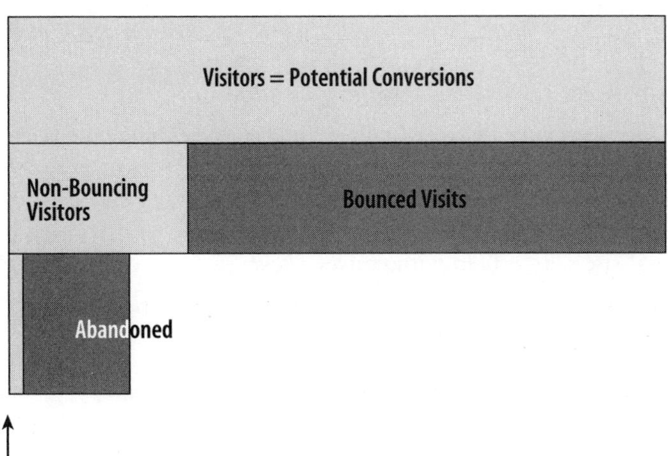

Figure 1.1 U.S. Conversion rates average 2–3 percent 2005–2007

Source: the e-tailing group, April 2007

If you are implementing web analytics for the first time, then you want to gain an insight into the initial visitor metrics to ascertain your traffic levels and visitor distribution. Examples of first-level metrics include the following:

- How many daily visitors you receive
- Your average conversion rate (sales, registration, download, etc.)
- Your top visited pages
- The average visit time on site and how often visitors come back
- The average visit page depth and how this varies by referrer
- The geographic distribution of visitors and what language setting they are using
- How "sticky" your pages are: Do visitors stay or simply bounce off (single-page visits)?

If your website has an e-commerce facility, then you will also want to know the following:

- What revenue your site is generating and where these customers are coming from
- What your top-selling products are—and their average order value

These metrics enable you to draw a line in the sand as the starting point from which you can increase your knowledge. Be warned, though, Google Analytics gives you statistics so readily that the habit of checking them can become obsessive! Hence, as you move deeper into your analysis, you will start to ask more complicated questions of your data. For example:

- What is the value of a visitor?
- What is the value of the web page?
- How does ROI differ between new versus returning visitors?
- How do visits and conversions vary by referring medium type or campaign source?
- How does bounce rate vary by page viewed or referring source?
- How is my site engaging with visitors?
- Does the use of internal site search help with conversions?
- How many visits and how much time does it take for a visitor to become a customer?

All of these questions can be answered with Google Analytics reports.

Decisions Web Analytics Can Help You Make

Knowledge without action is meaningless. The purpose of web analytics is to give you the knowledge from which you can make informed decisions about changing your online strategy—for the better.

In terms of benchmarks, it is important that any organization spends time planning its *key performance indicators (KPIs)*. KPIs provide a distillation of the plethora of website visitor data available to you as clear, actionable information. Simply put, KPIs represent the key factors, specific to your organization, that measure success.

Google Analytics gives you the data from which KPIs are built and in some cases can provide a KPI directly. For example, saying "we had 10,000 visitors this week" is providing a piece of data. A KPI based on this could be "our visitor numbers are up 10 percent month on month"—that is an indicator saying things are looking good. Most KPIs are ratios that enable you to take action, and the job of an analyst is to build these specific to your organization. I discuss building KPIs in detail in Chapter 10.

Using KPIs, typical decisions you can make include those shown in Table 1.1.

While engaging in this process to improve your website's performance, consider the changes as part of a continuous process—not a one-hit fix. That is, think in terms of the AMAT acronym:

- Acquisition of visitors
- Measurement of performance
- Analysis of trends
- Testing to improve

▶ Table 1.1 Typical Decisions Based on KPIs

Observation	Action
Blog visitors show different behavioral patterns from potential customers.	Segment these visitors to view the difference.
Goal conversions are higher for foreign language visitors than for those with US-English.	Investigate the potential for conducting business in additional languages.
Internal site search is being actively used by 70 percent of visitors. However, conversions are lower for this segment.	Investigate the quality of site search results.
Forum visitors are driving goal conversions (PDF downloads), but it is paid-search visitors who are driving transactions.	Acquire more forum visitors to drive branding, reach, and goal conversions. Acquire more paid-search visitors to provide further revenue growth.

The ROI of Web Analytics

Google Analytics is a free data collection and reporting tool. However, analyzing, interpreting, and making website changes all require a resource outlay at your end. The amount of investment you make in web analytics, therefore, depends on how significant your website is to your overall business.

How Much Time Should You Spend on This?

A great phrase often heard from Jim Stern at his eMetrics conference series (www.emetrics.org) is "What is the ROI of measuring your ROI?" In other words, how much time and effort should you spend on data measurement and analysis, considering that the vast majority of people performing this job role also have other responsibilities, such as webmaster, online marketer, offline marketer, content creator—even running a business. After all, you need to focus on delivering for your visitors and generating revenue or leads from your website.

The key to calculating this is understanding the value of your website in monetary terms—either direct as an e-commerce website or indirect from lead generation or advertisement click-throughs. Marketers are smart but they are not fortune-tellers. Purchasing clicks and doing nothing to measure their effectiveness is like scattering seeds in the air. Even highly paid expert opinions can be wrong. Moreover, content that works today can become stale tomorrow. Using web analytics, you can ascertain the impact your work has and what that is worth to your organization.

Table 1.2 demonstrates a before-and-after example of what making use of web analytics data can achieve. In this theoretical case, the target was to grow the online conversion rate by 1 percent, using an understanding of visitor acquisition and onsite factors such as checkout funnel analysis, exit points, bounce rates, and engagement metrics. By achieving this increase, the values of total profit, *P*, and ROI, *R*, shown in the last two rows of the table, put the analysis into context—that is, profit will rise by $37,500 and return on investment quadruples to 50 percent—without any increase in visitor acquisition costs.

▶ **Table 1.2** Economic Effect of a 1% Increase in Conversion Rate

Measure	Symbol	Calculation	Before	After
Visitors	v		100,000	100,000
Cost per visit	c		$1.00	$1.00
Cost of all visits	c_T	$v \times c$	$100,000	$100,000
Conversion rate	r		3%	4%
Conversions	C	$r \times v$	3,000	4,000
Revenue per conversion	V		$75	$75
Total revenue	T	$V \times C$	**$225,000**	**$300,000**
Non-marketing profit margin	m		50%	50%
Non-marketing costs	n	$m \times T$	$112,500	$150,000
Marketing costs	c_T	$v \times c$	$100,000	$100,000
Total profit	P	$T - (n + c_T)$	**$12,500**	**$50,000**
Total marketing ROI	R	P / c_T	**13%**	**50%**

Note: The excel spreadsheet of Table 1.2 is available at: www.advanced-web-metrics.com/scripts

To calculate how much time you should spend on web analytics in your organization, try a similar calculation; then ask your boss (or yourself) how much time such an increase in revenue buys you. As a guide, I have worked with clients for whom the time from web analytics implementation, initial analysis, forming a hypothesis, testing, interpretation, and presenting the results—that is, the before and after—takes six months. Of course, the compounded impact of your work will last much longer, so the actual lifetime value is always higher than this calculation suggests.

How Web Analytics Helps You Understand Your Web Traffic

As discussed earlier, viewing the 80-plus reports in Google Analytics can at first appear overwhelming—there is simply too much data to consume in one go. Of course, all of this data is relevant, but some of it will be more relevant to you, depending on your business model. Therefore, once you have visitor data coming in and populating your reports, you will likely want to view a smaller subset—the key touch points with your potential customers. To help you distill visitor information, Google Analytics can be configured to report on goal conversions.

Identifying goals is probably the single most important step of building a website—it enables you to define success. Think of goal conversions as specific, measurable actions that you want your visitors to complete before they leave your website. For example, an obvious goal for an e-commerce site is the completion of a transaction—that is, buying something. However, not all visitors will complete a transaction on their first visit, so another useful e-commerce goal is quantifying the number of people who add an item to the shopping cart whether they complete the purchase or not—in other words, how many begin the shopping process.

Regardless of whether you have an e-commerce website or not, your website has goals. A goal is any action or engagement that builds a relationship with your visitors, such as the completion of a feedback form, a subscription request, leaving a comment on a blog post, downloading a PDF whitepaper, viewing a special offers page, or clicking a `mailto:` link. As you begin this exercise, you will probably realize that you actually have many website goals.

With goals clearly defined, you simplify the viewing of your visitor data and the forming of a hypothesis. The at-a-glance key metrics are your goal conversions. For example, knowing instantly how many, and what proportion, of your visitors convert enables you to promptly ascertain the performance of your website and whether you should do something about it or relax and let the computers continue to do the work for you.

Where to Get Help

Google itself provides a number of self-help resources that you can tap into:

Google Analytics Help Center An online searchable manual and reference guide:

`www.google.com/support/googleanalytics`

Continues

Where to Get Help *(Continued)*

Analytics Help Group A Google Group with a threaded message-board system. Members are Google Analytics users although Google Support staff occasionally participate:

`www.google.com/analytics/analyticshelp`

Conversion University Going beyond the standard reporting, advanced topics, and methodologies:

`www.conversionuniversity.com`

Conversion University channel on YouTube:

`www.youtube.com/view_play_list?p=7A545E796C2CFA72`

Google Blog Official Google Analytics News blog where you can find latest product updates, what's new, events, Conversion University, Help Center, and more:

`http://analytics.blogspot.com`

Official Authorized Partners If you are investing in web analytics yet cannot afford full-time resources in house, a global network of third-party Google Authorized Analytics Consultants (GAAC) is available.

GAAC partners are independent of Google and have a proven track record in their field, providing paid-for professional services such as strategic planning, custom installation, onsite or remote training, data analysis, and consultation:

`www.google.com/analytics/support_partner_provided.html`

Official book website and blog from Brian Clifton:

`www.advanced-web-metrics.com`

Summary

In Chapter 1, you have learned the following:

The kinds of information you can obtain from analyzing traffic on your site This includes visitor volumes, top referrers, time on site and depth on site to conversion rates, page stickiness, visitor latency, frequency, revenue, and geographic distribution—to name a few.

The kinds of decisions that web analytics can help you with For example, web analytics can help you determine whether blog visitors have a positive impact on your website's reach and conversions, which visitor acquisition channels work best and to what extent these should be increased or decreased, whether site search is worth the investment, or if overseas visitors would be better served with more localized content.

The ROI of web analytics Knowing how much time and effort to invest in web analytics, without losing site of your objectives, will keep you focused on improving your organization's bottom line.

How web analytics helps you understand your web traffic By focusing metrics around goal-driven web design, you concentrate not only your own efforts, but also that of your visitors around clear calls-to-action. This simplifies the process of forming a hypothesis from observed visitor patterns.

Available Methodologies

Web analytics can be incredibly powerful and insightful—an astonishing amount of information is available when compared to any other forms of traditional marketing. The danger, however, is taking web analytics reports at face value; and this raises the issue of accuracy.

The key to successfully utilizing the volume of information collected is to get comfortable with your data—what it can tell you and what it can't and the limitations therein. This requires an understanding of the data collection methodologies. Essentially, there are two common techniques: page tags and server logfiles. Google Analytics is a page tag technique.

2

In this chapter, you will learn about the following:

How web visitor data is collected
The relative advantages of page tags and logfiles
The role of cookies in web analytics
The accuracy limitations of web traffic information
How to think about web analytics in relation to user privacy concerns

Page Tags and Logfiles

Page tags collect data via the visitor's web browser. This information is usually captured by JavaScript code (known as *tags* or *beacons*) placed on each page of your site. Some vendors also add multiple custom tags to collect additional data. This technique is known as *client-side data collection* and is used mostly by outsourced, hosted vendor solutions.

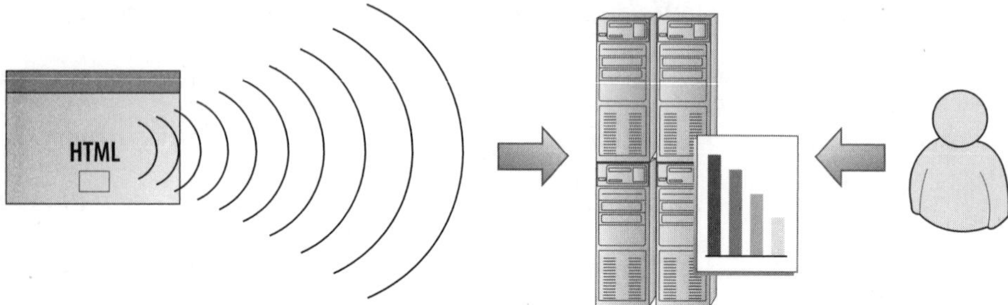

Figure 2.1 Schematic page tag methodology: Page tags send information to remote data collection servers. The analytics customer views reports from the remote server.

Note: Google Analytics is a hosted page tag service.

Logfiles refer to data collected by your web server independently of a visitor's browser. This technique, known as *server-side data collection*, captures all requests made to your web server, including pages, images, and PDFs, and is most frequently used by stand-alone software vendors.

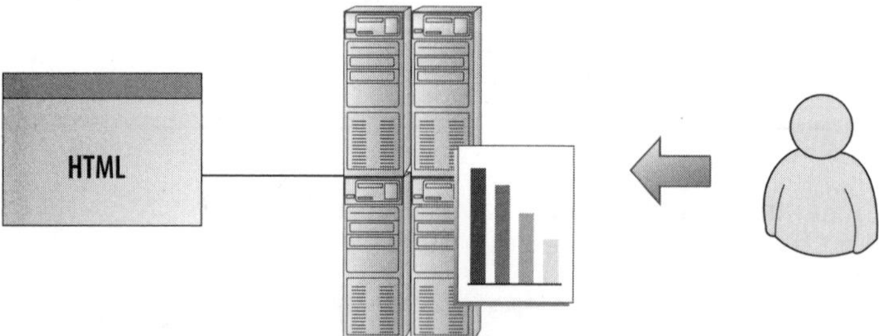

Figure 2.2 Schematic logfile methodology: The web server logs its activity to a text file that is usually local. The analytics customer views reports from the local server.

In the past, the easy availability of web server logfiles made this technique the one most frequently adopted for understanding the behavior of visitors to your site. In fact, most Internet service providers (ISPs) supply a freeware log analyzer with their web hosting accounts (Analog, Webalizer, and AWStats are some examples). Although this is probably the most common way people first come into contact with web analytics, such freeware tools are too basic when it comes to measuring visitor behavior and are not considered further in this book.

In recent years, page tags have become more popular as the method for collecting visitor data. Not only is the implementation of page tags easier from a technical point of view, but data management needs are significantly reduced because the data is collected and processed by external servers (your vendor), saving website owners from the expense and maintenance of running software to capture, store, and archive information.

Note that both techniques, when considered in isolation, have their limitations. Table 2.1 summarizes the differences. A common myth is that page tags are technically superior to other methods, but as Table 2.1 shows, that depends on what you are looking at. By combining both, however, the advantages of one counter the disadvantages of the other. This is known as a *hybrid* method and some vendors can provide this.

Note: Google Analytics can be configured as a hybrid data collector—see "Backup: Keeping a Local Copy of Your Data," in Chapter 6.

Other data collection methods

Although logfile analysis and page tagging are by far the most widely used methods for collecting web visitor data, they are not the only methods. Network data collection devices (packet sniffers) gather web traffic data from routers into black-box appliances. Another technique is to use a web server application programming interface (API) or loadable module (also known as a plugin, though this is not strictly correct terminology). These are programs that extend the capabilities of the web server—for example, enhancing or extending the fields that are logged. Typically, the collected data is then streamed to a reporting server in real time.

▶ **Table 2.1** Page Tag versus Logfile Data Collection

Methodology	Advantages	Disadvantages
Page tags	• Breaks through proxy and caching servers—provides more accurate session tracking • Tracks client-side events—e.g., JavaScript, Flash, Web 2.0 • Captures client-side e-commerce data—server-side access can be problematic • Collects and processes visitor data in nearly real time • Allows program updates to be performed for you by the vendor • Allows data storage and archiving to be performed for you by the vendor	• Setup errors lead to data loss—if you make a mistake with your tags, data is lost and you cannot go back and reanalyze • Firewall can mangle or restrict tags • Cannot track bandwidth or completed downloads—tags are set when the page or file is requested, *not* when the download is complete • Cannot track search engine spiders—robots ignore page tags
Logfile analysis software	• Historical data can be reprocessed easily • No firewall issues to worry about • Can track bandwidth and completed downloads—and can differentiate between completed and partial downloads • Tracks search engine spiders and robots by default • Tracks mobile visitors by default	• Proxy and caching inaccuracies—if a page is cached, no record is logged on your web server • No event tracking—e.g., no JavaScript, Flash, Web 2.0 tracking • Requires program updates to be performed by your own team • Requires data storage and archiving to be performed by your own team • Robots multiply visit counts

As you can see, the advantages of one data collection method cancel out the disadvantages of the other. However, freeware tools aside, the page tagging technique is by far the most widely adopted method because of its ease of implementation and low IT overhead.

Cookies in Web Analytics

Page tag solutions track visitors by using cookies. *Cookies* are small text messages that a web server transmits to a web browser so that it can keep track of the user's activity on a specific website. The visitor's browser stores the cookie information on the local hard drive as name–value pairs. *Persistent cookies* are those that are still available when the browser is closed and later reopened. Conversely, *session cookies* last only for the duration of a visitor's session (visit) to your site.

For web analytics, the main purpose of cookies is to identify users for later use—most often with an anonymous visitor ID. Among many things, cookies can be used to determine how many first-time or repeat visitors a site has received, how many times a visitor returns each period, and how much time passes between visits. Web analytics aside, web servers can also use cookie information to present personalized web pages. A returning customer might see a different page than the one a first-time visitor would view, such as a "welcome back" message to give them a more individual experience or an auto-login for a returning subscriber.

The following are some cookie facts:

- Cookies are small text files, stored locally, that are associated with visited website domains.
- Cookie information can be viewed by users of your computer, using Notepad or a text editor application.
- There are two types of cookies: first-party and third-party. A first-party cookie is one created by the website domain. A visitor requests it directly by typing the URL into his or her browser or following a link. A third-party cookie is one that operates in the background and is usually associated with advertisements or embedded content that is delivered by a third-party domain not directly requested by the visitor.
- For first-party cookies, only the website domain setting the cookie information can retrieve the data. This is a security feature built into all web browsers.
- For third-party cookies, the website domain setting the cookie can also list other domains allowed to view this information. The user is not involved in the transfer of third-party cookie information.
- Cookies are not malicious and can't harm your computer. They can be deleted by the user at any time.
- Cookies are no larger than 4KB.
- A maximum of 50 cookies are allowed per domain for the latest versions of IE7 and Firefox 2. Other browsers may vary (Opera 9 currently has a limit of 30).

Note: Google Analytics uses first-party anonymous cookies only.

Getting Comfortable with Your Data and Its Accuracy

When it comes to benchmarking the performance of your website, web analytics is critical. However, this information is only accurate if you avoid common errors associated with collecting the data—especially comparing numbers from different sources.

Unfortunately, too many businesses take web analytics reports at face value. After all, it isn't difficult to get the numbers. The harsh truth is that web analytics data can never be 100 percent accurate, and even measuring the error bars can be difficult.

So what's the point?

Despite the pitfalls, error bars remain relatively constant on a weekly, or even a monthly, basis. Even comparing year-by-year behavior can be safe as long as there are no dramatic changes in technology or end-user behavior. As long as you use the same yardstick, visitor number trends will be accurate. For example, web analytics data may reveal patterns like the following:

- 30 percent of my traffic came from search.
- 50 percent of traffic came to page *x.html*.
- We increased conversions by 20 percent last week.
- Pageviews at our site increased 10 percent during March.

With these types of metrics, marketers and webmasters can determine the direct impact of specific marketing campaigns. The level of detail is critical. For example, you can determine if an increase in pay-per-click advertising spending—for a set of keywords on a single search engine—increased the return on investment during that time period. As long as you can minimize inaccuracies, web analytics tools are effective for measuring visitor traffic to your online business.

Next, I'll discuss in detail why such inaccuracies arise, so you can put this information into perspective. The aim is for you to arrive at an acceptable level of accuracy with respect to your analytics data. Recall from Table 2.1 that there are two main methods for collecting web visitor data—logfiles and page tags—and both have limitations.

Issues Affecting Visitor Data Accuracy for Logfiles

1. One IP address registers as one person.

   Generally, a logfile solution tracks visitor sessions by attributing all hits from the same IP address and web browser signature to one person. This becomes a problem when ISPs assign different IP addresses throughout the session. A U.S.-based comScore study (www.comscore.com/request/cookie_deletion_white_paper.pdf) showed that a typical home PC averages 10.5 different IP addresses per month. Those visits will be counted as 10 unique visitors by a logfile analyzer. This issue is becoming more severe, as most Web users have identical web browser signatures (currently Internet Explorer). As a result, visitor numbers are often vastly over-counted. This limitation can be overcome with the use of cookies.

2. Cached pages are not counted.

   Client-side caching means a previously visited page is stored on a visitor's computer. In this case, visiting the same page again results in that page being served locally from the visitor's computer, and therefore the visit is not recorded at the web server.

Server-side caching can come from any web accelerator technology that caches a copy of a website and serves it from their servers to speed up delivery. This means that all subsequent site requests come from the cache and not from the site itself, leading to a loss in tracking. Today, most of the Web is in some way cached to improve performance. For example, see Google's cache description at www.google.com/intl/en/help/features.html#cached.

3. Robots multiply figures.

Robots, also known as spiders or web crawlers, are most often used by search engines to fetch and index pages. However, other robots exist that check server performance—uptime, download speed, and so on—as well as those used for page scraping, including price comparison, e-mail harvesters, competitive research, and so on. These affect web analytics because a logfile solution will also show all data for robot activity on your website, even though they are not real visitors.

When counting visitor numbers, robots can make up a significant proportion of your pageview traffic. Unfortunately, these are difficult to filter out completely because thousands of homegrown and unnamed robots exist. For this reason, a logfile analyzer solution is likely to over-count visitor numbers, and in most cases this can be dramatic.

4. Logfiles capture mobile users.

All is not lost for logfile analyzers. A mobile web audience study by comScore for January 2007 (http://www.comscore.com/press/release.asp?press=1432) showed that in the U.S., 30 million (or 19 percent) of the 159 million U.S. Internet users accessed the Internet from a mobile device.

Mobile Web audience statistics

The comScore study for January 2007 also showed 19 percent of U.K. Internet users accessing the Internet from a mobile device (5.7 million of the 30 million who access via a PC). The most popular sites accessed on U.K. mobiles were the BBC.com and Sky.com, attracting 2.3 million and 1.2 million unique visitors respectively.

Other reports show 28% of mobile phone owners around the world access the Internet on a wireless handset, up from 25% at the end 2004. [IPSOS, April, 2006]

In 2004, 36% of mobile phone users browsed the Internet or downloaded e-mail. That figure rose to 56% in 2005. In Japan 92% of users went online via their mobile devices. [A.T.Kearney, April, 2006]

For the vast majority of commercial of websites, the number of pageviews from mobile phones is currently very small in comparison with normal computer access. However, this number will continue to grow in the coming years. In fact, Japan and many parts of Asia are currently experiencing an explosive growth in mobile Internet access. As most mobile phones do not yet understand JavaScript or cookies, logfile tools are able to track visitors who browse using their phones—something page tag solutions cannot do. The next generation of mobile phones is already increasing mobile pageview volume. Some can be tracked by JavaScript and cookies, such as the iPhone. However, maybe a superior tracking method will evolve for tracking mobile visitors.

Issues Affecting Visitor Data from Page Tags

1. Setup errors cause missed tags.

The most frequent error by far observed for page tagging solutions comes from its setup. Unlike web servers, which are configured to log everything delivered by default, a page tag solution requires the webmaster to add the tracking code to each page. Even with an automated content management system, pages can and do get missed.

In fact, evidence from analysts at MAXAMINE (www.maxamine.com) who used their automatic page auditing tool has shown that some sites claiming that all pages are tagged can actually have as many as 20 percent of pages missing the page tag—something the webmaster was completely unaware of. In one case, a corporate business-to-business site was found to have 70 percent of its pages missing tags. Missing tags equals no data for those pageviews.

2. JavaScript errors halt page loading.

Page tags work well, provided that JavaScript is enabled on the visitor's browser. Fortunately, only about one to three percent of Internet users have disabled JavaScript on their browsers, as shown in Figure 2.3. However, the inconsistent use of JavaScript code on web pages can cause a bigger problem: Any errors in other JavaScript on the page will immediately halt the browser scripting engine at that point, so a page tag placed below it will not execute.

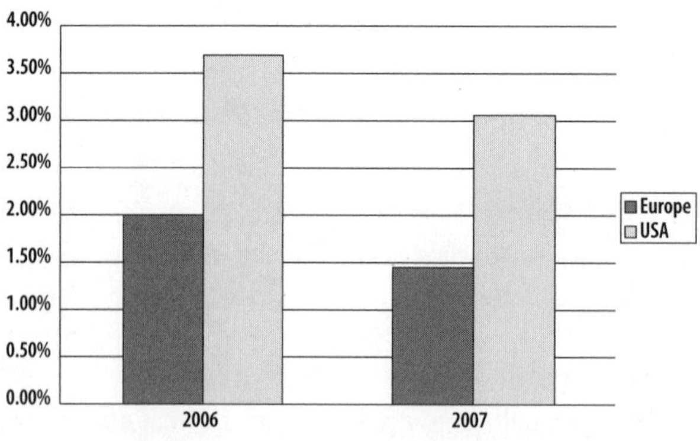

Source: 1,000,000,000 visits across multiple industry web properties using IndexTools (www.visualrevenue.com/blog—Dennis R. Mortensen)

Figure 2.3 Percentage of Internet users with JavaScript-disabled browsers

Page Tag Implementation Study

The following data is from over 10,000 websites, whose page tags were validated. The page tags checked are from a variety of web analytics vendors. (Thanks to Stephen Kirby of MAXAMINE for this information.)

Summary:

- The more frequently a website's content changes, the more prone the site is to missing page tags. In the following image, website content was updated on January 14; by mistake, the updated pages did not include page tags.

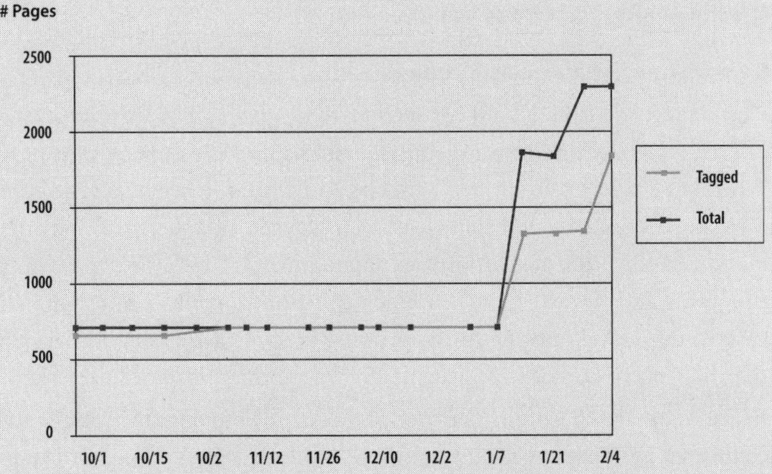

- Large websites very rarely achieve 100 percent tagging accuracy, as shown in the following chart.

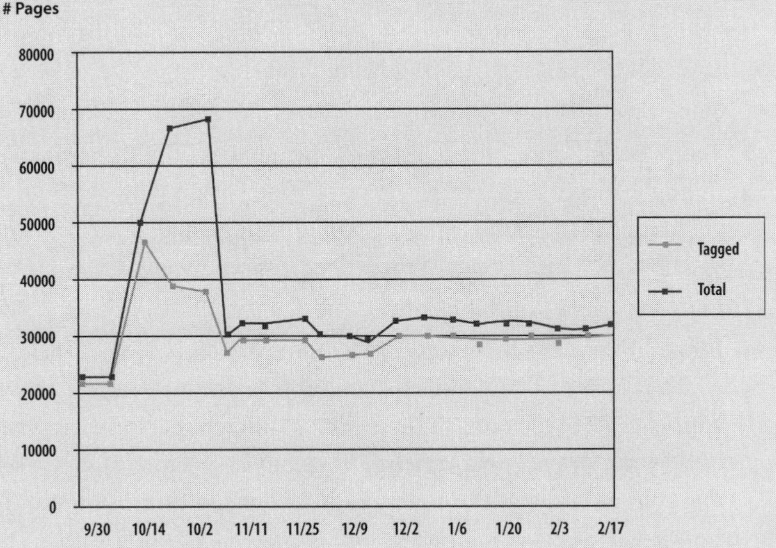

3. Firewalls block page tags.

   Corporate and personal firewalls can prevent page tag solutions from sending data to collecting servers. In addition, firewalls can also be set up to reject or delete cookies automatically. Once again, the effect on visitor data can be significant. Some web analytics vendors can revert to using the visitor's IP address for tracking in these instances, but mixing methods is not recommended. As discussed previously in Issues Affecting Visitor Data Accuracy for Logfiles (comScore report), using visitor IP addresses is far less accurate than simply not counting such visitors. It is therefore better to be consistent with the processing of data.

Issues Affecting Visitor Data When Using Cookies

1. Visitors can reject or delete cookies.

   Cookie information is vital for web analytics because it identifies visitors, their referring source, and subsequent pageview data. The current best practice is for vendors to process first-party cookies only. This is because visitors often view third-party cookies as infringing on their privacy, opaquely transferring their information to third parties without explicit consent. Therefore, many anti-spyware programs and firewalls exist to block third-party cookies automatically. It is also easy to do this within the browser itself. By contrast, anecdotal evidence shows that first-party cookies are accepted by more than 95 percent of visitors.

   Visitors are also becoming savvier and often delete cookies. Independent studies conducted by Belden Associates (2004), JupiterResearch (2005), Nielsen//NetRatings (2005) and comScore (2007) concluded that cookies are deleted by at least 30 percent of Internet users in a month.

2. Users own and share multiple computers.

   User behavior has a dramatic effect on the accuracy of information gathered through cookies. Consider the following scenarios:

   - Same user, multiple computers

     Today, people access the Internet in any number of ways—from work, home, or public places such as Internet cafes. One person working from three different machines still results in three cookie settings, and all current web analytics solutions will count each of these user sessions as unique.

   - Different users, same computer

     People share their computers all the time, particularly with their families, which means that cookies are shared too (unless you log off or switch off your computer each time it is used by a different person). In some instances, cookies are deleted deliberately. For example, Internet cafes are set up to do this automatically at the end of each session, so even if a visitor uses that cafe regularly and works from the same machine, the web analytics solution will consider that visitor a different and new visitor every time.

3. Latency leaves room for inaccuracy.

   The time it takes for a visitor to be converted into a customer (latency) can have a significant effect on accuracy. For example, most low-value items are either instant purchases or are purchased within seven days of the initial website visit. With such a short time period between visitor arrival and purchase, your web analytics solution has the best possible chance of capturing all the visitor pageview and behavior information and therefore reporting more accurate results.

   Higher-value items usually mean a longer consideration time before the visitor commits to becoming a customer. For example, in the travel and finance industries, the consideration time between the initial visit and the purchase can be as long as 90 days. During this time, there's an increased risk of the user deleting cookies, reinstalling the browser, upgrading the operating system, buying a new computer, or dealing with a system crash. Any of these occurrences will result in users being seen as new visitors when they finally make their purchase. Offsite factors such as seasonality, adverse publicity, offline promotions, or published blog articles or comments can also affect latency.

4. Data collection may be skewed by offline visits.

   It is important to factor in problems unrelated to the method used to measure visitor behavior but which still pose a threat to data accuracy. High-value purchases such as cars, loans, and mortgages are often first researched online and then purchased offline. Connecting offline purchases with online visitor behavior is a long-standing enigma for web analytics tools. Currently, the best practice way to overcome this limitation is to use online voucher schemes that a visitor can print and take with them to claim a free gift, upgrade or discount at your store. If you would prefer to receive online orders, provide similar incentives, such as web-only pricing, free delivery if ordered online, etc.

   Another issue to consider is how your offline marketing is tracked. Without taking this into account, visitors that result from your offline campaign efforts will be incorrectly assigned or grouped with other referral sources, and therefore skew your data. How to measure offline marketing is discussed in detail in Chapter 11.

Comparing Data from Different Vendors

As shown earlier, it is virtually impossible to compare the results of one data collection method with another. The association simply isn't valid. However, given two comparable data collection methods—both page tags—can you achieve consistency? Unfortunately, even comparing vendors that employ page tags has its difficulties.

Factors that lead to differing vendor metrics include the following:

1. First-party versus third-party cookies

There is little correlation between the two because of the higher blocking rates of third-party cookies by users, firewalls, and anti-spyware software. For example, the latest versions of Microsoft Internet Explorer block third-party cookies by default if a site doesn't have a compact privacy policy (see www.w3.org/P3P).

2. Page tags: Placement considerations

Page-tag vendors often recommend that their page tags be placed just above the </body> tag of your HTML page to ensure that the page elements, such as text and images, load first. This means that any delays from the vendor's servers will not interfere with your page loading. The potential problem here is that repeat visitors, those more familiar with your website navigation, may navigate quickly, clicking on to another page before the page tag has loaded to collect data.

This was investigated in a study by Stone Temple Consulting (www.stonetemple.com/articles/analytics-report-august-2007-part2.shtml). The results showed that the difference between a tracking tag placed at the top of a page and one placed at the bottom accounted for a 4.3 percent difference in unique visitor traffic for the same vendor's tool. Their hypothesis for the cause was the 1.4 second delay between loading the top of the page and the bottom page tag. Clearly, the longer the delay, the greater the discrepancy will be.

In addition, non-related JavaScript placed at the top of the page can interfere with JavaScript page tags that have been placed lower down. Most vendor page tags work independently of other JavaScript and can sit comfortably alongside other vendor page tags—as shown in the Stone Temple Consulting report in which pages were tagged for five different vendors. However, JavaScript errors on the same page will cause the browser scripting engine to stop at that point and prevent any JavaScript below it, including your page tag, from executing.

3. Did you tag everything?

Many analytics tools require links to files such as PDFs, Word documents, or executable downloads or outbound links to other websites, to be modified in order to be tracked. This may be a manual process whereby the link to the file needs to be modified. The modification represents an event or action when it is clicked, which sometimes is referred to as a *virtual pageview*. Comparing different vendors requires this action to be carried out several times with their specific codes (usually with JavaScript). Take into consideration that whenever pages have to be coded, syntax errors are a possibility. If page updates occur frequently, consider regular website audits to validate your page tags.

4. Pageviews: A visit or a visitor?

   Pageviews are quick and easy to track; and because they only require a call from the page to the tracking server, they are very similar among vendors. The challenge is differentiating a visit from a visitor; and because every vendor uses a different algorithm, no single algorithm results in the same value.

5. Cookies: Taking time out

   The allowed duration of timeouts—how long a web page is left inactive by a visitor—varies among vendors. Most page-tag vendors use a visitor-session cookie timeout of 30 minutes. This means that continuing to browse the same website after 30 minutes of inactivity is considered to be a new repeat visit. However, some vendors offer the option to change this setting. Doing so will alter any data alignment and therefore affect the analysis of reported visitors. Other cookies, such as the ones that store referrer details, will have different timeout values. For example, Google Analytics referrer cookies last six months. Differences in these timeouts between different web analytics vendors will obviously be reflected in the reported visitor numbers.

6. Page-tag codes: Ensuring security

   Depending on your vendor, your page tag code could be hijacked, copied, and executed on a different or unrelated website. This contamination results in a false pageview within your reports. By using filters, you can ensure that only data from your domains is reported.

7. PDF files: A special consideration

   For page-tag solutions, it is not the completed PDF download that is reported, but the fact that a visitor has clicked on a PDF file link. This is an important distinction, as information on whether or not the visitor completes the download—for example a 50-page PDF file—is not available. Therefore, a click on a PDF link is reported as a single event or pageview.

Note: The situation is different for logfile solutions. When viewing a PDF file within your web browser, Adobe Reader can download the file one page at a time, as opposed to a full download. This results in a slightly different entry in your web server logfile, showing an HTTP status code 206 (partial file download). Logfile solutions can treat each of the 206 status code entries as individual pageviews. When all the pages of a PDF file are downloaded, a completed download is registered in your logfile with a final HTTP status code of 200 (download completed). Therefore, a logfile solution can report a completed 50-page PDF file as one download and 50 pageviews.

8. E-commerce: Negative transactions

All e-commerce organizations have to deal with product returns at some point, whether it's because of damaged or faulty goods, order mistakes, or other reasons. Accounting for these returns within web analytics reports is often forgotten about. For some vendors, it requires the manual entry of an equivalent negative purchase transaction. Others require the reprocessing of e-commerce data files. Whichever method is required, aligning web visitor data with internal systems is never bulletproof. For example, the removal or crediting of a transaction usually takes place well after the original purchase, and therefore in a different reporting period.

9. Filters and settings: Potential obstacles

Data can vary when a filter is set up in one vendor's solution but not in another. Some tools can't set up the exact same filter as another tool, or they apply filters in a different way or at a different point in time during data processing.

Consider for example a page level filter to exclude all error pages from your reports. Visit metrics such as time on site, page depth etc. may or may not be adjusted for the filter depending on the vendor. This is because some vendors treat page level metrics separately to visitor level metrics.

10. Process frequency

This is best illustrated by example: Google Analytics does its number-crunching to produce reports hourly. However, because it takes time to collate all the logfiles from all of the data-collecting servers around the world, reports are three to four hours behind the current time. In most cases, it is usually a smooth process, but sometimes things go wrong. For example, if a logfile transfer is interrupted, then only a partial logfile is processed. Because of this, Google collects and reprocesses all data for a 24-hour period at the day's end. Other vendors may do the same, so it is important not to focus on discrepancies that arise on the current day.

Note: This is the same reason why you should not panic if you note "missing" data from your reports—for example, no data showing for the period 10 a.m. to 11 a.m. These should be picked up during the data reprocessing that takes place at the end of the day. If you have waited more than 24 hours and the data is still missing, contact the Google Analytics support team at http://www.google.com/support/googleanalytics/bin/request.py

11. Goal conversion versus pageviews: Establishing consistency

Using Figure 2.4 as an example, assume that five pages are part of your defined funnel (click stream path) with the last step (page 5) being the goal conversion (purchase). During checkout, a visitor goes back up a page to check a delivery charge (label A) and then continues through to complete payment. The visitor is so happy with the simplicity of the entire process, she then purchases a second item using exactly the same path during the same visitor session (label B).

Depending on the vendor you use, this process can be counted differently, as follows:

- Twelve funnel page views, two conversions, two transactions
- Ten funnel page views (ignoring step A), two conversions, two transactions
- Five funnel page views, two conversions, two transactions
- Five funnel page views, one conversion (ignoring step B), two transactions

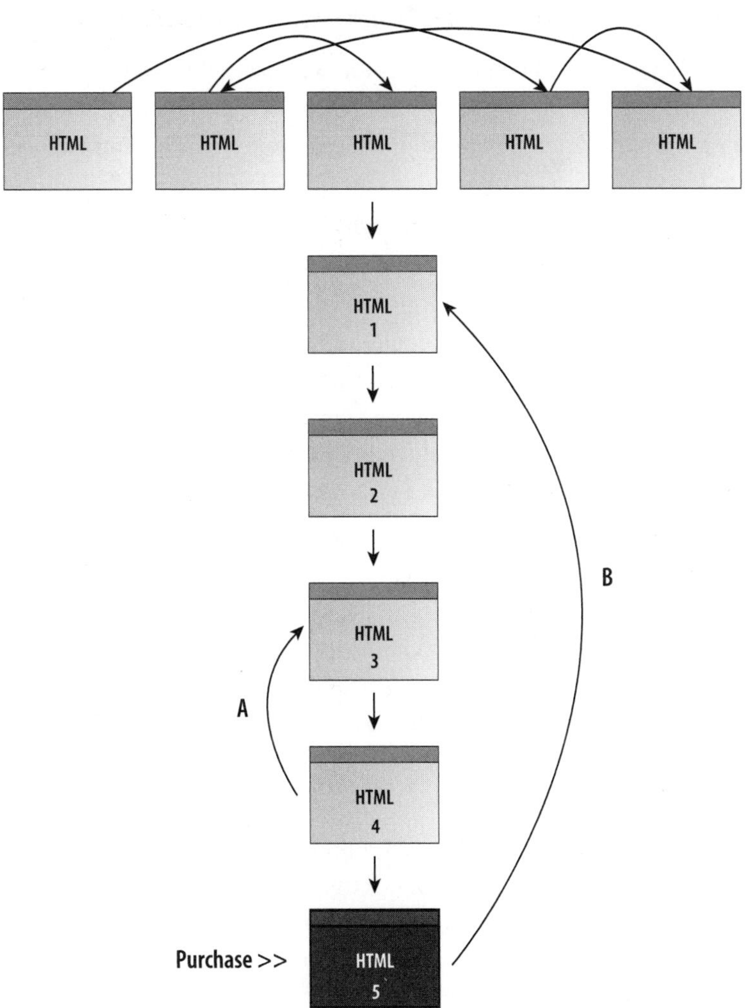

Figure 2.4 A visitor traversing a website, entering a five-page funnel, and making two transactions

Most vendors, but not all, apply the last rationale to their reports. That is, the visitor has become a purchaser (one conversion); and this can only happen once in the session, so additional conversions (assuming the same goal) are ignored. For this to be valid, the same rationale must be applied to the funnel pages. In this way, the data becomes more visitor-centric.

Note: In the example of Figure 2.4, the total number of pageviews equals twelve and would be reported as such in all pageview reports. It is the funnel and goal conversion reports that will be different.

Unparallel Results: Why PPC Vendor Numbers Do Not Match Web Analytics Reports

If you are using pay-per-click (PPC) networks, you will typically have access to the click-through reports provided by each network. Quite often, these numbers don't exactly align with those reported in your web analytics reports. This can happen for the following reasons:

1. Tracking URLs: Missing PPC click-throughs

   Tracking URLs are required in your PPC account setup in order to differentiate between a non-paid search engine visitor click-through and a PPC click-through from the same referring domain—Google.com or Yahoo.com, for example. Tracking URLs are simple modifications to your landing page URLs within your PPC account and are of the form `www.mysite.com?source=adwords`. Tracking URLs forgotten during setup, or sometimes simply assigned incorrectly, can lead to such visits incorrectly assigned to non-paid visitors.

2. Clicks and visits: Understanding the difference

   Remember that PPC vendors, such as Google AdWords, measure clicks. Most web analytics tools measure visitors that can accept a cookie. Those are not always going to be the same thing when you consider the effects on your web analytics data of cookie blocking, JavaScript errors, and visitors who simply navigate away from your landing page quickly—before the page tag collects its data. Because of this, web analytics tools tend to slightly underreport visits from PPC networks.

3. PPC: Important account adjustments

   Google AdWords and other PPC vendors automatically monitor invalid and fraudulent clicks and adjust PPC metrics retroactively. For example, a visitor may click your ad several times (inadvertently or on purpose) within a short space of time. Google AdWords investigates this influx and removes the additional click-throughs and charges from your account. However, web analytics tools have no access to these systems and so record all PPC visitors. For further information on how Google treats invalid clicks, see http://adwords.google.com/support/bin/topic.py?topic=35.

4. Keyword matching: Bid term versus search term

The bid terms you select within your PPC account and the search terms used by visitors that result in your PPC ad being displayed can often be different: think "broad match." For example, you may have set up an ad group that targets the word "shoes" and solely relies on broad matching to match all search terms that contain the word "shoes." This is your bid term. A visitor uses the search term "blue shoes" and clicks on your ad. Web analytics vendors may report the search term, the bid term, or both.

5. Google AdWords: A careful execution

Within your AdWords account, you will see data updated hourly. This is because advertisers want and need this to control budgets. Google Analytics imports AdWords cost data once per day, and this is for the date range minus 48 to 24 hours from 23:59 of the previous day—so AdWords cost data is always at least 24 hours old.

Why the delay? Because it allows time for the AdWords invalid click and fraud protection algorithms to complete their work and finalize click-through numbers for your account. Therefore, from a reporting point of view, the recommendation is to not compare AdWords visitor numbers for the current day. This recommendation holds true for all web analytics solutions and all PPC advertising networks.

Note: Although most of the AdWords invalid click updates take place within 24 hours, it can take longer. For this reason, even if all other factors are eliminated. AdWords click-throughs within your PPC account and those reported in your web analytics reports may never match exactly.

6. Third-party ad tracking redirects: Weighing in the factors

Using third-party ad tracking systems—such as Atlas Search, Blue Streak, DoubleClick, Efficient Frontier, and SEM Director, for example—to track click-throughs to your website means your visitors are passed through redirection URLs. This results in the initial click being registered by your ad company, which then automatically redirects the visitor to your actual landing page. The purpose of this two-step hop is to allow the ad tracking network to collect visitor statistics independently of your organization, typically for billing purposes. As this process involves a short delay, it may prevent some visitors from landing on your page. The result can be a small loss of data and therefore failure to align data.

In addition, redirection URLs may break the tracking parameters that are added onto the landing pages for your own web analytics solution. For example, your landing page URL may look like this:

```
http://www.mysite.com/?source=google&medium=ppc&campaign=08
```

When added to a third-party tracking system for redirection, it could look like this:

```
http://www.redirect.com?http://www.mywebsite.com?source=google&medium=
ppc&campaign=08
```

The problem occurs with the second question mark in the second link, because you can't have more than one in any URL. Some third-party ad tracking systems will detect this error and remove the second question mark and the following tracking parameters, leading to a loss of campaign data.

Some third-party ad tracking systems allow you to replace the second ? with a # so the URL can be processed correctly. If you are unsure of what to do, you can avoid the problem completely by using encoded landing-page URLs within your third-party ad tracking system, as described at the following:

```
http://www.w3schools.com/tags/ref_urlencode.asp
```


Data Misinterpretation: Lies, damn lies, and statistics

The following are not accuracy issues. However, the reference to Mark Twain is simply to point out that data is not always so straightforward to interpret. Take the following two examples:

1. New visitors plus repeat visitors does not equal total visitors.

   A common misconception is that the sum of the new plus repeat visitors should equal the total number of visitors. Why isn't this the case? Consider a visitor making his first visit on a given day and then returning on the same day. They are both a new and a repeat visitor for that day. Therefore, looking at a report for the given day, two visitor types will be shown, though the total number of visitors is one.

   It is therefore better to think of visitor types in terms of "visit" type—that is, the number of first-time visits plus the number of repeat visits equals the total number of visits.

2. Summing the number of unique visitors per day for a week does not equal the total number of unique visitors for that week.

   Consider the scenario in which you have 1,000 unique visitors to your website blog on a Monday. These are in fact the only unique visitors you receive for the entire week, so on Tuesday the same 1,000 visitors return to consume your next blog post. This pattern continues for Wednesday through Sunday.

If you were to look at the number of unique visitors for each day of the week in your reports, you would observe 1,000 unique visitors. However you cannot say that you received 7,000 unique visitors for the entire week. For this example, the number of unique visitors for the week remains at 1,000.

Accuracy Summary and Recommendations

Clearly, web analytics is not 100 percent accurate, and the number of possible inaccuracies can at first appear overwhelming. However, as the preceding sections demonstrated, you can get comfortable with your implementation and focus on measuring trends, rather than precise numbers. For example, web analytics can help you answer the following questions:

- Are visitor numbers increasing?
- By what rate are they increasing (or decreasing)?
- Have conversion rates gone up since beginning PPC advertising?
- How has the cart abandon rate changed since the site redesign?

If the trend showed a 10.5 percent reduction, for example, this figure should be accurate, regardless of the web analytics tool that was used.

When all the possibilities of inaccuracy that affect web analytics solutions are considered, it is apparent that it is ineffective to focus on absolute values or to merge numbers from different sources. If all web visitors were to have a login account in order to view your website, this issue could be overcome. In the real world, however, the vast majority of Internet users wish to remain anonymous, so this is not a viable solution.

As long as you use the same measurement for comparing data ranges, your results will be accurate. This is the universal truth of all web analytics.

Here are 10 recommendations for web analytics accuracy:

1. Select the data collection methodology based on what best suits your business needs and resources.
2. Be sure to select a tool that uses first-party cookies for data collection.
3. Don't confuse visitor identifiers. For example, if first-party cookies are deleted, do not resort to using IP address information. It is better simply to ignore that visitor.
4. Remove or report separately all non-human activity from your data reports, such as robots and server performance monitors.
5. Track everything. Don't limit tracking to landing pages. Track your entire website's activity, including file downloads, internal search terms, and outbound links.

6. Regularly audit your website for page tag completeness. Sometimes site content changes result in tags being corrupted, deleted, or simply forgotten.
7. Display a clear and easy-to-read privacy policy (required by law in the European Union). This establishes trust with your visitors because they better understand how they're being tracked and are less likely to delete cookies.
8. Avoid making judgments on data that is less than 24 hours old, because it's often the most inaccurate.
9. Test redirection URLs to guarantee that they maintain tracking parameters.
10. Ensure that all paid online campaigns use tracking URLs to differentiate from non-paid sources.

These suggestions will help you appreciate the errors often made when collecting web analytics data. Understanding what these errors are, how they happen, and how to avoid them will enable you to benchmark the performance of your website. Achieving this means you're in a better position to then drive the performance of your online business.

Privacy Considerations for the Web Analytics Industry

With the huge proliferation of the Web, people are now more aware of privacy issues, concerns, and obligations. In my opinion, this is a step forward—the industry needs an informed debate about online privacy. So far, the discussion has been fairly basic, with people talking about online privacy as a single entity and using the example of the web analytics industry as proof of loss of privacy—for example, many people complain that tracking their visit to a website is an invasion of their privacy that they did not consent to. However, there are actually two privacy issues that web users and website owners should be aware of:

Non–Personally Identifiable Information (non-PII) This is anonymous aggregate data that cannot be used to identify or deduce demographic information. It is best illustrated by example. Suppose you wish to monitor vehicle traffic close to a school so that you can predict and improve the safety and efficiency of the surrounding road structure. You might stand on a street corner counting the number of vehicles, their type (car, van, truck, bus, etc.), time of day, and how long it takes for them to pass the school gates. This is an example of non-personal information—there is nothing in this aggregate data that identifies the individual driver or owner of each vehicle. Incidentally, nor can you identify whether the same vehicle is repeatedly driving around the school in a circle.

As you can see, this is a great way to collect data to improve things for all people involved (school pupils, residents, shop owners, and drivers) without any interference of privacy. This example is directly analogous to using the Web. By far, the vast majority of Web users who are surveyed claim they are happy for their non-personal information to be collected and used to improve a website's effectiveness and ultimately their user experience.

Personally Identifiable Information (PII) Taking the previous non-PII example further, suppose the next day you started to collect vehicle license plate details, or stopped drivers to question them on their driving habits, or followed them home to determine whether they were local residents or not. These are all examples of collecting personal data—both asked-for data such as their name, age, and address, as well as non-volunteered information that can be discovered, such as gender and license plate details.

Collecting personally identifiable information clearly has huge privacy implications and is regulated by law in most democratic countries. Collecting data in this way would mean that all drivers would need to be explicitly informed that data collection was occurring and offered the choice of not driving down the street. They can then make an informed decision as to whether they wish to take part in the study or not. Again, this is analogous to using the Web—asking the visitor to opt-in to sharing their personal information.

Note: On the internet, IP addresses are classed as personally identifiable information.

The current issue with regard to privacy on the Web is that many users are confused as to what form of tracking, if any, is taking place when they visit a website. This is reflected in the fact that on large, high-traffic websites for which I have worked (1–50 million visitors per month), the number of pageviews for the privacy policy statement were consistently and considerably less that 0.01 percent of the total.

Even when viewing privacy statements, the public is cynical. Often, these statements tend to be written in a legal language that is difficult to understand, they change without notice, and they primarily appear to be there to protect the website owner rather than the privacy of the visitor.

Regardless of the public's confusion or apathy about website privacy, it is your responsibility as a website owner to inform visitors about what data collection practices are occurring when a visitor views your website. In fact, within the European Union, law requires it.

Google's entry into the web analytics market and their strong stance on end-user privacy, transparency, and accountability is adding clarity to this whole area. As a best-practice illustration of a clear privacy statement, take a look at the Information Commissioner's Office (the U.K. independent authority to protect personal information) as an example:

```
www.ico.gov.uk/Global/privacy_statement.aspx
```

Summary

In Chapter 2, you have learned the following:

- How web visitor data is collected, the relative advantages of page tags and log-file tools, as well as why page tagging has become the de facto standard.
- The role of cookies in web analytics, what they contain, and why they exist, including the differences between first-party and third-party cookies.
- The accuracy limitations of web traffic information in terms of collecting web visitor data, its interpretation, and comparing numbers from different vendors.
- How to think about web analytics in relation to end-user privacy concerns and your responsibilities as a website owner to respect your visitors' privacy.

Where Google Analytics Fits

Understanding how Google Analytics data collection works is a great way to recognize what can be achieved with web analytics reporting. Don't worry—this is not an engineering book, so technicalities are kept to a minimum. However, it is important to know what can and cannot be accomplished, as this knowledge will help you spot erroneous data that may show up in your reports.

As well as a discussion of the key features and capabilities of Google Analytics, included in this chapter is a description of Urchin software—a separate web analytics tool from Google. Understanding the differences between Urchin and Google Analytics will help you make an informed decision when you are considering data collection tools.

3

In this chapter, you will learn about the following:

The key features of Google Analytics

How Google Analytics works

The Google Analytics approach to user privacy

What Urchin software is

The differences between Google Analytics and Urchin

Key Features and Capabilities of Google Analytics

This is not an exhaustive list, but it highlights some key features you can find in Google Analytics:

1. Multiple language interfaces and support

Google Analytics can display reports in 25 languages, though this number is continually growing. Languages include Czech, Chinese, Danish, Dutch, English (US), English (UK), Filipino, Finnish, French, German, Hungarian, Italian, Indonesian, Japanese, Korean, Malaysian, Norwegian, Polish, Portuguese (Brazil), Portuguese (Portugal), Russian, Spanish, Swedish, Taiwanese, and Turkish.

In addition to the display of reports in multiple languages, all documentation is internationalized and each language is directly supported by Google staff.

2. High scalability

The Google Analytics target audience can be compared to that of online advertising—just about everyone! Clients range from a few pageviews per day to some of the best-known brands and most highly trafficked sites on the web—that is, sites receiving more than 1 billion pageviews per day.

3. Features applicable for enterprise and small business users

I started my career running my own business of web professionals, so I understand the analytic needs of a small company. Now, having worked at Google for a number of years, I am familiar with the other end of the spectrum—working with some of the largest organizations in the world. What still amazes me to this day is just how similar both large and small companies are in their analytics requirements—from understanding what is happening on their website and how to interpret the data to what action to take to improve matters, small and large organizations face the same challenges.

Both users express an understanding of the need for measurement, yet also fear data overload when combined with other aspects of the business and their job. Both also expect the collection and reporting of data to be at the smaller end of their investment budget, with professional services the key to unlocking their online business potential.

4. Two-click integration with AdWords

If you manage a pay-per-click campaign, you know what a chore tagging your landing page URLs can be—each one has to have at least one campaign variable appended to differentiate visitors that click through from non-paid search results. In addition, you will want to import your AdWords cost and impression data. Google Analytics achieves this with two check boxes. As a result, all your AdWords landing page URLs are tagged, and cost data is imported automatically each day.

5. Full campaign reporting—not just AdWords

   Google Analytics enables you to track and compare all your visitors—from non-paid organic search, paid ads (pay-per-click, banners), referrals, e-mail newsletters, affiliate campaigns, links from within digital collateral such as PDF files, and any other search engine or medium that forwards a visitor to your website. You can even get a handle on your offline marketing campaigns.

6. Funnel visualization

   Funnels are paths visitors take before achieving a goal conversion. An obvious conversion is an e-commerce purchase, for which the funnel is the checkout process. However, others exist, such as a registration sign-up process or feedback form. By visualizing the visitor path, you can discover which pages result in lost conversions and where your would-be customers go.

7. Customized dashboards

   A dashboard is the first section you see when viewing your reports. Here you can place and organize your key report selections for an at-a-glance comparison. Dashboard reports are copies from the main sections of your Google Analytics. Up to 12 reports can be changed and reordered at any time, on a per-user basis.

8. Site overlay report

   Site overlay is a graphical way of looking at the popularity of links on your pages. You view your key metrics overlaid on top of your web page links. It's an easy to view snapshot of which links are working for you.

9. Map overlay reports

   Similar to site overlay, map overlay is a graphical way of presenting data that reflects where visitors are connecting from around the world when viewing your website. Based on IP address location databases, they show your key metrics overlaid on top of a world, regional, or country map. This provides a clear representation of which parts of the world visitors are connecting from, down to city level.

   Geo-IP information has improved dramatically in recent years—mainly due to improvements in online credit card fraud detection. The database used in Google Analytics is the same as that used for geo-targeting ads in your AdWords campaigns. Data can be as accurate as a 25-mile (40 km) radius. However, sometimes location details are not available and this is reported as "(not set)" in your reports.

10. Cross-segmentation

    Cross segmentation is the terminology used for cross-referencing, or correlating, one set of data against another. If you are familiar with MS Excel, cross-segmentation is analogous to pivot tables. An example of cross-segmentation might be displaying the geo location report for California and then cross-segmenting to display which search engines these visitors are coming from. As another example, suppose you want to determine, for U.K. visitors, the most frequently used keywords to find your site. That would be a cross-reference of U.K. visitors against keywords.

11. Data export and scheduling

Report data can be manually exported in a variety of formats, including XLS, CSV, PDF, or the open-source XML. You may also schedule any report (even cross segmented) to be e-mailed to you and your colleagues automatically, for up to 25 e-mail addresses. For example, you may want to e-mail your web designer the list of error pages generated by your website each week.

12. Date range comparison

In addition to showing side-by-side date range comparisons within the same browser window, Google Analytics has a unique "timeline window" method for selecting date ranges without losing sight of long-term trends. For example, you can select a date range that shows a visitor number spike you were previously unaware of.

13. E-commerce reporting

You can trace transactions to campaigns and keywords, get loyalty and latency metrics, and identify your revenue sources. Similarly, you can drill down on this information on a per-product basis.

14. Site search reporting

For complex websites (i.e., a large number of pages), internal site search is an important part of the site navigation system and in many cases is critical for providing a positive user experience. A dedicated report section enables you to monetize the value of your internal site search engine, comparing with those visitors who do not search. In addition, you can discover which pages lead to visitors performing a search, as well as list the post-search destination pages.

15. Event tracking

This report shows you events displayed separately from pageviews. For example, if your website incorporates Flash elements or embedded video, you will want to see how users interact with these separately from your pageview reports. Any Flash element, Ajax content, file downloads, and even load times can be reported on in this way.

Did You Know...?

- Google Analytics can distinguish visitors from any source—for example, any search engine, any pay-per-click advertising network (such as AdWords, Yahoo Search Marketing, Microsoft adCenter, Miva), e-mail campaigns, banner ads, affiliates, etc.

- In addition to tracking standard pageviews, Google Analytics can track error pages, file downloads, clicks on mailto links, partial form completion, and exit links. See Chapter 7 for further details.
- Unreadable dynamic URLs can be converted into human-readable form. For example:

  ```
  www.mysite.com/home/product?rid=191045&scid=184282
  ```

  can be converted to

  ```
  www.mysite.com/products/menswear/shirts/white button down
  ```

  See Chapter 7 for further details.
- Google retains your data (for free) for at least 25 months, so you can go back and perform year-by-year comparisons.
- In building a relationship with your organization, a visitor may use multiple referrers before converting. In this way, all referrers are tracked. However, for a conversion only the last referrer is given the credit.

  For example, consider the following scenario: A visitor first views a banner ad on the Web and clicks through to your site. The visitor does not convert on that first visit but returns later that day after performing a keyword query on a search engine. Still not convinced that they are ready to purchase (or convert into a lead), the visitor leaves your website. Later in the week, a friend of the visitor recommends via e-mail a review article published on a blog. Happy with the review, the same visitor clicks the link from the blog article directly to your website. On this third visit, a purchase is made. For this scenario, Google Analytics credits the conversion to the blog website and its URL will be listed in your reports.

  However, there is one exception to this rule: when the last referrer is "direct." A direct visit means the visitor typed your website address directly into his or her browser or used a bookmark to arrive on your website. In that case, the penultimate referrer is given credit. For example, using the preceding scenario, if the purchaser bookmarks your website and then later returns to make a repeat purchase by selecting the bookmark, credit for that conversion will still be given to the referring blog.
- You can use a regular expression (regex) to filter URL data into visitor segments. Maximum regex length is 256 characters. See Chapter 7 for further details.
- You can change or append the recognized search engines list. For example, by default all Google search engine properties are grouped under a single search engine referrer—"google." However, you may wish to split google.co.uk, google.fr, google.de, and others from google.com. This can be achieved by a simple modification of the page tag. See Chapter 9 for further details.
- If you have an existing web analytics solution, you can run Google Analytics alongside it by appending the Google Analytics page tag to your pages. This

way, you can evaluate Google Analytics or even enhance existing data you may already be collecting.

- You can track visitor data into multiple Google Analytics accounts. For example, tracking at a regional or country level as well as having an aggregate account for all visits. See Chapter 6 for further details.

How Google Analytics Works

From Chapter 2 you gained an understanding of data collection techniques and the role that cookies play in web analytics, but how does Google Analytics work? This is best illustrated with the schematic shown in Figure 3.1. By this method, all data collection, processing, maintenance, and program upgrades are managed by Google as a hosted service. Figure 3.1 explains:

1. Visitors arrive at your website via many different routes, including search engines, e-mail marketing, referral links (other websites), embedded links (PDF, DOC, XLS, etc.), or directly by typing the address into a browser's address bar. Whatever the route, when the visitor views one of your pages with the Google Analytics JavaScript page tag, this information plus other visitor data (e.g., page URL, timestamp, unique ID, screen resolution, color depth) is collected and a set of cookies are created to identify the visitor.
2. The Google Analytics JavaScript page tag sends this information to Google data collection servers via a call of a transparent, 1×1-pixel GIF image at google-analytics.com. The entire process takes a fraction of a second.
3. Each hour, Google processes the collected data and updates your Google Analytics reports. However, because of the methodology and the huge quantity of data involved, reports are displayed three hours in arrears; and this may sometimes be longer—though not more than 24 hours.

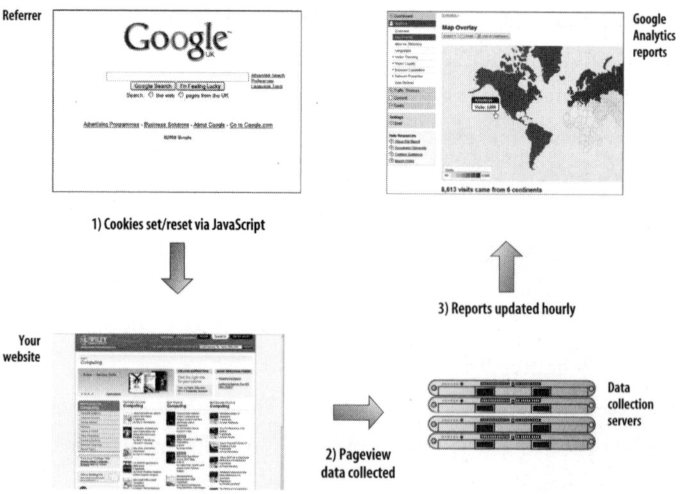

Figure 3.1 Schematic diagram of how Google Analytics works

Google Analytics and User Privacy

All Google Analytics reports contain aggregate non-personally-identifiable information. That said, three parties are involved in the Google Analytics scenario: Google, an independent website, and a visitor to that website. Google has designed its privacy practices to address each of these participants by requiring each website that uses Google Analytics to abide by the privacy provisions in the terms of service, specifically section 7 (see www.google.com/analytics/tos.html):

> You will not (and will not allow any third party to) use the Service to track or collect personally identifiable information of Internet users, nor will You (or will You allow any third party to) associate any data gathered from Your website(s) (or such third parties' website(s)) with any personally identifying information from any source as part of Your use (or such third parties' use) of the Service. You will have and abide by an appropriate privacy policy and will comply with all applicable laws relating to the collection of information from visitors to Your websites. You must post a privacy policy and that policy must provide notice of your use of a cookie that collects anonymous traffic data.

Note: The content of section 7 of tos.html may vary depending on which country you operate in. Ensure you view the most relevant Terms of Service by selecting from the drop down menu at the top of the page.

The Google Analytics cookies collect standard Internet log data and visitor behavior information in an anonymous form. They do not collect any personal information such as addresses, names, or credit card numbers. The logs include standard log information such as IP address, time and date stamp, browser type, and operating system. The behavior information includes generic surfing information, such as the number of pages viewed, language setting, and screen resolution settings in the browser, and can include information about whether or not a goal was completed by the visitor to the website. The website can define the goal to mean different things, such as whether or not a visitor downloaded a PDF file, completed an e-commerce transaction, or visited more than one page, and so on. Note that Google Analytics does not track a user across multiple unrelated sites, and it uses different cookies for each website.

Note: IP addresses are only stored until processing has determined the geo-location of a visit, then they are discarded.

Google Analytics prepares anonymous and statistical reports for the websites that use it. As you will see, such reports include different information views and show data such as geographic location (based on generic IP-based geo-location codes), time of visit, and so on. These reports are anonymous and statistical. They do not include any information that could identify an individual visitor—for example, they do not include IP addresses.

Common questions asked by potential Google Analytics clients include the following:

- What does Google do with the data it collects?
- Who at Google sees the analytics data?
- How securely is data kept?
- As a website owner, what is my obligation to data privacy?

I answer these questions from my own perspective after working at Google for a number of years.

- What does Google do with the data it collects?

  Google Analytics is a tool specifically targeted at advertisers (and potential advertisers) who want to gain a better understanding of their website traffic. In fact, it is one of many tools that make up what I refer to as an advertiser's toolkit. Others include Google Trends, Webmaster Central, Product Search (formally Froogle), Google Maps, Website Optimizer, and Checkout. Google Analytics provides advertisers with the transparency and accountability they need in order to have confidence in the pay-per-click, online auction model. Essentially, a happy advertiser is good for business.

  Keep in mind that the Google AdWords auction model prevents anyone from interfering with the pricing of ads. The system is completely transparent, so it would be ludicrous for Google to artificially adjust bids—destroying a business overnight. On the Web, the competition is always only one click away.
- Who at Google sees the analytics data?

  Google Analytics data, as with all data at Google, is accessed on a strict need-to-know basis—for example by support staff and maintenance engineers. If, as a client, you want Google staff to look at your reports—for example, to provide help with managing an AdWords campaign—then you must request this from your Google Account Manager or via the Google Analytics Help Center (www.google.com/support/googleanalytics/). All internal Google access to your reports is monitored for auditing purposes.
- How secure is the analytics data?

  Data security and integrity is paramount for continued end-user confidence in all Google services. As such, Google Analytics data is subject to the same rigorous security checks and audits as all other Google products. Of course, one can never be 100 percent certain of security in any organization, but Google employs some of the best industry professionals in the world to ensure that its systems remain secure.

- As a site owner, what is my obligation to data privacy?

  In addition to Google's commitment to data privacy and integrity, owners of websites that use Google Analytics also have an obligation to visitor privacy. In fact, this is true for any analytics solution. For Google Analytics, the terms of service state that you will not associate any data gathered from your website with any personally identifiable information. You will, of course, also need to comply with all applicable data protection and privacy laws relating to your use of Google Analytics, and have in place (in a prominent position on your website) an appropriate privacy policy.

  These are commonsense best-practice approaches to owning a website and collecting visitor information about its usage. However, I recommend that you view your obligations as a website owner from the Terms of Service link at the bottom of any page on the Google Analytics website (www.google.com/analytics). To ensure that you read the most relevant terms for your location, select the region that most closely matches your own from the country drop-down menu at the top of the page.

  A best-practice illustration of a clear privacy statement can be viewed at www.ico.gov.uk/Global/privacy_statement.aspx. The ICO is the website of the Information Commissioner's Office—the U.K. independent authority regarding the protection of personal information.

What Is Urchin?

Although this book is about using Google Analytics to measure your visitor traffic, it is worth mentioning that Google has two web analytics products: Google Analytics and Urchin software.

Urchin is the software company and technology that Google acquired in April 2005 and that went on to become Google Analytics—a free web analytics service that uses the resources at Google. Urchin software is a downloadable web analytics tool that runs on a local server (Unix or Windows). Typically, this is the same machine as your web server. The Urchin tool creates reports by processing web server logfiles—including hybrid ones—which combine logfile information with page tag information. This hybrid approach is the most accurate of the common web analytics data collection methods available—as discussed in Chapter 2.

Urchin is essentially the same technology as Google Analytics—the difference is that resources for Urchin log storage and data processing are provided by your organization. As Table 2.1 showed, logfile tools can report on information that page tag solutions alone cannot provide. Urchin software provides complementary reports that Google Analytics cannot provide:

- Error page/status code reports
- Bandwidth reports
- Login name reports—standard Apache .htaccess or any authentication that logs usernames in the logfile
- Visitor history report—tracking individual visitors (anonymously)

Note: As discussed in Chapter 9, it is possible to configure your website to report error pages within Google Analytics. However, for Urchin software no additional changes are required to track error pages. These are tracked by default by your web server, so Urchin reports on them out of the box.

Google Analytics versus Urchin

With two analytics products from Google to choose from, how do you determine which one of these is right for your organization? As you may have guessed from the title of this book, Google Analytics is perfect for most organizations, for two very simple reasons:

- Google Analytics is a free service, whereas Urchin software is a licensed product and therefore must be purchased.
- Google Analytics removes a large part of the IT overhead usually associated with implementing a web analytics tool. That is, the data collection, storage, program maintenance, and upgrades are conducted for you by Google. For Urchin software, these become your responsibility.

The second point is not trivial. In fact, in my experience, the IT overhead of implementing tools was the main reason why web analytics remained a niche industry for such a large part of its existence. Maintaining your own logfiles has an overhead, mainly because web server logfiles get very large, very quickly.

Note: As a guide, every 1,000 visits produces approximately 4 MB of log info. Therefore, 10,000 visits per month is ~500 MB per year. If you have 100,000 visits per month, that's 5 GB per year, and so on. Those are just estimates—for your own site, these could easily be double the estimate.

Urchin also requires disk space for its processed data (stored in a proprietary database). Though this will always be a smaller size than the raw collected numbers, storing and archiving all this information is an important task because if you run out of disk space, you risk file or database corruption from disk write errors. This kind of file corruption is almost impossible to recover from.

As an aside, if you maintain your own visitor data logfiles, the security and privacy of collected information (your visitors) also becomes your responsibility.

Why, then, might you consider Urchin software at all? Urchin software does have some real advantages over Google Analytics. For example, data is recorded and stored by your web server, rather than streamed to Google, which means the following:

- Urchin can keep and view data for as long as you wish.

  Google Analytics currently commits to keeping data for a minimum of 25 months.

Note: To date, Google has made no attempt to remove data older than 25 months.

- Urchin allows your data to be audited by an independent third party. This is usually important for publishers who sell advertising space so they can verify visitor numbers to provide credibility for advertisers (trust in their rate card).

  Google Analytics does not pass data to third parties.
- Urchin can reprocess data as and when you wish—for example, to apply a filter retroactively.

  Google Analytics currently does not reprocess data retroactively.
- Urchin works behind the firewall—that is, it's suitable for intranets.

  Google Analytics cannot run behind a closed firewall.
- Urchin stores data locally in a proprietary database and includes tools that can be used to access the data outside of a web browser, allowing you to run ad hoc queries.

  Google Analytics stores data in remote locations within Google datacenters around the world in proprietary databases and does not provide direct access to the stored data for ad hoc queries.

Criteria for Choosing between Google Analytics and Urchin

- If you have an intranet site behind a firewall that blocks Internet activity, then the decision is easy—use Urchin software, as Google Analytics is a hosted solution that needs access to the Internet in order to work.
- If you are unable to page tag—for example, on WML sites used for mobile phones—use Urchin.

- If you are measuring the success (or not) of your website—its ability to convert and the effectiveness of online marketing—then select Google Analytics, as it is much easier to implement, has stronger AdWords integration, and is virtually maintenance-free.
- If you are a hosting provider wishing to offer visitor reports to thousands of customers, consider Urchin, as it has a command-line interface that can be scripted to create and modify multiple website reports at once.
- Use both together if you need the flexibility of maintaining your own site visitor data—and you have the resources to manage it. Combining Google Analytics with Urchin software gives you the best of both worlds—the advanced features of Google Analytics (free) and the flexibility of Urchin. Chapter 6 discusses how you can configure your page tags to stream data to both Google Analytics and Urchin simultaneously.

Summary

In Chapter 3, you have learned the following:

- The key features and capabilities of Google Analytics, which will enable you to ascertain what it can do for you and whether it is suitable for the analytics needs of your organization
- How Google Analytics works from a non-technical perspective, so that you can understand how data is collected and processed by Google
- How seriously Google Analytics takes its responsibility for visitor data—both in terms of Google Analytics users and website visitors
- What Urchin is, how it compares with Google Analytics, and what criteria you should consider when selecting an analytics product from Google

Using Google Analytics Reports

Part II is a user guide that walks you through using Google Analytics reports to understand website visitor behavior. Rather than describe every report, I've highlighted the key areas of the user interface as well as how to find your way around the information presented. I deliberately focus on the most important aspects you need to know first in order for you to enjoy the process of discovering more of its capabilities and going deeper into the data. This chapter emphasizes ten fundamental reports that can help you answer your most burning questions.

II

In Part II, you will learn about the following:

4

Using the Google Analytics Interface

The Google Analytics user interface makes use of the latest developments in Web 2.0/Ajax technology to construct report data in a highly accessible format. For example, rather than use a side menu to navigate through different reports (though that is available), the user is encouraged to drill into the data itself.

This process is intuitive, because the data is shown in context, with visit data displayed alongside conversion and e-commerce data. Instead of determining which related navigation item to click, you simply click the link within the viewed data. This is a difficult process to describe on paper, but by the end of this chapter you will be quickly and efficiently gaining insights into your data.

In this chapter, you will learn about the following:

Discoverability and the context of data
How to navigate your way around the plethora of information
Comparing date periods
Hourly reporting
Scheduling e-mail exports of data
The value of cross segmentation
How to assess the value of a page

Discoverability

A common complaint from users of other web analytics tools is that the vast quantity of data generated is often overwhelming, resulting in users getting lost and unable to decipher the information. As a result, a great deal of effort went into building the Google Analytics report interface to make it as intuitive to use as possible. In addition to data being very accessible, the user interface, shown in Figure 4.1, enhances *discoverability*. By this I mean how easy is it for you to ascertain whether the report you are looking at is good news, bad news, or indifferent to your organization. In other words, Google Analytics simplifies the process of turning raw data into useful information so that you can either reward your team, fix something, or change your benchmarks.

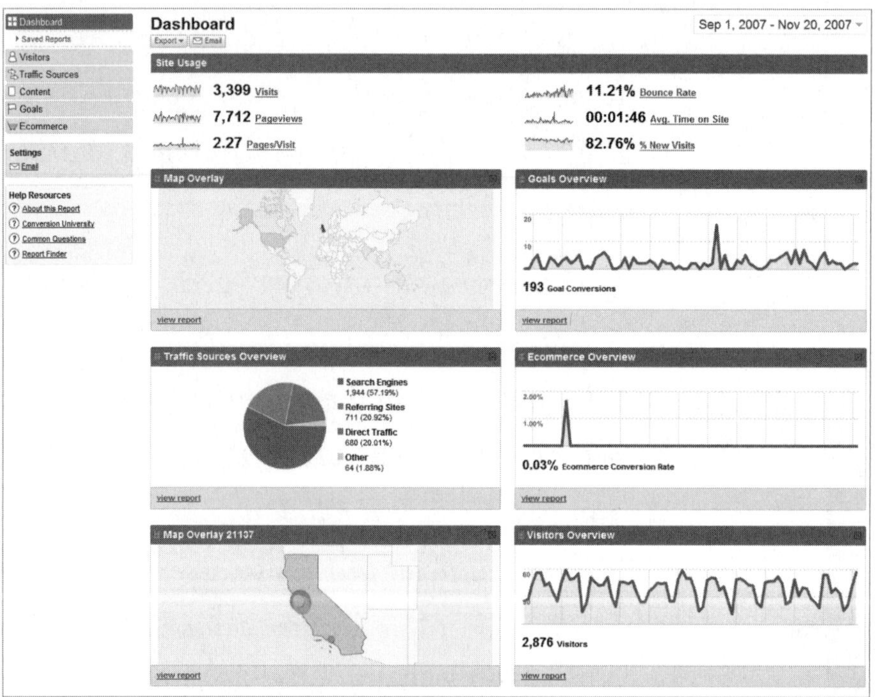

Figure 4.1 The initial report screen of Google Analytics

The Google Analytics drill-down interface is intended to be intuitive. It differs from other web analytics tools, which have a menu-driven style of navigation. You can select menu-driven navigation if you prefer it, but the Google Analytics interface makes it much easier to explore your data in context—that is, within the data, so that you do not waste your time navigating back and forth between reports to answer your questions. In addition, links within the reports suggest related information; and fast, interactive segmentation enables you to reorganize data on-the-fly. Short narratives, scorecards, and *sparklines* summarize your data at every level. Moreover, to help you understand, interpret, and act on data relationships, context-sensitive Help and Conversion University articles are available in every report.

Note: A sparkline is a mini-image (thumbnail) of graphical data that enables you to put numbers in a temporal context without the need to display full charts. For example, the following screen shot shows an array of numbers that on their own would be meaningless. However, the sparkline graphics show these in context by illustrating the trends over the time period selected. It's a neat and condensed way conveying a lot of information.

3,399 Visits
7,712 Pageviews
2.27 Pages/Visit

Navigating Your Way Around: Report Layout

An example report is shown in Figure 4.2.

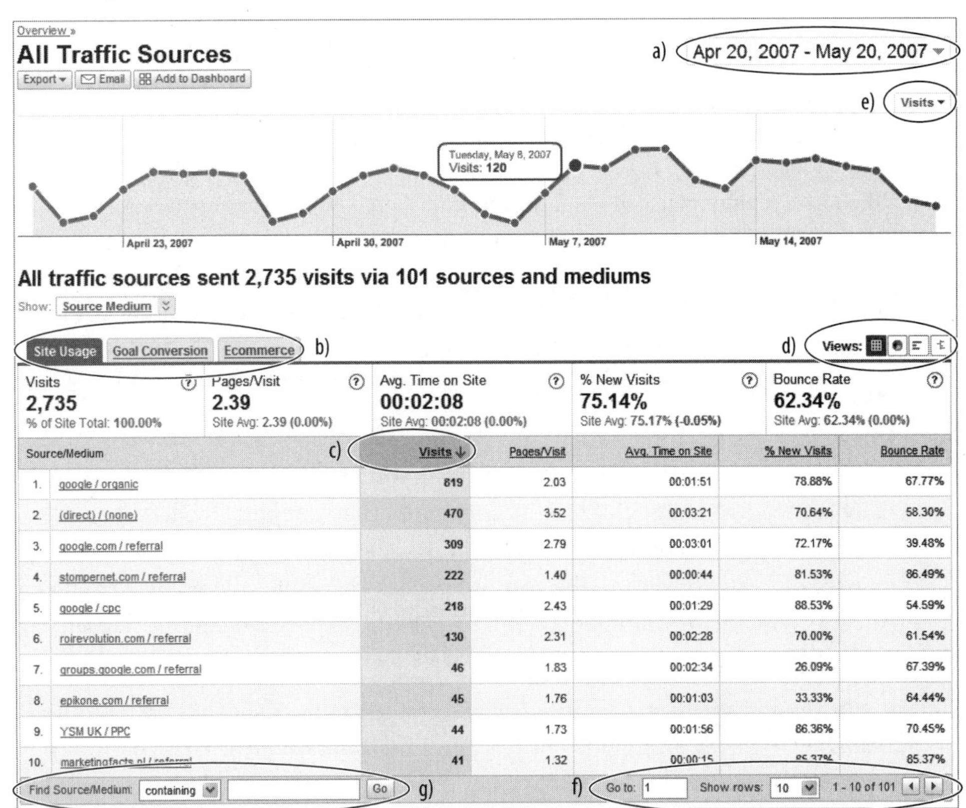

	Source/Medium	Visits ↓	Pages/Visit	Avg. Time on Site	% New Visits	Bounce Rate
1.	google / organic	819	2.03	00:01:51	78.88%	67.77%
2.	(direct) / (none)	470	3.52	00:03:21	70.64%	58.30%
3.	google.com / referral	309	2.79	00:03:01	72.17%	39.48%
4.	stompernet.com / referral	222	1.40	00:00:44	81.53%	86.49%
5.	google / cpc	218	2.43	00:01:29	88.53%	54.59%
6.	roirevolution.com / referral	130	2.31	00:02:28	70.00%	61.54%
7.	groups.google.com / referral	46	1.83	00:02:34	26.09%	67.39%
8.	epikone.com / referral	45	1.76	00:01:03	33.33%	64.44%
9.	YSM UK / PPC	44	1.73	00:01:56	86.36%	70.45%
10.	marketingfacts.nl / referral	41	1.32	00:00:15	[illegible]	85.37%

a) the default date range
b) the three tabbed views found in most reports
c) the default table sort order
d) the four different data representation views
e) the different values to chart
f) the inline filter for searching within a report
g) viewing different or more rows of data

Figure 4.2 A typical Google Analytics report

The following list describes each of the elements highlighted in Figure 4.2:

a) Changing the date range

By default, when you view reports, you view the last month of activity. That is, assuming today is day $x+1$ of the month, the default date range for reports is from day x of the previous month to day x of the current month. By default, the current day is not included, as this skews calculated averages.

b) Tabbed layout

A common feature of most Google Analytics reports is the tabbed layout. Visitor information is grouped into three areas: Site Usage, Goal Conversion, and Ecommerce. As you might guess, Site Usage reflects visitor activity relevant to that report. From here, if you wish to view how many of those visitors convert, click either the Goal Conversion or Ecommerce tabs.

Note: The Ecommerce tab does not show if you have not configured Google Analytics to collect e-commerce data.

Of course, ideally you will want all this data (Site Usage, Goal Conversion, Ecommerce) viewable in one long, continuous row. However, that would never fit into your browser so neatly; therefore, simply export your data into CSV format—or one of the other supported formats (PDF, XLM, or TSV)—and view this using MS Excel or another compatible application. The export contains the data from all the tabs. For more detailed information, see the section "Scheduled Export of Data."

c) Changing the sort order

For any particular report or visitor segment you may be viewing, you will initially see the Site Usage chart with concomittent table report. By default, tables are sorted by the second column entry in descending order; usually this is the number of *visits*. To reverse the sort order, click the Visits column header entry. Alternatively, sort on another column by clicking the desired column header.

d) Changing the data view

If you would rather see data in a pie chart than a table, the data view option available in mostly all reports enables you to select a different view to display your data: table (default), pie chart, bar chart, or delta. The delta view compares the displayed metric to the site average (or the second date range if selected—see Figure 4.5).

e) Changing what data is plotted

The trend graph on top of every report enables you to change which data element is charted over time. Each report section has its own plot alternatives. Examples to select from may include visitors, visits, pageviews, conversion rate, revenue, ROI, bounce rate, average time on site, and so on. Moving your cursor over the chart highlights particular data points showing the date and corresponding value.

f) Increasing the table rows displayed

By default, each report in Google Analytics (with the exception of the Dashboard and Site Overlay reports) shows a data over time chart, with the corresponding data tabulated below it. The initial 10 rows of the table are shown. To scroll through to other table rows, or to increase the number of rows displayed, select one of the options at the bottom right corner of the table.

g) Using inline filters

Often, high-traffic websites contain so much data that expanding the number of table rows or scrolling through them in batches is not an efficient way to find information—for example, finding error pages that may be relatively small in number but clearly significant to the user experience. The inline filter offers a quick way to do this. The filter applies to the data in the first table column and includes all data in that column, not just the data displayed.

For example, Figure 4.2 shows the referral Source Medium report. When the inline filter is blank, no filter is applied. To apply a filter, type a keyword in the filter text box (label f) and select either "containing" or "excluding" from the drop-down option. For example, you could try the keyword "referral" and select "containing." This would result in a table including only data rows in which the word "referral" is present in the source/medium data (see Figure 4.3.) Try different examples and combinations to become familiar with this.

Preview Google Analytics features

As with all web-based software applications, the best way to get to know its capabilities is to see it in action. With the Google Analytics report interface, you can do this quickly. You can see an initial preview of some of its capabilities at http://64.233.179.110/analytics/tour/index_en-US.html. (The walkthrough is in English, with other languages shown as subtitles.)

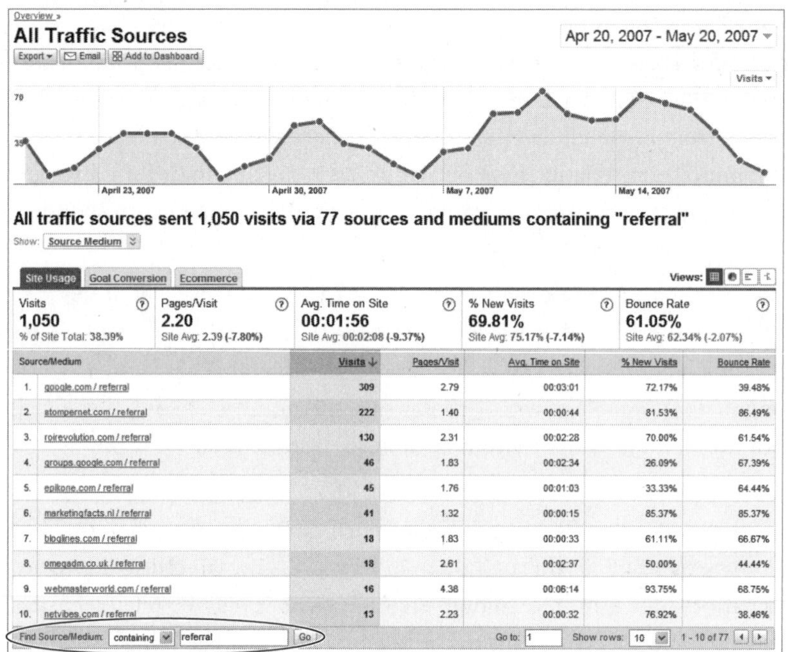

	Source/Medium	Visits	Pages/Visit	Avg. Time on Site	% New Visits	Bounce Rate
1.	google.com / referral	309	2.79	00:03:01	72.17%	39.48%
2.	stompernet.com / referral	222	1.40	00:00:44	81.53%	86.49%
3.	roirevolution.com / referral	130	2.31	00:02:28	70.00%	61.54%
4.	groups.google.com / referral	46	1.83	00:02:34	26.09%	67.39%
5.	epikone.com / referral	45	1.76	00:01:03	33.33%	64.44%
6.	marketingfacts.nl / referral	41	1.32	00:00:15	85.37%	85.37%
7.	bloglines.com / referral	18	1.83	00:00:33	61.11%	66.67%
8.	omegadm.co.uk / referral	18	2.61	00:02:37	50.00%	44.44%
9.	webmasterworld.com / referral	16	4.38	00:06:14	93.75%	68.75%
10.	netvibes.com / referral	13	2.23	00:00:32	76.92%	38.46%

Figure 4.3 Using the inline filter to show only referral traffic

The inline filter also works with partial matches and regular expressions.

Regular Expression Overview

For partial matches, say you only wanted to view referrals from the website www.roirevolution.com (row 3). Using inline filtering as shown in Figure 4.3, you could enter the partial keyword **roi**. This will match all entries that have the letters *roi* in them.

For regular expressions, you could uniquely view the entry for stompernet.com (refer to Figure 4.3) by using **stomp.+t**. In English, that expression is equivalent to "find entries with the letters *stomp* (in order), followed by one of more of any character, followed by the letter *t*."

Perl regular expressions are used to match or capture portions of a field using characters, numbers, wildcards, and meta-characters. They are often used for text manipulation tasks. A list of common wildcards and meta-characters is shown below.

Regular expressions are best understood by example. To show matches, the example test string used is "the quick brown fox uses his brain to build bridges to allow him to jump over all the lazy dogs. Although some dogs are also smart."

Regular Expression Overview *(Continued)*

Wildcards

The following list explains the most common wildcards:

. match any single character

example: br..n matches brown brain, brain

* match zero or more of the previous item

example: br* matches brown, brain, build, bridges

+ match one or more of the previous item

example: br+ matches brown, brain, bridges but not build

? match zero or one of the previous item

example: al? matches Although, also but not all, allow

Meta-characters

The following list explains how to use meta-characters:

() remember contents of parenthesis as item

Used when combining with a second regular expression

[] match one item in this list

example: [aeiou]+ can be used to find words that contain a vowel

- create a range in a list

example: [a-z]+ can be used to find letters, [0-9]+ can be used to find numbers

| or

example: (al|all)+ matches Although, also, all, allow

^ anchor to the beginning of the field

example: ^the matches only once at the beginning of the test string

$ anchor to the end of the field

example: the$ has no matches, as the test string does end with "the"

\ escape any of the above meta-characters

example: \.$ matches the full stop at the end of the test string only

Continues

Regular Expression Overview *(Continued)*

Tips for Regular Expressions

1. Make the regular expression as simple as possible. Complex expressions take longer to process or match than simple expressions.
2. Avoid the use of .* if possible because this expression matches everything zero or more times and may slow processing of the expression. For instance, if you need to match all of the following:

   ```
   index.html, index.htm, index.php, index.aspx, index.py, index.cgi
   ```
   use
   ```
   index\.(h|p|a|c)+.+
   ```
   not
   ```
   index.*
   ```
3. Try to group patterns together when possible. For instance, if you wish to match a file suffix of .gif, .jpg, and .png, use
   ```
   \.(gif|jpg|png)
   ```
   not
   ```
   \.gif|\.jpg|\.png
   ```
4. Be sure to escape the regular expression wildcards or meta-characters if you wish to match those literal characters. Common ones are periods in filenames and parentheses in text.
5. Use anchors whenever possible (^ and $, which match either the beginning or end of an expression), as these speed up processing.

Selecting and Comparing Date Ranges

When looking at your Google Analytics reports, one of the first things you will probably wish to change is the time period to view. By default, when you log in you will see the last month's worth of web visitor activity. Perhaps, however, you only want to focus on the current day's activities. In that case, click the Date Range drop-down list (see Figure 4.4a) and select the current day. You can also enter the date manually by using the fill-in fields provided. In this respect, the date range selector works like any other calendar tool.

- To select an entire calendar month, click the month name.
- To select an entire week (Monday–Sunday), click the rounded ends of a particular week.

Note that the default "Comparison" value is set to "Site." This means all report metrics shown will be compared to your overall site averages. For example, if you viewed the report of visits referred by search engines, the average time on site for these visits will be compared to the average time on site for all visits.

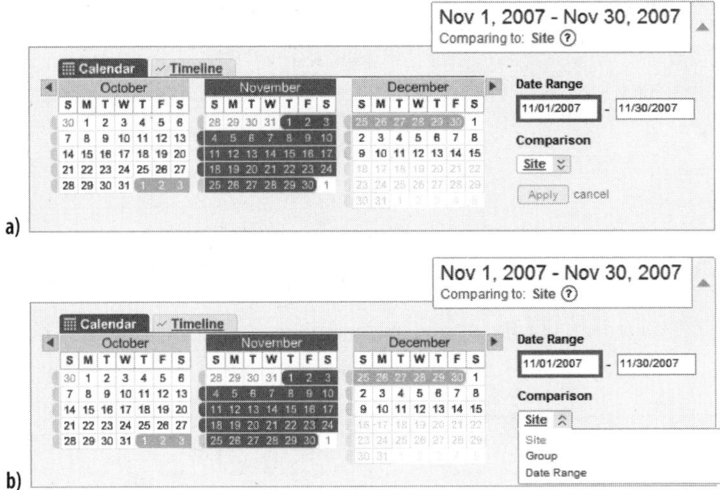

Figure 4.4
Selecting a date range

To compare the current date range data with any other date range, change the comparison drop down menu to "Date Range," as shown in Figure 4.4b. By default, Google Analytics will select a date range to compare. For example, if your first date range is the current day, the previous day will be automatically selected as the comparison. If your first date range is the last 30 days of data, the previous 30 days will be selected by default, and so forth. You can overwrite the second date range as required.

Another comparison option is "Group." Choosing this enables you to compare your selected data with the non-selected data. For example, if you view the report of visits referred by search engines, the average time on site for these visits will be compared to the average time on site for visits that were not referred by a search engine.

All comparison data is shown within the same browser window. Positive data changes—that is, an increase over the previous period—are shown in green, whereas negative changes are shown in red, as shown in Figure 4.5. The exception to this is when viewing bounce rates. In this case, a decrease in bounce rate would be green and an increase would be red, to reflect that a decrease in bounce rate is desirable.

Note: Care should be taken when viewing chart data for different date ranges. By default, Google Analytics will select a suitable second date range for you—previous 30 days, for example. However, this might not always align with the first date range—for example, Mondays may not align with Mondays. When comparing date ranges, always attempt to align days of the week. For example, compare Monday–Friday of this week with Monday–Friday of the previous week.

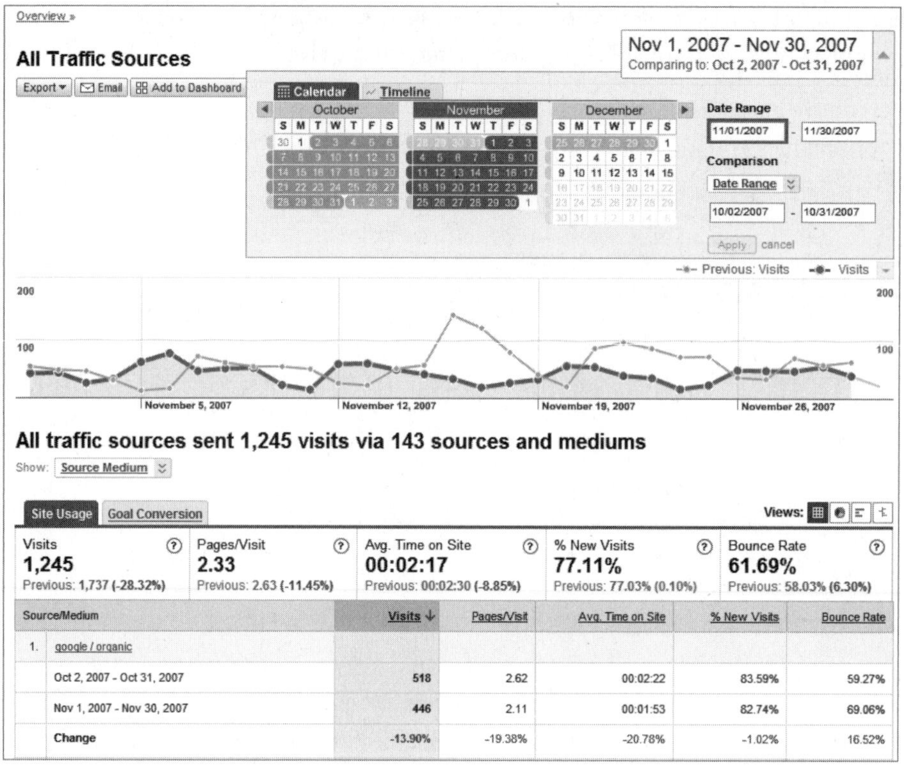

Figure 4.5 Comparing two date ranges

An alternative way to select your date range is to use the *timeline sliders*, as shown in Figure 4.6. The Timeline view enables you to make informed decisions regarding what date range to select because you can see the visitor totals before selecting it. In theory, if you can see that large peak in mid-August, for example, then you are much more likely to select it for comparison. Without that information, you may select a different range and miss a key event on your website. The timeline slider bars enable you to make this comparison—you drag the data window to the area you wish to investigate and expand or contract the window boundaries as desired.

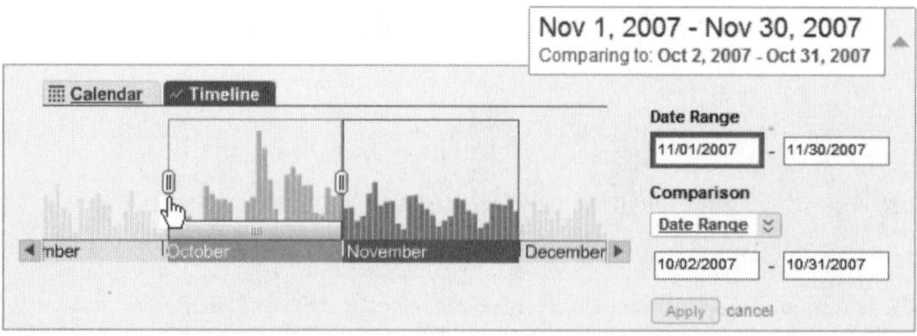

Figure 4.6 Timeline selection

Hourly Reporting

The Visitors > Visitor Trending section and the Ecommerce > Total Revenue section of the reports have an additional feature: data can be viewed over an hourly time frame. This report enables you to track at what times of the day visitor traffic arrives on your site, midnight to midnight (see Figure 4.7). Knowing what times of the day are most productive for you provides powerful insight for scheduling campaigns or downtime—for example, the starting and stopping of ads, changing your keyword buys, viral marketing events, and the best time to perform web server maintenance.

Of course, care should be taken when interpreting this report if you are receiving significant visitors from different time zones—for example, U.S. versus European time zones. If this is your situation, consider segmenting your visitors into separate profiles using a geographical filter before interpreting these reports. See Chapter 8 for more information.

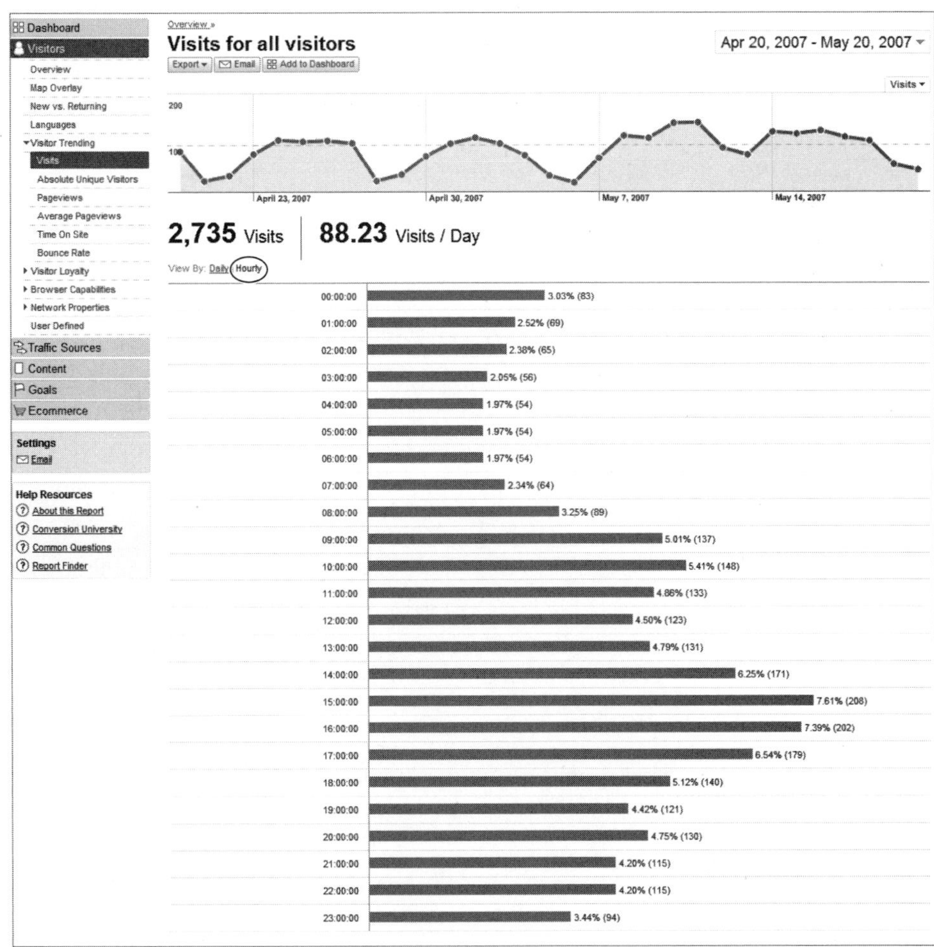

Figure 4.7 Hourly reporting of visitors

Scheduled Export of Data

Data export is available in four industry standard formats: PDF, XML, CSV, and TSV. Select Export from the top of each report to have your data exported in PDF (for printable reports), CSV or TSV (to import into Excel or another spreadsheet application), or XML (for importing into third-party applications). See Figure 4.8.

Manually exporting data is great for manipulating data further or creating one-off reports to present to your team. Once you have chosen which reports are important to your stakeholders, you will probably wish to have these sent to them via e-mail—either ad hoc or scheduled on a regular basis. To do this, chose the Email link next to the Export link (refer to Figure 4.8). Reports can be scheduled to be sent daily, weekly, monthly, or quarterly, as per Figure 4.9.

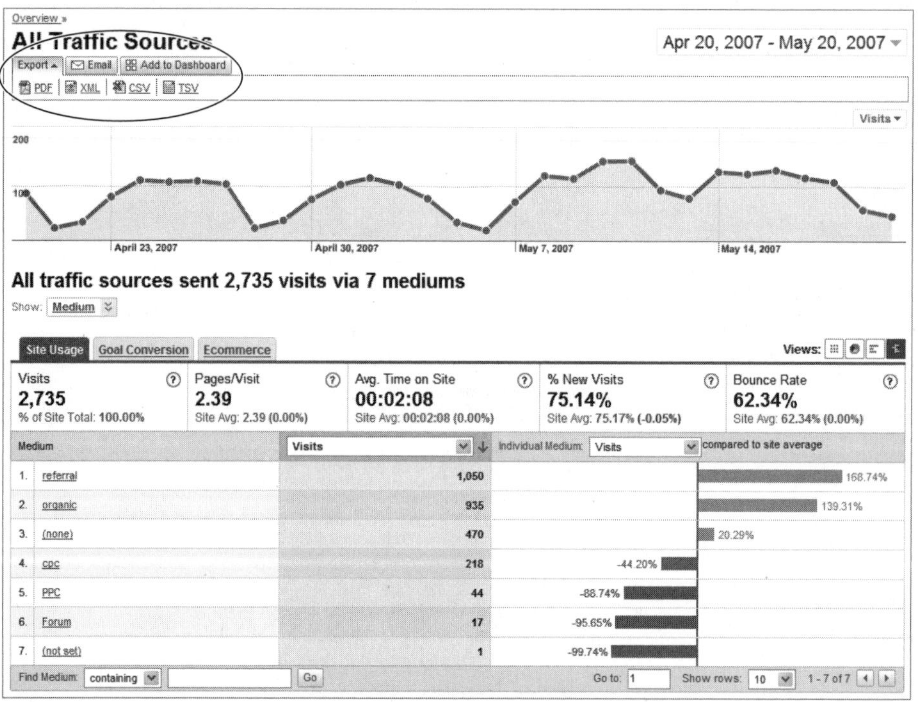

Figure 4.8 Data export options

Note: All times are local to Mountain View, California (Google headquarters). Although the exact time is not specified, a daily report sent in the morning will actually be sometime in the afternoon for European customers.

Figure 4.9 Scheduling a report for e-mail export

If you wish to group a set of reports into an existing e-mail schedule, use the Add to Existing link, as shown in Figure 4.10.

Set Up Email: Goals Overview
Back to report
Send Now
Schedule
Add to Existing
GA Experts eComm & Map overlay report (sent quarterly)
Reports: Ecommerce Overview Map Overlay
Recipients: bclifton@google.com
Attachment: pdf
Daily Visitor Overview report (sent quarterly)
Reports: Visitors Overview
Recipients: theboss@mysite.com , brian@omegadm.co.uk
Attachment: pdf
GA-Experts Errors (sent weekly)
Reports: Top Content
Recipients: david@omegadm.co.uk , brian@omegadm.co.uk
Attachment: pdf
Add Report

Figure 4.10 Adding a report to an existing e-mail schedule

Note: E-mail schedule settings are saved on a per user or profile combination. Therefore, two different users for the same profile can set their own e-mail schedules.

Cross-Segmentation

Cross-segmentation, also known as *cross-referencing*, is one of the key factors that define a web analytics application as enterprise class. Google Analytics has a host of cross-segmenting options available in most of its reports. Cross-segmentation is analogous to pivot tables in Microsoft Excel. It enables you to compare one set of data against another. Cross-segmentation takes place whenever you select an item from the Segment drop-down menu within a report.

Figure 4.11 illustrates the following example: Show me only U.K. visitors who used an organic (non-paid) search engine to reach my website—that is, cross-referencing U.K. visitors against a referral source.

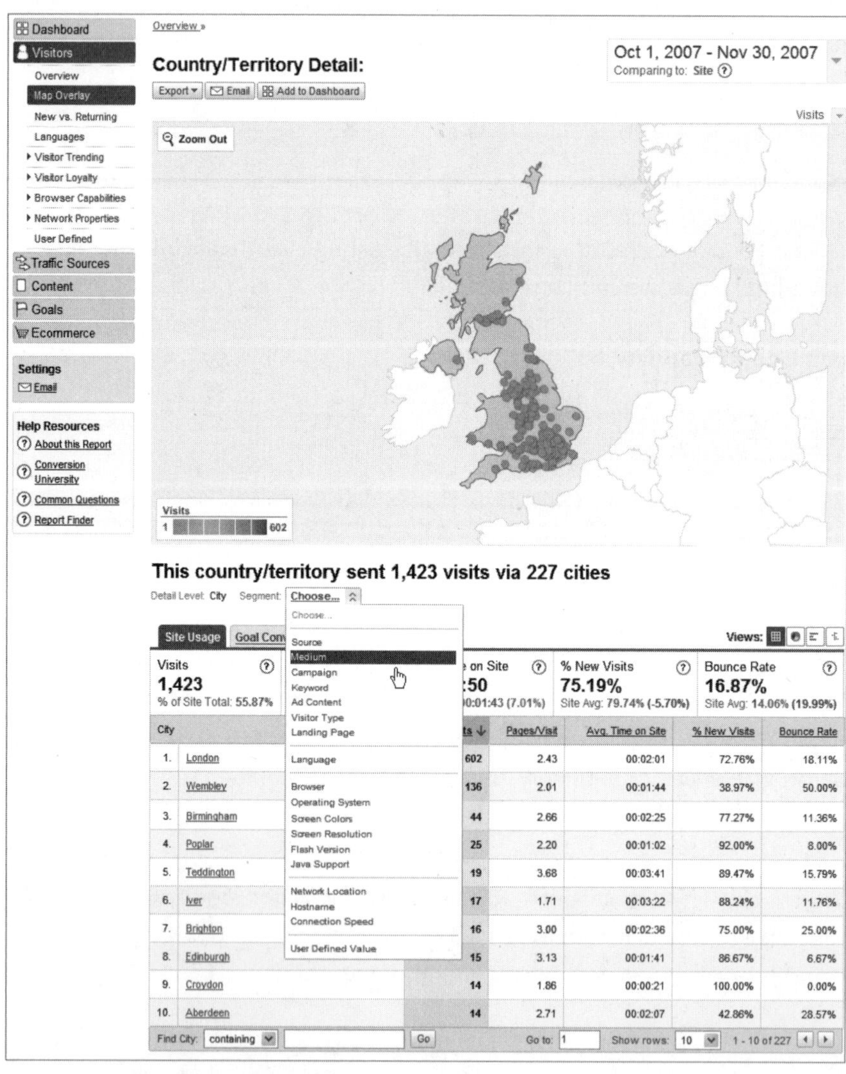

Figure 4.11 Example of cross-segmentation

Cross-segmenting your data is a powerful way for you to understand your visitor personas—both geographics and demographics—and is discussed in greater detail in Chapter 8.

Summary

In this chapter we have reviewed the Google Analytics interface, particularly in relation to discovering information. By understanding the report layout, you will quickly become accustomed to drilling down into the data, investigating whether a number or trend is good, bad, or indifferent for your organization. Sparklines, cross-segmentation, inline filters, regular expressions, changing graphing and data views, re-sorting table data, data export formats, and e-mail scheduling should all now be familiar concepts and terminology to you.

In Chapter 4, you have learned about the following:

- The different ways you can view data with chart options and data views
- The different ways you can select and compare date ranges and how to make use of the timeline feature to select periods of interest—such as data peaks or troughs
- The role of inline filters and the use of regular expressions to refine displayed data to a specific page or group of pages
- How to drill down and focus on particular visitor segments using the cross-segmentation menus
- How to schedule the e-mailing of reports in different file formats

Top 10 Reports Explained

At my last count, Google Analytics had over 80 default reports—and when you take into consideration cross-segmentation options, the number grows exponentially. Clearly, no one person is going to look at all those reports on a regular basis—nor should you try to. Being overwhelmed with data is not my idea of fun. My approach is to first understand the key areas of your website and what is happening from a visitor's point of view.

In this chapter, I focus on ten important first-level reports that can give you that initial understanding. Of course, my report selection may not reflect the information most important to you—every website is different in some way. Once you initially understand the drivers or blocking points for your visitors, you can focus on more detail and build your own list of top reports.

In this chapter, you will learn about the following:

The dashboard overview
The top 10 reports
Content reports

The Dashboard Overview

Before delving into specific reports, I want to discuss the dashboard view—as this is not really a report in itself. The Google Analytics dashboard is the first screen displayed when you log in to view your reports (refer to Figure 4.1 in Chapter 4). This is the overview or summary area where you can place a chart or table copied from the main body of the Google Analytics reports. From here, if you notice a significant change, you can click through to go to the detailed report section.

You can also change the selection of reports on your dashboard at any time, with a maximum of 12. To add to the dashboard, navigate to a report and click the Add to Dashboard link at the top of the page, as highlighted in Figure 4.8. When viewing the dashboard, you can move the report's placement by dragging and dropping it into the desired position.

Try the following exercise as an example. Suppose a key market for you is California, and at the current time being able to log in to Google Analytics and immediately view the data from California visitors is a key requirement. From the Visitors menu, select Map Overlay. From the displayed map, drill into the area of the map as required (see Figure 5.1), and then click Add to Dashboard.

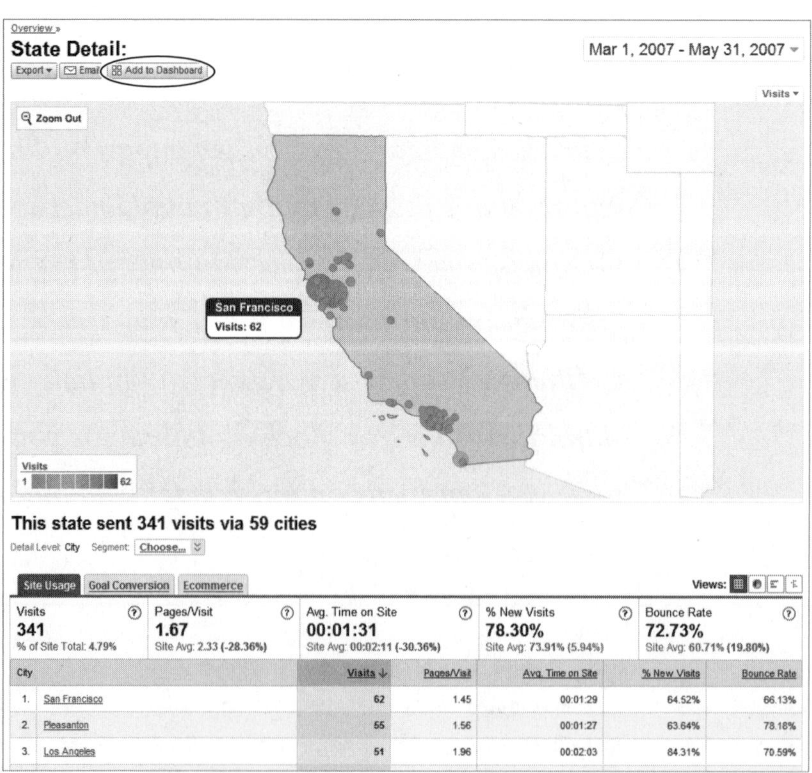

Figure 5.1 Visitor Map Overlay for California

Now click the Dashboard link at the top of the side menu. Your map overlay of California will be displayed as the last item on the report page. Drag and drop the map overlay into the top position (or any desired position). From now on, each time you log in to Google Analytics and view your reports, the first item displayed in your dashboard will be the map overlay of visitors from California.

Once you have your key reports set on your dashboard, consider scheduling an e-mail export of this to senior management. Click the Email button at the top of the dashboard report and set it accordingly. I recommend this be scheduled weekly at most, although monthly may be the optimal frequency in order for you to maintain interest—a key factor when disseminating information to people not directly involved with the performance of your website. You'll learn more about this in Chapter 10.

The Top 10 Reports

This section is not intended as a definitive list of the only reports you should look at. Rather, these are suggestions to take you beyond the initial visitor volume numbers that you will first see. Reviewing these reports for your organization will give you an understanding of visitor behavior before mapping your organization's stakeholders and determining what key performance indicators to use for benchmarking your website.

Visitors: Map Overlay

As shown in Figure 5.1, Map Overlay shows you where your visitors come from, enabling you to identify your most lucrative geographic markets. You can zoom in from world view to continent, regional, and country view, and along the way examine visitor statistics from that part of the world—right down to city level. Below the displayed map is the tabulated data for the selected region. For each location, you can cross-segment your visitors against other metrics, such as referral source, medium, language, and so on, as shown in Figure 5.2. For example, once you have found your location of interest, cross-segment to view which search engines are popular with your visitors there.

The displayed maps in this report are color coded by density—the darker the color, the higher the reported metric, such as more visits or revenue. A density key is shown in the bottom left corner and you can mouse over the regions, countries, or cities to view top-level metrics.

In addition to showing you which parts of the world your visitors are coming from and measuring whether they are relevant to your business—for example, are they converting—geographic information is extremely powerful for targeting your online marketing activities. For online marketing, Google AdWords (and other pay-per-click networks) enable you to geo-target your advertisements (see Figure 5.3). The Map Overlay report of Google Analytics can be used in two ways: It enables you to identify new locations for potential online campaigns, and it enables you to measure the effectiveness of existing geo-targeted campaigns.

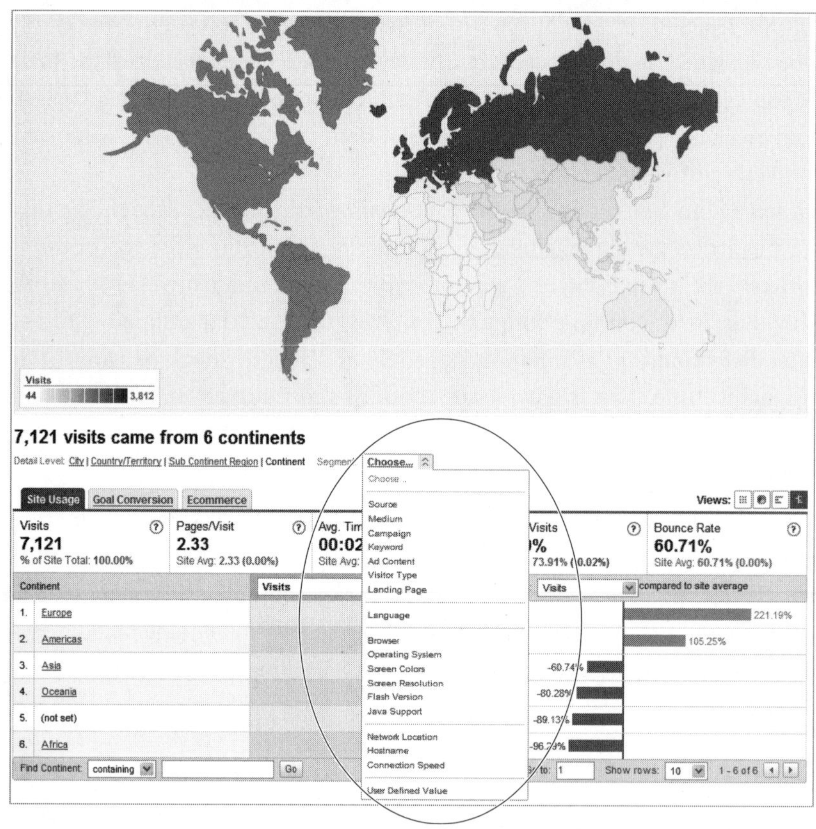

Figure 5.2 Cross-segmenting your geographic visitors

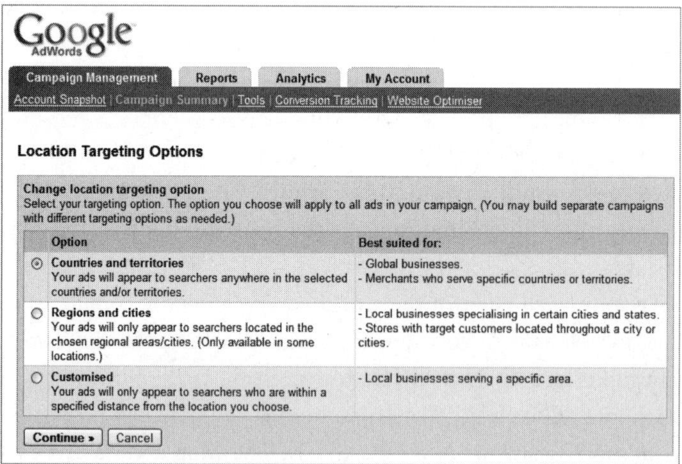

Figure 5.3 Geo-targeting options in AdWords

Visually stunning, the Map Overlay report is also an extremely powerful report—it gets across the information you need to know at a glance. Consider the two charts shown in Figure 5.4 for the same profile and date range. Figure 5.4a shows the visitor information, whereas Figure 5.4b shows the conversion data from the same visitors. As you can see, Europe gets by far the most visitors, yet visitors from Asia and the Americas provide proportionally much higher conversions.

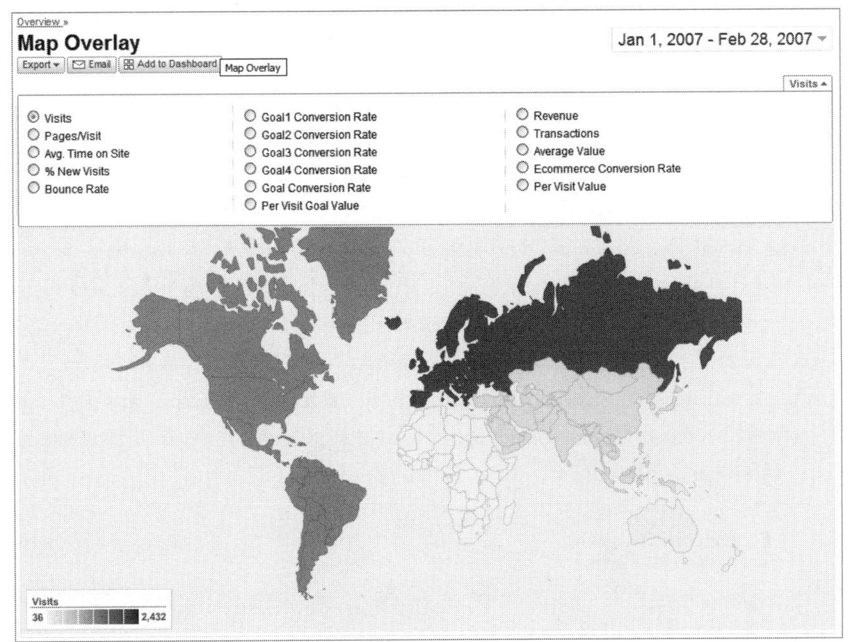

a)

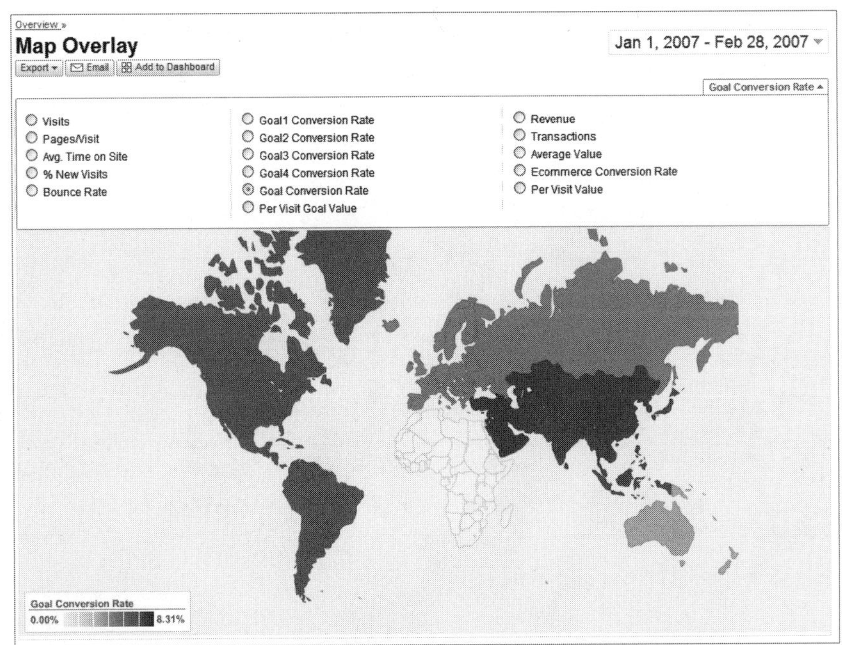

b)

Figure 5.4 a) Geographic density of visits; b) Geographic density of conversions for the same data set

Ecommerce: Overview Report

Even if you do not have an e-commerce facility on your website, you can still monetize your website by adding goal values. Either way, the e-commerce reports of Google Analytics enable you to identify revenue sources and trace transactions back to specific campaigns—right down to the keyword level. Individual product data can be viewed and grouped (shown as categories), as can loyalty and latency metrics.

Note: Monetizing a non-ecommerce website is discussed in detail in Chapter 11.

From the initial Ecommerce Overview report (see Figure 5.5), a wealth of information is provided for you to feast on. From here, any click-through takes you to a more detailed report. For example, click one of the top-performing products to view its individual report, and then cross-segment against other fields, such as referral source, campaign name, keywords, and so on. These are the details that are driving visitor transactions. Such information is critical for a successful product-by-product search engine marketing initiative.

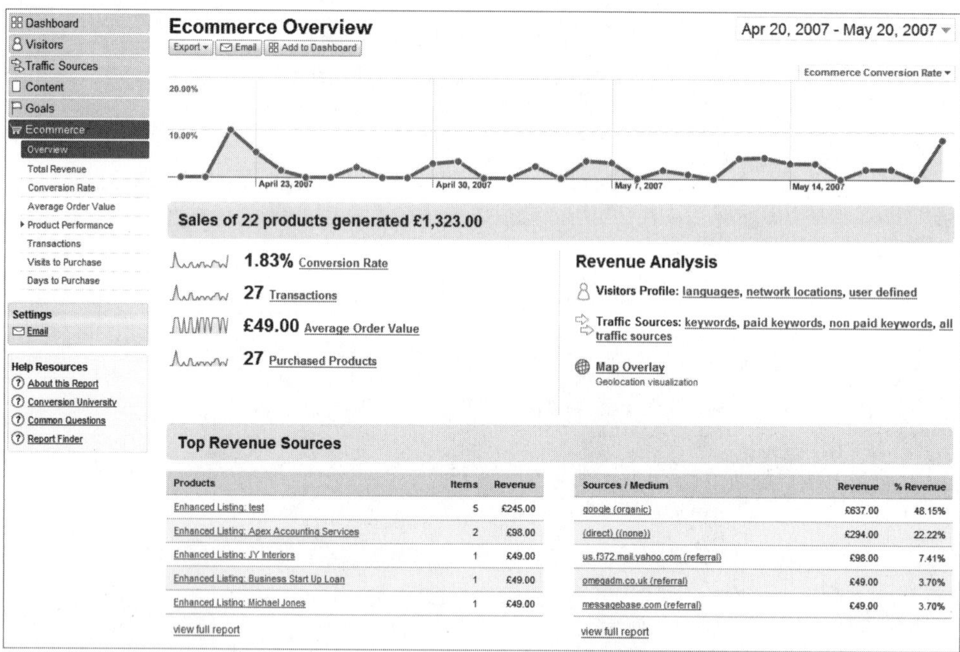

Figure 5.5 A typical e-commerce report

Goals: Overview Report

As discussed throughout this book, goal reporting (conversions) is an important measurement for your organization. Regardless of whether you have an online retail facility or not, measuring goal conversions is the de facto way to ascertain whether your website is engaging to your visitors.

In addition to measuring your goal conversion rate, you can also monetize these rates by applying a goal value. Figure 5.6 shows the Goals Overview report with monetized goal values. Clicking any of the links within this report enables you to view each goal report in further detail.

Within the Goals reporting section, the Goal Verification report enables you to view the specific URLs that trigger the reporting of a goal. This is particularly useful when a wildcard is used to define the goal—for example, *.pdf. In this case, the Goal Verification report will list all the PDF downloads that trigger the reporting of that defined conversion.

Also within this section, the Reverse Goal Path report considers the last three steps (pages) visitors took before reaching a goal. This is an excellent place to look for visitor paths that could be considered for funnel analysis. Funnel analysis is discussed next.

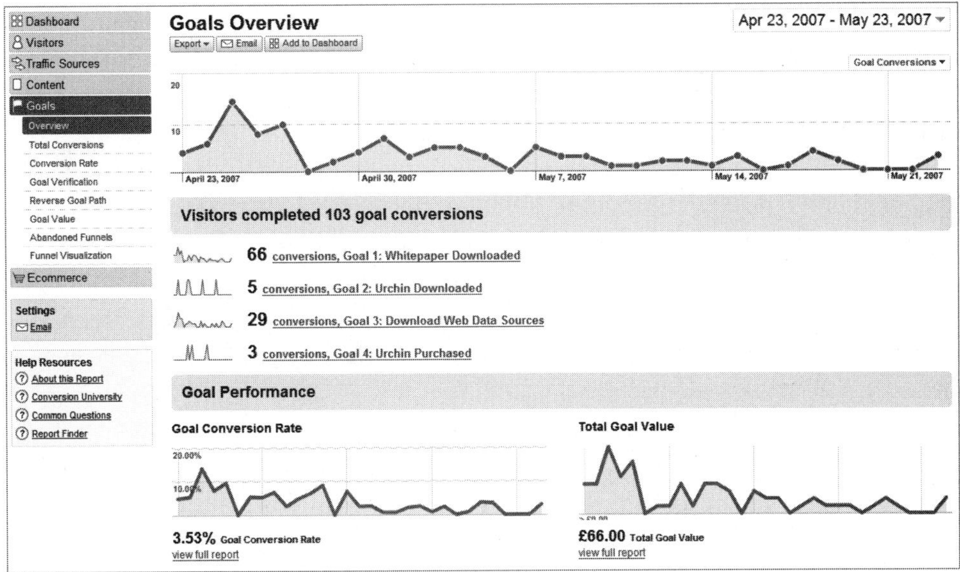

Figure 5.6 Goals Overview report

What is a conversion?

It is important to clarify that a goal is synonymous with conversion in this context. Say, for example, one of your website goals is *.pdf—that is, the download of any PDF file. A visitor arrives on your website and downloads five PDF files. Google Analytics will count this as one goal conversion (not five, as you may expect). The rationale for this is that visitors can only convert once during their session, which makes sense.

To view the total number of PDF downloads and which files they were, you can either view the Goals > Goal Verification report or, to cross-segment the data, go to the Content > Top Content report and use the inline filter to display only .pdf files, as shown in the following figure:

Goals: Funnel Visualization Report

Funnel analysis (sometimes referred to as *path analysis*) is a subsection of the Goals reports. Some goals have clearly defined paths that a visitor takes to reach the goal. An obvious example is an e-commerce checkout process; others include newsletter sign-ups, registration subscriptions, reservation systems, and brochure requests. Not all goals have a defined path; but if yours do, then it is useful to visualize how your visitors traverse them (or not) to reach the goal. The Funnel Visualization report does just that, and an example is shown in Figure 5.7.

The pages of a funnel a visitor is expected to pass through (as defined by your configuration) to reach the goal is the central section highlighted in Figure 5.7—in this example, to download a whitepaper. The tables to the left of the central funnel are entrance pages into the funnel. A well-defined funnel should have the vast majority of visitors passing downwards, not in from the side, into a minimum number of funnel steps. The tables to the right are exit pages out of the funnel steps—that is, where visitors go to when they leave the funnel page. The exit pages listed can be to other pages within your website, or the visitor leaving the site completely.

Funnel visualization enables you to assess how good your funnel pages are at persuasion—that is, how good are they at getting visitors to proceed to the next step, getting closer to approaching conversion. A funnel with pages optimized for persuasion and conversions should have a minimal number of exit points (pages to the right of the funnel), thereby leading to a high conversion rate. A detailed funnel analysis is considered in the section "Identifying Poor Performing Pages" in Chapter 11.

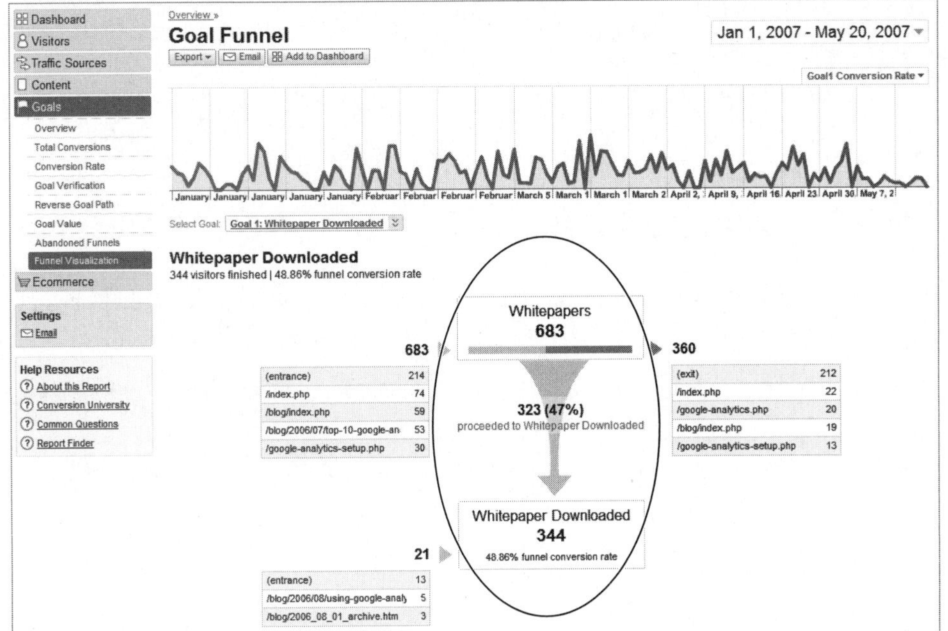

Figure 5.7 Funnel Visualization report for a two-step funnel

Traffic Sources: AdWords Reports

As you might expect from a product by Google, Google Analytics integrates tightly with Google AdWords—and undoubtedly in the future there will be further integration with other Google products. Within the Traffic Sources report is a dedicated subsection for AdWords data. Figure 5.8 shows the two AdWords reports available—AdWords Campaigns and Keyword Positions. These two reports are populated by data imported directly from your AdWords account, assuming you have one and have configured it to be imported into your Google Analytics account—more on this in Chapter 6.

The power of combining your AdWords account data with Google Analytics is illustrated in Figure 5.8—that is, when you wish to drill down into the data. For example, clicking the campaign name takes you to the Ad Group level of data with the same column headings. Clicking an ad group provides further detail, showing the actual keywords used by AdWords visitors to find your website—as shown in Figure 5.9.

One item you may have noticed in the detail of Figure 5.9 is the keyword (content targeting). This is the term used to describe visitors from the content network of AdWords. The Google content network comprises websites, news pages, and blogs that partner with Google to display targeted AdWords ads. The partner uses AdSense to manage this. At this time it is not possible to view the actual keyword matching that AdSense has performed.

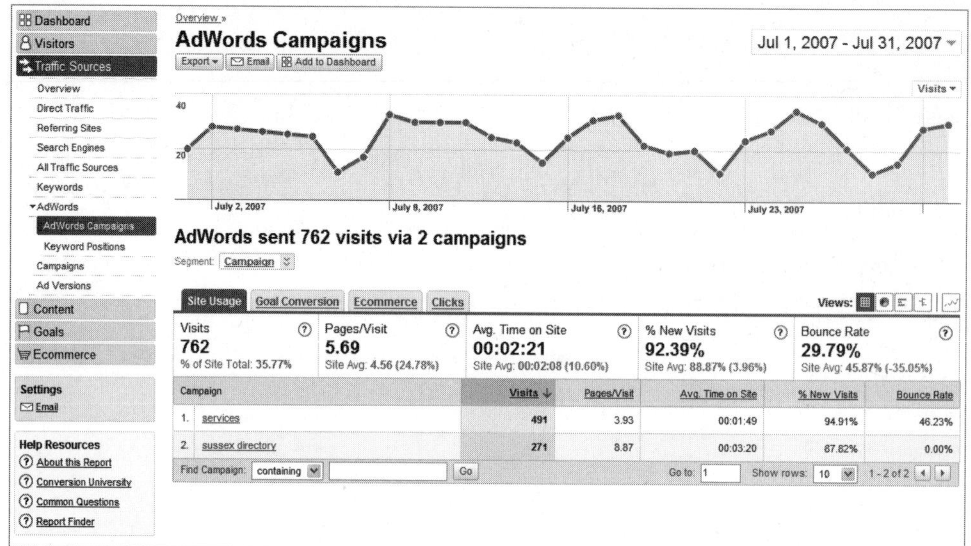

Figure 5.8 AdWords Campaign report showing the visitor summary

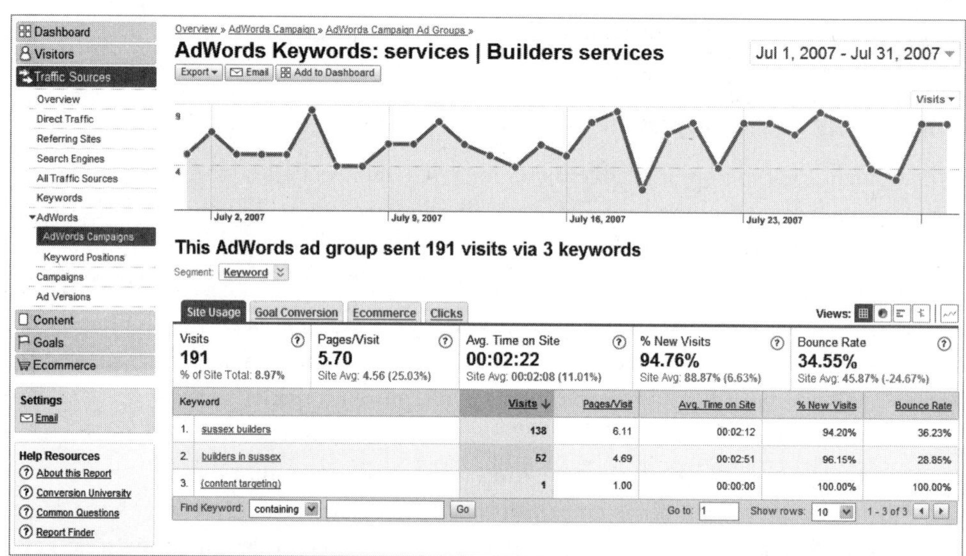

Figure 5.9 AdWords Keywords report detail obtained by drilling down through the campaign links of Figure 5.8

Note: Further information on AdSense and the content network can be found at www.google.com/adsense and http://adwords.google.com/select/afc.html

From the AdWords Campaigns report you can click the tabbed view to see how your campaigns, ad groups, and keywords convert (Goal Conversion tab) or purchase (Ecommerce tab). The last tab on this row is unique to this report (Clicks) and its contents are shown in Figure 5.10. The data in the Clicks report is imported directly from your AdWords account with the exception of the last three columns, which are calculated from your website revenue—both monetized goals and e-commerce revenue. Apart from the cost data, you should keep a close eye on your ROI and margin data. Chapter 11 looks at interpreting this data in more detail.

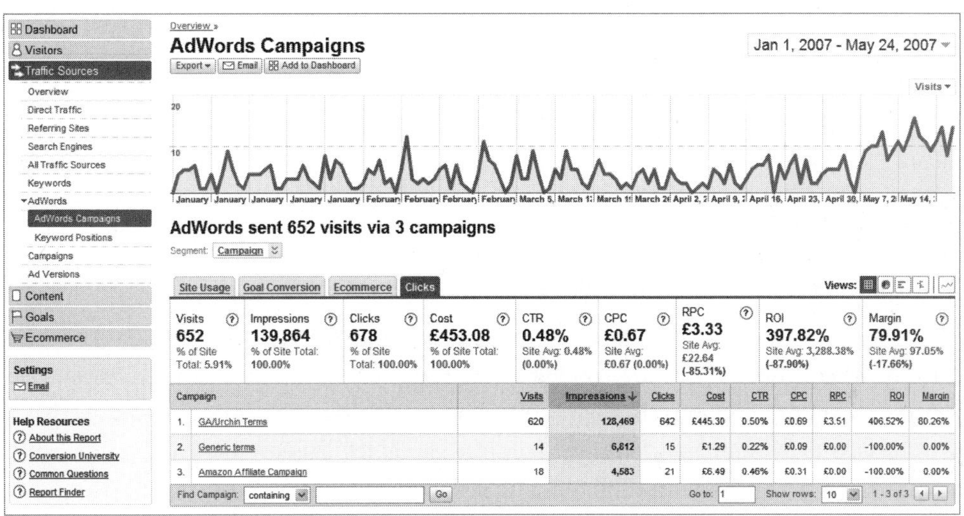

Figure 5.10 AdWords Campaigns report—showing the click detail

Another powerful feature of this report is the capability to cross-segment your AdWords data. Although you can do this in many other reports, I illustrate it here with the example shown in Figure 5.11. This shows the AdWords e-commerce report, at Ad Group level, cross-segmented by landing page URL—in English this means: for your purchased keywords, which landing pages did visitors use?

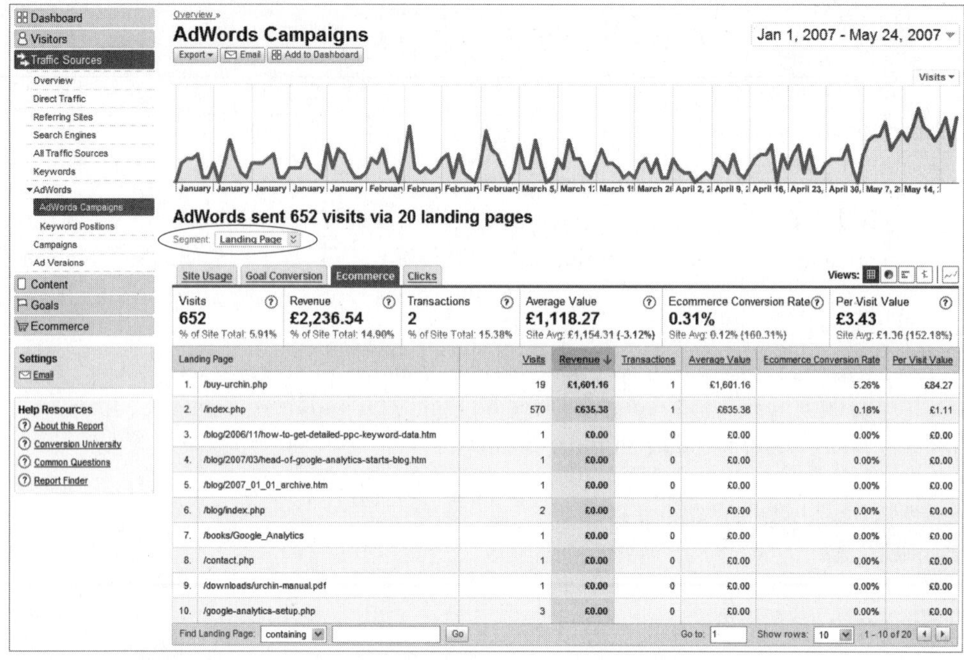

Figure 5.11 AdWords cross-segmentation by landing page URL

Traffic Sources: Source and Medium Report

Source Medium sounds like a delicacy from your local cafe! In fact, it is a powerful indicator of where your visitors are coming from. The source denotes the referral site—that is, the domain of another website that links to you and that a visitor clicked to arrive at your website. Common referral sources include search engines (paid and non-paid), a link from a partner organization, affiliate websites, blog articles, e-mail click-throughs, or forum posts—in fact, the referral can literally be from anywhere on the Internet. For visitors who type your web address directly into their browser (or use their browser's bookmarks or favorites folder), the label `direct` is listed as the source.

Medium refers to the online channel used by the visitor. The following values are medium labels:

- **organic** Label applied to visits from non-paid search engines
- **cpc** Label applied to visits from Google AdWords (cost per click)
- **referral** Label applied to visits from a link on another website
- **(none)** Medium label for direct visitors—those who type in your web address directly or use their browser's bookmarks or favorites folder

For both source and medium labels, it is possible to set your own values—as discussed in the section "Online Campaign Tracking" in Chapter 7. For example, within an e-mail message to potential customers, if you tag a link that points back to your website, you will see how many visitors arrive as a result of that e-mail link, including their paths and conversions.

For the example shown in Figure 5.12 (note that "Show: Medium" has been selected for the segment), in addition to the standard medium labels of organic, cpc, referral, and (none), the mediums PPC, Forum, Email, ppc, and Web are also shown. These non-standard values come from applying tagged landing page URLs—as described in Chapter 7. For this particular example, whenever the website owner left a link—be it on another site or within e-mails—it was done so with these medium labels appended to the link.

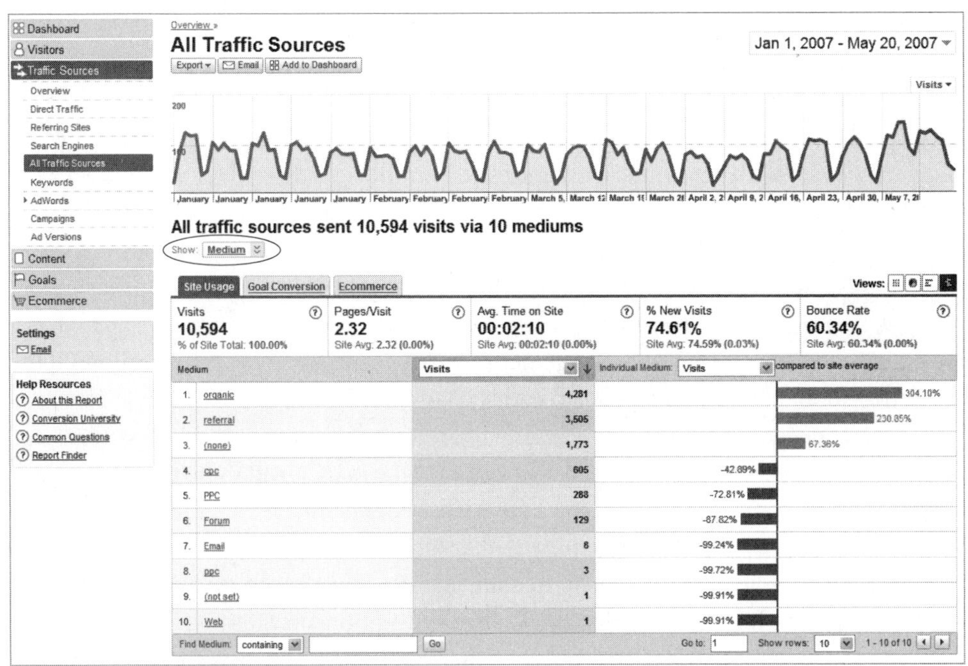

Figure 5.12 Referral medium report

You can further drill down into a specific medium by clicking its link to reveal the source detail, as shown in Figure 5.13.

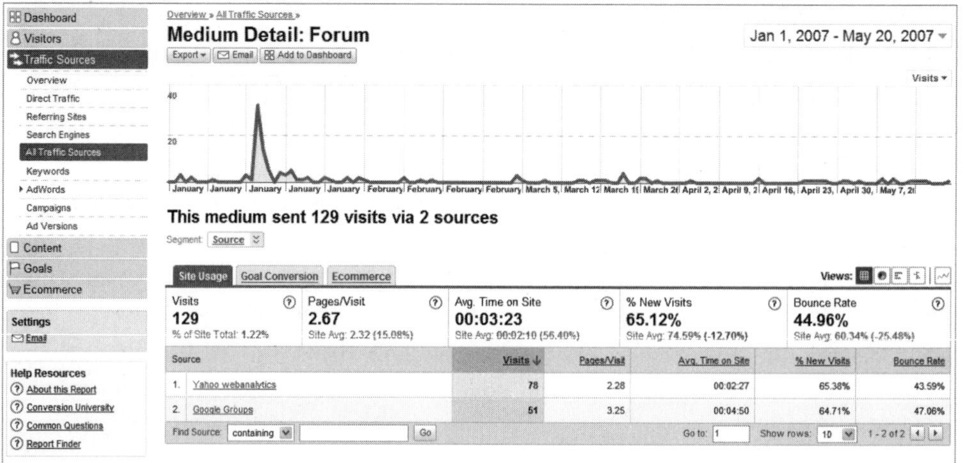

Figure 5.13 Referral medium detail report showing the sources

Content: Top Content Report

Knowing which pages are popular on your site is an obvious first place to look when assessing your website's performance. In addition to common per-page metrics such as pageviews, time on site, bounce rate (single-page visits) and percentage of visits that leave on this page (% Exit), an additional column is labeled $Index. This is a measure of the value of a page, and it is computed from goal and e-commerce values. The higher the $Index value, the higher the importance of that page in generating conversions. The calculation of $Index is discussed later in this chapter.

The Top Content report is much more than just a hit counter for successful pageviews. It can provide valuable insight into visitor behavior. Consider the report shown in Figure 5.14. Notice in this example I have used the inline filter to exclude blog visitors. Why? Because it was suspected for this site that blog visitors would exhibit very different behavior from those visitors likely to complete the goal conversions defined. That is, visitors not viewing the blog area spend slightly less time on the site (–20.45% compared to the site average for all visitors), are much less likely to bounce away from the site after only one pageview (–18.63%), are much less likely to exit (–22.58%), but, incredibly, are much more likely to convert as shown by the high $Index (+782.98%).

Note: Because the differences in behavior between an average blog reader and an average non-blog reader are so great, it would make sense to segment all reports by this criterion. Segmenting visitors into different profiles is discussed in Chapter 8.

You can drill down and investigate page properties in greater detail by clicking the page links. This enables you to perform navigational analysis and cross-segmentation against other metrics. For example, Figure 5.15 shows the navigational analysis of the page /index.php (the website home page). This shows how visitors arrived on that page and where they went to afterwards.

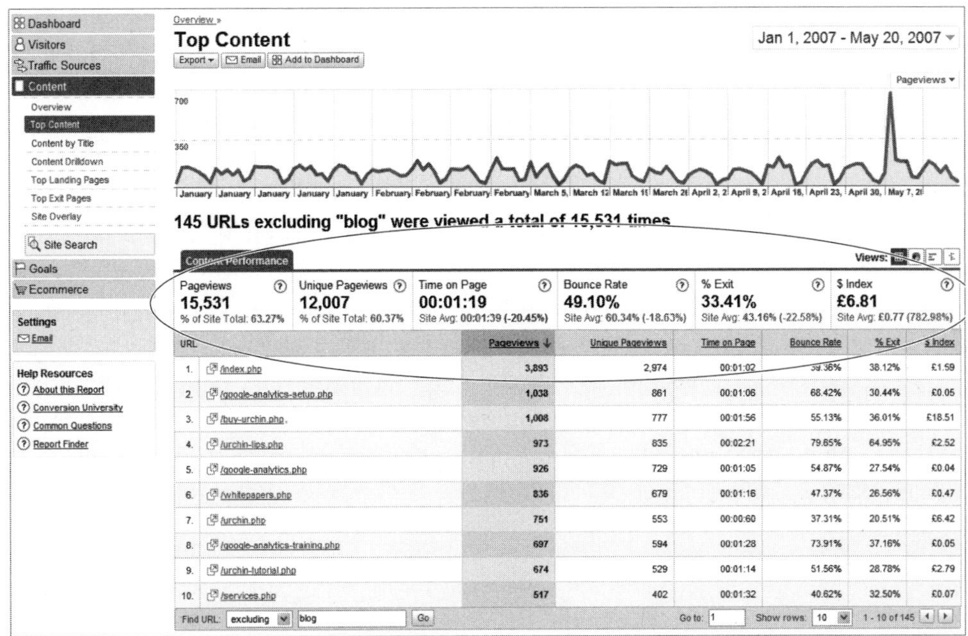

Figure 5.14 Top Content report, with blog visitors excluded

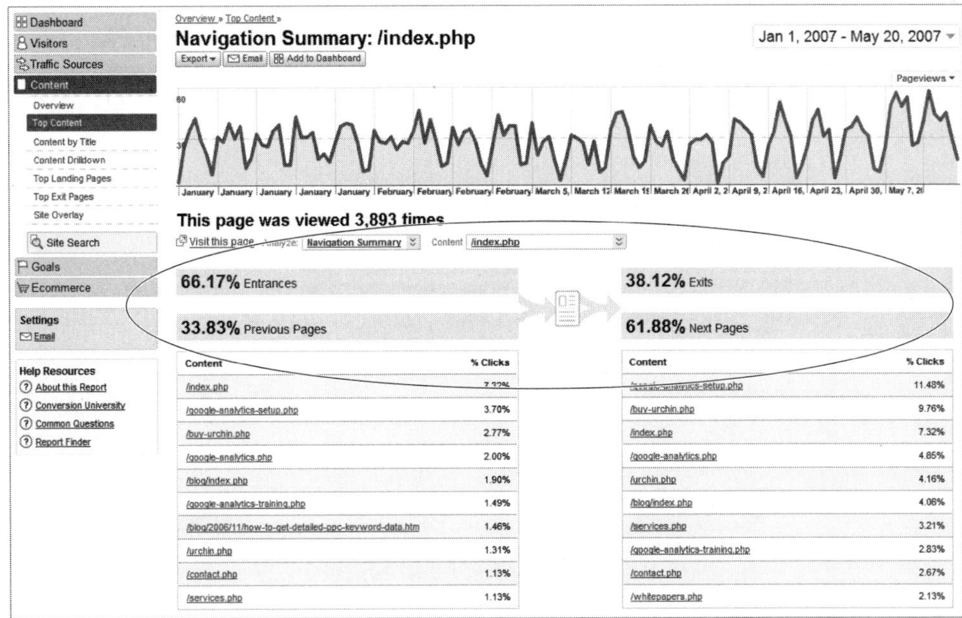

Figure 5.15 Navigation Summary

Content: Site Overlay Report

Site Overlay loads a page from your website and then overlays it with the key metrics for each link on that page. It's an excellent visual way to see which links on your website drive traffic, conversions, transactions, and revenue (see Figure 5.16). The default view is to display the number of clicks received for each link on a page using a small bar chart under the link—mouse over the bar chart to see the corresponding pop-up metrics. The view can be easily changed using the Displaying drop-down menu at the top of the report.

As shown in Figure 5.16, you can see that the Free Whitepapers link from the side menu on the left is driving the most goal values for this page (£44.00). As this is a link pointing directly to a defined goal page, this would be expected. However, what is interesting in this example is that the Jump Start side menu link is also driving significant goal revenue, as indicated by the graphical bar below Jump Start links.

Click any of your links to navigate through to that page and view its site overlay statistics. Chapter 9 describes how you can use the Site Overlay report to differentiate links that point to the same URL. For example, if the Free Whitepapers link highlighted in Figure 5.16 is also present elsewhere on the same page, you can spot which location works best.

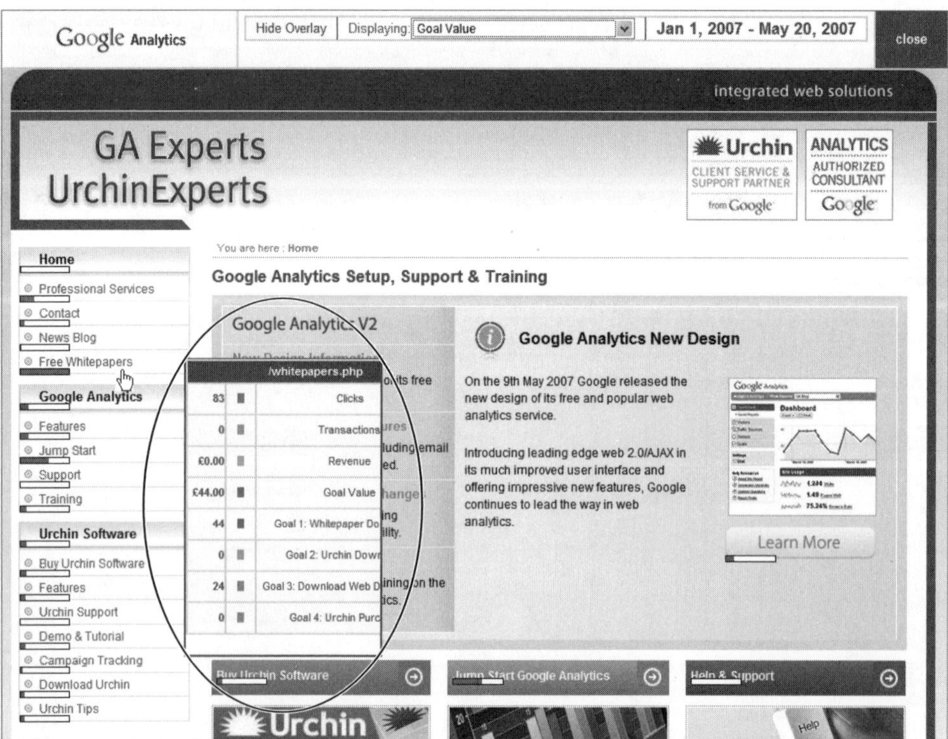

Figure 5.16 Site Overlay report

> **Current limitations of site overlay**
>
> In order for site overlay to work correctly, the page referenced by each link must exist as an HREF element on the page being viewed. That is, if you use the function `trackPageView()` for generating virtual pageviews (as described in Chapter 7), the Site Overlay report will not work. Nor will site overlay work for pages containing Flash content.
>
> Another example is the submission of forms. A submit button or form tag does not contain an HREF element. Therefore, if you have a goal conversion configured as a form submission, the Site Overlay report will not show this as part of the metrics.

Traffic Sources: AdWords Positions Report

This is a unique report, not found in any other web analytics tool, and an extremely powerful report it is, too. The AdWords Positions report tells you what position your AdWords ad was in when the visitor clicked on it. In addition, you can drill down and view how your ad conversion rate, bounce rate, per-visit goal value, number of transactions, revenue, and other metrics vary by position, using the Position breakdown drop-down menu.

In Figure 5.17, the left side of the report table lists the AdWords keywords you have bid on during the specified time frame. Selecting one of these options changes the view on the right to a schematic screen shot of the Google search engine, with the positions your ad was shown at, and the number of visits received while in that position. This emulates what the positions would look like on the Google search engine results page.

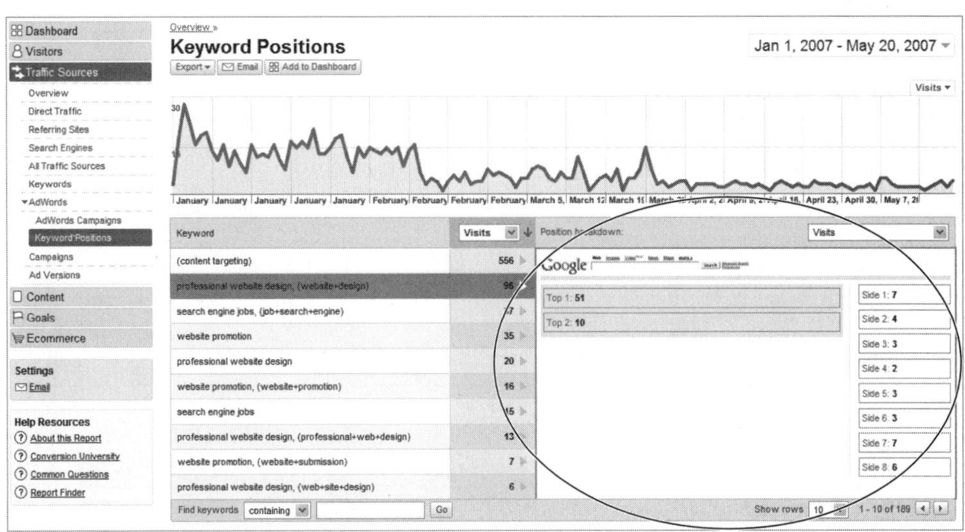

Figure 5.17 AdWords Keyword Positions report

You might expect that the higher your position in the AdWords auction model, the more visitors you receive. Figure 5.18 illustrates the data showing just that—an expected long-tail chart (this figure was created by exporting the report data into MS Excel).

However, long-tail charts are not always the case. Figure 5.19 shows a different keyword selected from the same report. As you can see, positions 3, 5, and 9 are more popular. With this information you may consider the use of the Position Preference feature in your AdWords account. Position Preference is an AdWords option that enables you to set where you would like your ad to rank among all ads shown on a given search engine results page. For instance, from Figure 5.19, you may prefer your ads to appear only when they rank between positions 3 and 9. By enabling Position Preference in your AdWords account, the AdWords system will attempt to make your ad appear in the positions you set—though no position is guaranteed. For more information on Position Preference in AdWords, see

`http://adwords.google.com/support/bin/answer.py?hl=en&answer=31788`

The data shown in Figures 5.18 and 5.19 reflects only visits. However, any of the other segments listed in the Position breakdown menu can be selected and compared (see Figure 5.20). Viewing how conversions, transactions, and revenue vary by ad position springs to mind as interesting reports to check.

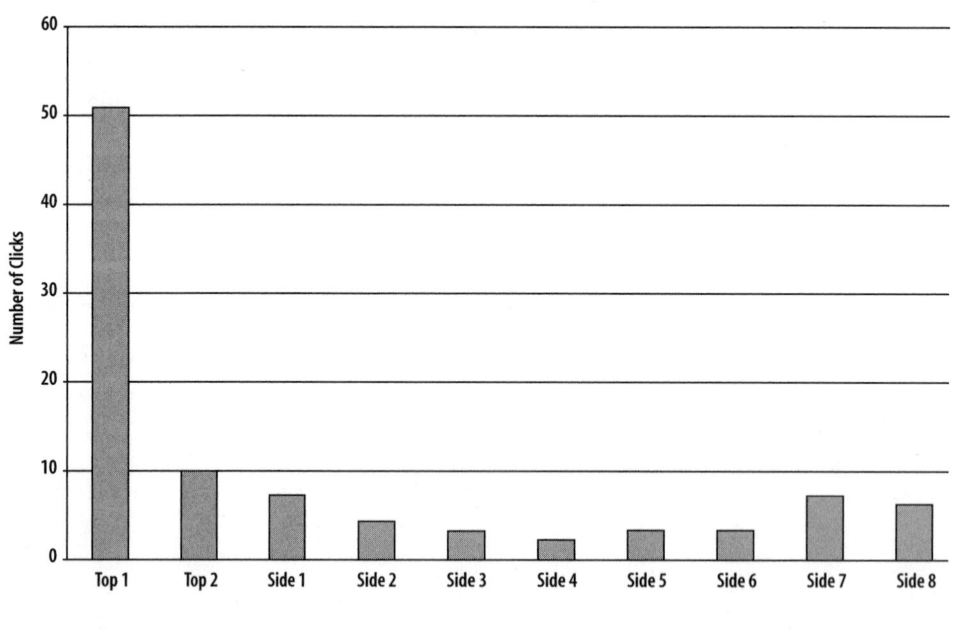

Figure 5.18 Number of clicks by AdWords position—export 1

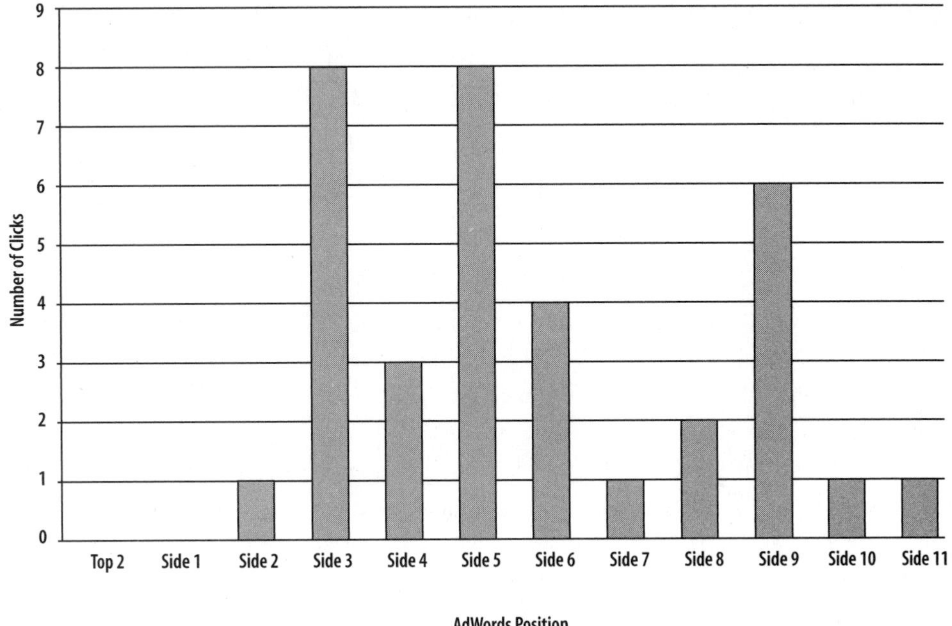

Figure 5.19 Number of clicks by AdWords position—export 2

Position breakdown:
Visits
Visits
Pages/Visit
Avg. Time on Site
% New Visits
Bounce Rate
Goal1 Conversion Rate
Goal2 Conversion Rate
Goal3 Conversion Rate
Goal4 Conversion Rate
Goal Conversion Rate
Per Visit Goal Value
Revenue
Transactions
Average Value
Ecommerce Conversion Rate
Per Visit Value
Google
Top 1: **51**
Top 2: **10**

Figure 5.20 Segmenting the AdWords Positions report

Note: As per other Google Analytics reports, the data shown in the AdWords Keyword Positions report is based on visitors with cookies. Therefore, the numbers may not match the totals viewed in your AdWords account reports, as AdWords can only track clicks. For a more detailed explanation of discrepancies between AdWords and Google Analytics reports, see Chapter 2.

Site Search Usage

The Site Search reports contained in the Content section of Google Analytics are dedicated to understanding the usage of your internal search engine (if you have one). For large, complex websites with thousands, and in some cases hundreds of thousands, of product pages, having an internal site search engine is critical for a successful visitor experience—no navigational system can perform as well as a good internal search engine in these cases.

At the very least, site search reports are a form of market research—every time visitors enter a keyword into your search box, they are telling you exactly what they want to find on your website. Marketers can use this information to better target campaigns. Product managers can use this as a feedback mechanism for designing new features or adding new products.

A report on the search terms used by visitors on your website is clearly powerful information for your organization. However, understanding where on your website a visitor reaches for the search box, what page they go to following a search, how long they stay on your site after conducting a search, whether they perform further search refinements, whether they are more likely to make a conversion, and whether their average order value is higher are also vital clues that can help you optimize the visitor experience.

The answers to all these questions can be found in the Content > Site Search section, as shown in Figure 5.21.

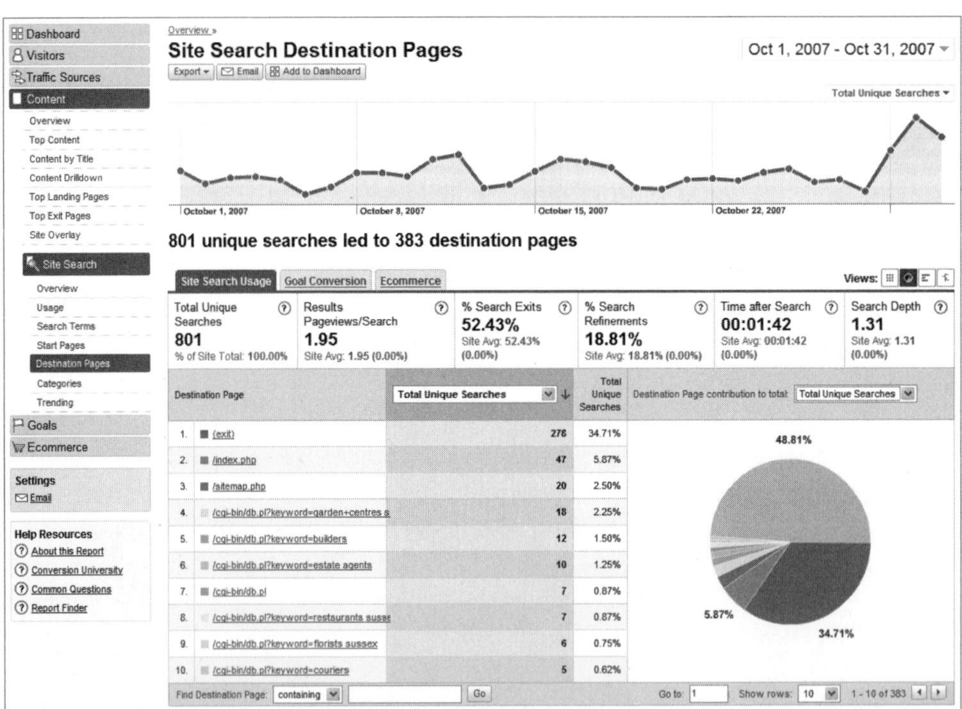

Figure 5.21 Site Search report showing which destination pages are visited following a search

Content Reports: $Index Explained

$Index is a per-page metric that you have seen throughout the Content reports section. As described earlier in this chapter, $index is a measure of the value of a page. The calculation of this metric is defined as follows:

$Index = (goal value + e-commerce revenue) / unique pageviews

$Index goes beyond a simple measurement of popularity by indicating how valuable a specific page is to you. Essentially, it is a way for you to prioritize the importance of pages on your website. For example, when you are optimizing your website content for user experience—that is, to improve conversion rates—you probably want to start by first looking at the pages with the highest $Index, as these have been shown to have the greatest impact.

To understand its significance, consider the following page paths that four different visitors take on a website. In these examples, the goal page is set as page D, and its goal value when reached is $10 (assuming no e-commerce revenue):

Page path 1: B > C > B > **D**

Page path 2: B > E > B > **D**

Page path 3: A > B > C > B > C > E > F > **D** > G

Page path 4: B > C > B > F

To calculate $Index for these pages, Google Analytics sets each unique page in a path that precedes the goal page (D) to have the same goal value ($10). That is, goal values are only attributed to the pages leading up to and including the goal page, not after. These goal values are assigned to a page only once per path. This may sound complicated as written, but actually the calculation is quite simple, as illustrated by Table 5.1.

Unique pageviews are used for the calculation to show how many times a page in a session contributes to the goal.

► **Table 5.1** Calculating $Index

Page	Goal value + revenue / unique pageviews	$Index
A	10/1	10
B	30/4	7.5
C	20/3	6.7
D	30/3	10
E	20/2	10
F	10/2	5
G	0/1	0

As you can see from Table 5.1, the highest value pages over all visitor sessions (highest $Index) are pages A, D, and E—whenever these pages are in a path, a goal conversion occurs. Second highest is Page C—its value is 7.5, as it occurs in most paths that contain a goal conversion. Page G never appears before a goal so there is no goal value for it.

The order of $Index values for pages on this example website is as follows:

(A, D, E) B C F G

With this in mind, if you were to perform page optimization testing, it would make sense to first work on pages A and E (page D is the goal page, and in this case it is the thank-you page so optimization is not required). You may also question the value of keeping page G—it appears to add no value to this website, as indicated by its zero $Index value. That's a good question that should be investigated.

Because $Index is so powerful at highlighting key pages that contribute a monetary value for your website, I recommend you always sort your Content reports by $Index to see how they factor into your success (see Figure 5.22).

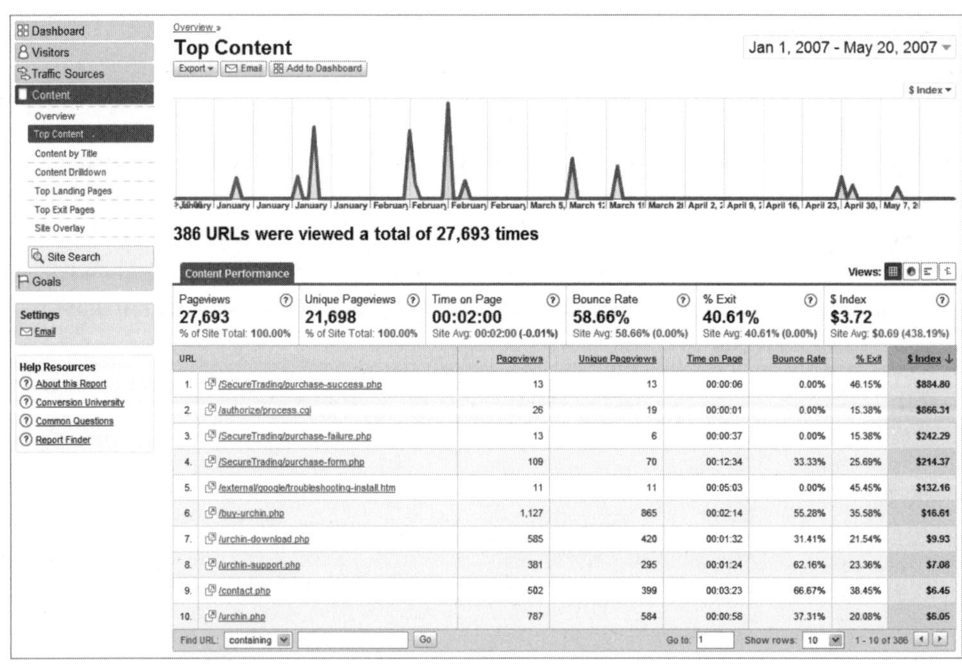

	URL	Pageviews	Unique Pageviews	Time on Page	Bounce Rate	% Exit	$ Index ↓
1.	/SecureTrading/purchase-success.php	13	13	00:00:06	0.00%	46.15%	$884.80
2.	/authorize/process.cgi	26	19	00:00:01	0.00%	15.38%	$866.31
3.	/SecureTrading/purchase-failure.php	13	6	00:00:37	0.00%	15.38%	$242.29
4.	/SecureTrading/purchase-form.php	109	70	00:12:34	33.33%	25.69%	$214.37
5.	/external/google/troubleshooting-install.htm	11	11	00:05:03	0.00%	45.45%	$132.16
6.	/buy-urchin.php	1,127	865	00:02:14	55.28%	35.58%	$16.61
7.	/urchin-download.php	585	420	00:01:32	31.41%	21.54%	$9.93
8.	/urchin-support.php	381	295	00:01:24	62.16%	23.36%	$7.08
9.	/contact.php	502	399	00:03:23	66.67%	38.45%	$6.45
10.	/urchin.php	787	584	00:00:58	37.31%	20.08%	$6.05

Figure 5.22 Listing $Index page values

The list of $Index values shown in Figure 5.22 could be considered your prioritization list for optimizing pages, and the power of this is illustrated by example: Notice row 3 of the table. The page /SecureTrading/purchase-failure.php is the failure page displayed when a purchaser incorrectly completes his or her payment details. It obviously has a high relevance to a successful order (high $Index) and shows a significant number of pageviews compared to /SecureTrading/purchase-success.php—the page displayed when payment is completed successfully.

The data clearly indicates that the owner of this website should investigate the payment form (/SecureTrading/purchase-form.php) to identify whether elements on that page are causing visitor confusion. For example, maybe date values are expected in U.S. format, which is not clear to a European visitor. Whatever the reason, the use of $Index has highlighted an opportunity to improve the efficiency of a page that provides significant revenue to the organization.

You can plot the trend of $Index over time for a specific page by clicking through on its page link and selecting the appropriate chart to display.

Note: $Index is independent of path route and path length. Using the preceding example, $Index for page B = 10 for paths 1, 2 and 3.

Summary

This chapter covered a selection of reports that I consider to be the top 10 reports (areas of interest) in Google Analytics. I have deliberately covered a small, though important, selection. This is simply because with all the various cross-segmentation and drill-down options available, going through all of these would be tedious and laborious. Rather, I have attempted to whet your appetite. It is hoped that seeing the advanced capabilities of Google Analytics has increased your interest in investigating further.

In Chapter 5, you have learned about the following:

- Using the dashboard as a place to save and organize your most important reports and key metrics
- Ten reports that can help you understand visitor behavior and that provide a starting point for further investigation and optimization
- How $Index can be used to evaluate the importance of a web page

Now that you are familiar with the user interface and report structure, it's time to get started with implementing Google Analytics on your own website. This is what the next chapter is all about.

Implementing Google Analytics

Part III provides a detailed description of everything to do in order to collect visitor data—from creating an account to installing the tracking code in a best practice manner.

Following this, we look at the configuration of goals, funnels, filters and visitor segmentation. Finally, Google Analytics Hacks is a work around chapter for when you have bespoke requirements.

If you are a webmaster or web developer, this section is for you. However, in keeping with this book's philosophy, the content is not aimed at programmers, so technicalities are kept to a minimum. You should, though, at least be familiar with HTML and JavaScript.

In Part III, you will learn to do the following:

Getting Started

This chapter is all about getting the basics right—creating an account in the right place (stand-alone or linked to AdWords), tagging your pages, becoming familiar with the concept of multiple profiles, and ensuring that you have AdWords visitors tracked and the concomitant impression and cost data for such visitors being imported. If you are an agency or hosting provider, you need to consider a couple of additional points, which are described in this chapter.

In this chapter, you will learn about the following:

Creating your Google Analytics account
Tagging your pages with the tracking code
Collecting data into multiple Google Analytics accounts
Creating a back up of your web traffic data to a local server
Using profiles in conjunction with accounts
Setting up agency client accounts
Linking Google Analytics and Google AdWords accounts
Common implementation questions

Creating Your Google Analytics Account

Opening a Google Analytics account and performing a base setup is very straightforward. An initial setup enables you to receive data that you can use to begin to understand your website traffic. The time required to do this varies depending on your expertise and familiarity with HTML, your website architecture, and the level of access you have to your web pages. Setting up one website can take as little as an hour or as long as a full working day.

However, it is important to manage your expectations. The initial collection of data is only the first step in understanding your visitor traffic. Configuring your Google Analytics account to your specific needs (see Chapters 7–9) is what will give you the most insight. Nonetheless, collecting the base data first will give you the initial information with which you can fine-tune your setup, so let's get the basic foundations right.

You can open a Google Analytics account in one of two ways. If you have an AdWords account, it makes sense to do it there, so that your cost data can be automatically imported. Click the Analytics tab at the top of your account area, as shown in Figure 6.1a. If you do not have an AdWords account, visit the stand-alone version at www.google.com/analytics/sign_up.html, as shown in Figure 6.1b. Both versions are identical, though the stand-alone version is limited to a maximum of five million pageviews per month—approximately 3,000 visitors per day. Obviously, Google wishes to encourage you to try their online advertising solutions!

If you use the stand-alone version, note that the e-mail address you use to create the account is a Google account. A Google account is simply a registered e-mail address for single sign-on for any Google hosted service. It gives you access to Google Analytics and other Google services such as AdWords, Gmail, Google Groups, personalized search, your personalized home page, and more. If you've used any of these services before, you already have a Google account.

AdWords users—a special case

If you have a Google AdWords account, it is important to create your Google Analytics account from within the AdWords interface. This enables you to quickly and easily link the two—that is, automatically import your AdWords cost data, and be able to log into Google Analytics via your AdWords account interface. You will also be able to log in via the stand-alone interface if you wish.

If you have created a stand-alone Google Analytics account first and then wish to link to your AdWords account, ensure that your AdWords Administrator e-mail address is also a Google Analytics Administrator. Then when you click on the Analytics tab within AdWords, you will be given the option to link your two accounts.

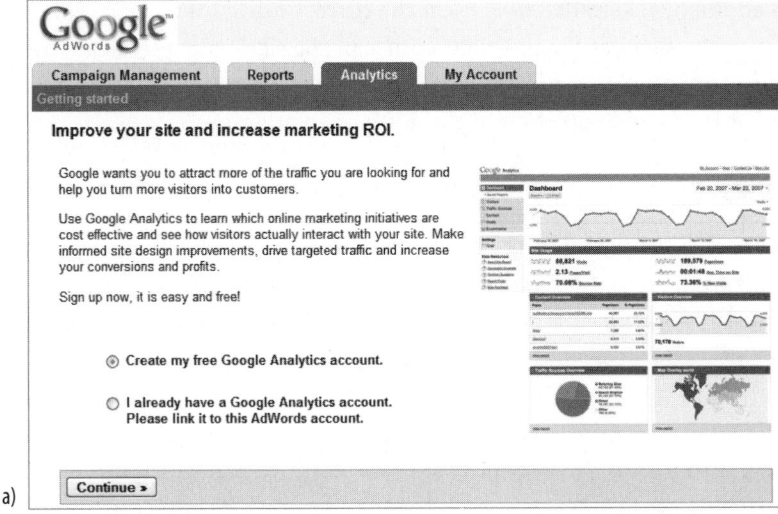

a)

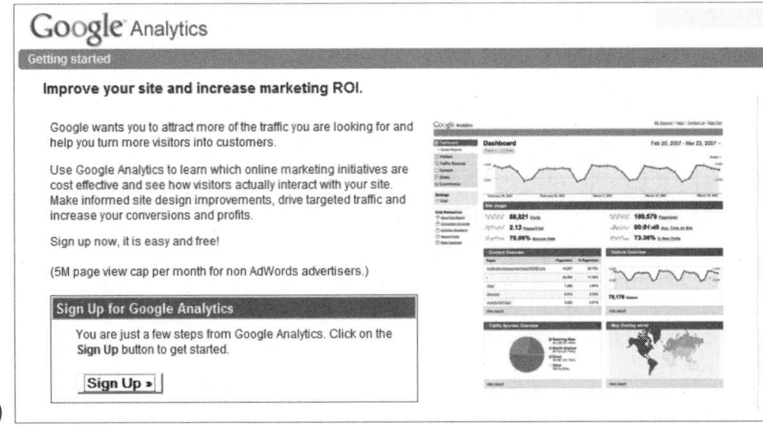

b)

Figure 6.1 Creating a Google Analytics account from (a) within AdWords or (b) via the stand-alone interface

Note: Note: Any e-mail address, such as your company e-mail address, can be registered and used as your Google account. The only requirement it that it must belong to an individual and not a mailing list. Further information is available at www.google.com/accounts.

Once you have your Google account, simply follow the instructions during the sign-up process. If you are using the stand-alone version and you have multiple Google accounts, choose the Google account you most frequently use. That way you will be automatically logged into Google Analytics if you have previously logged in to another

Google service. In addition, ensure that you select the correct region (the one closest to you) from the drop-down menu at the top-right corner of the sign-up page. This sets the language for the sign-up process and ensures that you are shown the correct Terms of Service that you agree to on completion of the account creation process.

Tagging Your Pages

The most important part of the sign-up process is the penultimate setup screen, which identifies your unique tag to be placed on all your pages. This is referred to as the Google Analytics Tracking Code (GATC). It is the use of this single tag to collect visitor data—the exact same tag for every page—that makes Google Analytics so easy to install.

The GATC

The GATC is simply a snippet of JavaScript that is pasted in your pages. The code is hidden to the visitor and acts as a beacon for collecting visitor information and sending it to Google Analytics data collection servers. The precise details of the GATC are unique to each Google Analytics account; an example is given in Figure 6.2.

There are three parts to the GATC:

- The call of a JavaScript file from Google servers

  The file ga.js contains the necessary code to conduct data collection. This file is approximately 18Kb in size, although once it is called it is cached by the visitor's browser and available for all subsequent pageviews.

- Your unique account ID, in the form UA-XXXX-YY

  This must be used exactly as quoted or your data will be sent to the wrong account. A filter to prevent this from happening to your account is detailed in Chapter 8.

- The call of the JavaScript routine `_trackPageview()`

  This is the workhorse of Google Analytics. Essentially, `_trackPageview()` collects the URL of the pageview a visitor loads in his or her browser, including associated parameters such as browser type, language setting, referrer, timestamp, etc. Cookies are then read and set and this information is passed back to Google data collecting servers (as described in the schematic of Figure 3.1).

If you have a relatively small website in terms of number of pages, you can simply copy and paste the GATC into your pages. Alternatively, you may have built your website using a HTML template or a content management system (CMS). If so, simply add the GATC to your master template or footer file. The recommended placement is just above the `</body>` tag at the bottom of the page. This will minimize any delay in page loading, as the ga.js file will be loaded last.

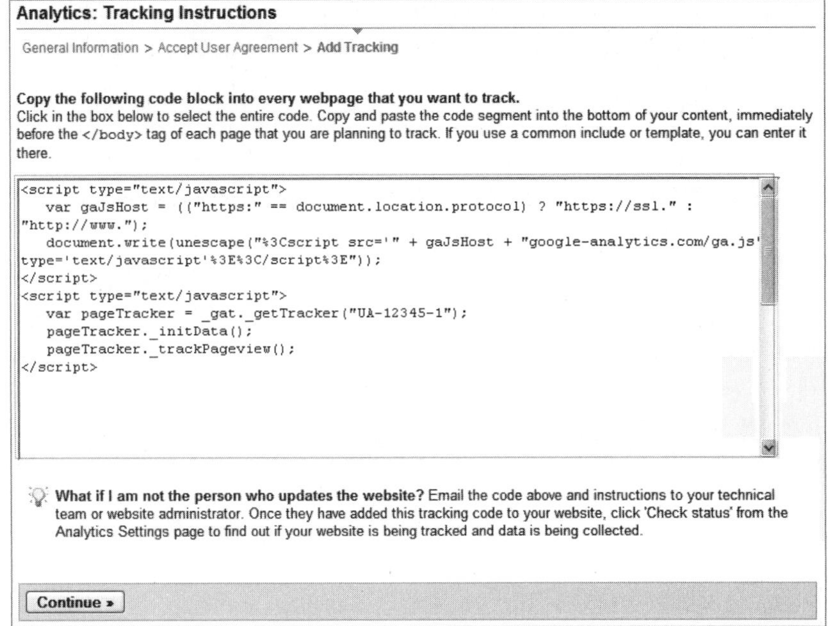

Analytics: Tracking Instructions
General Information > Accept User Agreement > Add Tracking
Copy the following code block into every webpage that you want to track.
Click in the box below to select the entire code. Copy and paste the code segment into the bottom of your content, immediately before the </body> tag of each page that you are planning to track. If you use a common include or template, you can enter it there.
<script type="text/javascript">
var gaJsHost = (("https:" == document.location.protocol) ? "https://ssl." : "http://www.");
document.write(unescape("%3Cscript src='" + gaJsHost + "google-analytics.com/ga.js' type='text/javascript'%3E%3C/script%3E"));
</script>
<script type="text/javascript">
var pageTracker = _gat._getTracker("UA-12345-1");
pageTracker._initData();
pageTracker._trackPageview();
</script>
What if I am not the person who updates the website? Email the code above and instructions to your technical team or website administrator. Once they have added this tracking code to your website, click 'Check status' from the Analytics Settings page to find out if your website is being tracked and data is being collected.
Continue »

Figure 6.2 Typical GATC to add to your pages

As you may have noticed in Figure 6.2, the first section of the code is there to ensure that the correct location of the ga.js file is loaded based on the protocol being used. Secure and encrypted pages use the HTTPS protocol. Non-encrypted pages use HTTP. If you are calling a secure page on your website, then you need to call a modified version of the GATC. The ga.js code does this for you automatically.

Migrating from urchin.js to ga.js

Prior to December 2007, the file referenced by the GATC was called urchin.js and contained different code to that of ga.js. If you are still using urchin.js, you should migrate to the newer ga.js code. To get your new tracking code, you'll need to have administrator access to the Google Analytics account. Follow these steps:

1. Log in into your Google Analytics account.
2. For each profile, click Edit.
3. Click the Check Status link.
4. Follow the on-screen instructions for adding the new tracking code (ga.js).

Once your pages are tagged, you should start to see data in your account within four hours. However, for new accounts, it can take up to 24 hours, so be patient at this stage!

Server-Side Tagging

If you do not have a content management system but use the Apache web server to host your pages, you can use the mod_layout module (similar in principal to a plug-in) to tag your pages for you as they are requested by visitors. Ask your development team or hosting provider to install the mod_layout loadable module from http://tangent.org.

Using this module enables you to tag your pages quickly and efficiently at the server side—the Apache web server automatically inserts the GATC on every page it serves. Note that this means exactly that, *every* page served, so you should add exclusions to those files where the GATC is not required, such as robots.txt, cgi-bin files, and so forth.

A full description of mod_layout is beyond the scope of this book, but an example configuration for your httpd.conf file is given in the following snippet. In this example, two file types are ignored (*.cgi and *.txt) and the file contents of utm_GA.html (the GATC content—as per Figure 6.2) are inserted just above the </body> tag of the HTML page being served:

```
#mod_layout directives
LayoutMergeBeginTag </body>
LayoutIgnoreURI *.cgi
LayoutIgnoreURI *.txt
LayoutHeader /var/www/html/mysite.com/utm_GA.html
LayoutMerge On
```

Warning: If your pages use the CAPTCHA method (http://en.wikipedia.org/wiki/CAPTCHA) of generating security images to protect your site from automated form submission, test that your security image still loads. If not, you may need to exclude the embedded file that calls the security image from mod_layout.

Collecting Data into Multiple Google Analytics Accounts

Your visitor data may be significant to several Google Analytics accounts—for example, if you have separate accounts for affiliates or you wish to share a subset of your data with agencies. Perhaps you have multiple websites set in different time zones and currencies linked to your AdWords account. If so, you will want to keep these as separate Google Analytics accounts; otherwise, you will have data alignment issues; for example, you have to choose one currency and time zone for all reports. Keeping Google Analytics accounts separate in this case makes sense, but you probably want an aggregate account as well, ignoring the alignment issues—that is, for top-level data reports.

In all cases, you can collect visitor data into multiple accounts by initiating more than one tracker object call on your pages, as highlighted in the following GATC:

```
<script type="text/javascript">
   var gaJsHost = (("https:" == document.location.protocol) ? "https://ssl."
: "http://www.");
   document.write(unescape("%3Cscript src='" + gaJsHost + "google-
analytics.com/ga.js' type='text/javascript'%3E%3C/script%3E"));
</script>
<script type="text/javascript">
   var firstTracker = _gat._getTracker("UA-12345-1");
   firstTracker._initData();
   firstTracker._trackPageview();

   var secondTracker = _gat.getTracker("UA-67890-1");
   secondTracker._initData();
   secondTracker._trackPageview();
</script>
```

Backup: Keeping a Local Copy of Your Data

Keeping a local copy of your Google Analytics data can be very useful for your organization. For example, Google currently commits to keeping Google Analytics data for up to 25 months, enabling you to compare annual reports. That is adequate for most users, but what if you wish to retain your data for longer?

By modifying the GATC with a single line of code, it is possible to send your web visitor data to Google Analytics collection servers and simultaneously log this data into your own web server log files. The following modified GATC code highlights the necessary change:

```
<script type="text/javascript">
   var gaJsHost = (("https:" == document.location.protocol) ? "https://ssl."
: "http://www.");
   document.write(unescape("%3Cscript src='" + gaJsHost + "google-
analytics.com/ga.js' type='text/javascript'%3E%3C/script%3E"));
</script>
<script type="text/javascript">
   var pageTracker = _gat._getTracker("UA-12345-1");
   pageTracker._setLocalRemoteServerMode();
   pageTracker._initData();
   pageTracker._trackPageview();
</script>
```

This is simple to achieve, as all web servers log their activity by default, usually in plain text format. Once implemented, open your logfiles to verify the presence of additional _utm.gif entries that correspond to the visit data as seen by Google Analytics. A typical Apache logfile line entry looks like the following:

```
86.138.209.96 www.mysite.com-[01/Oct/2007:03:34:02 +0100] "GET
/__utm.gif?utmwv=1&utmt=var&utmn= 2108116629 HTTP/1.1" 200 35
"http://www.mysite.com/pageX.htm" "Mozilla/4.0 (compatible; MSIE 6.0;
Windows NT 5.1; SV1; .NET CLR 1.1.4322)"
"__utma=1.117971038.1175394730.1175394730.1175394730.1; __utmb=1; __utmc=1;
__utmz=1.1175394730.1.1.utmcid=23|utmgclid=CP-Bssq- oIsCFQMrlAodeUThgA|
utmccn=(not+set)|utmcmd=(not+set)|utmctr=looking+for+site
```

Defining a logfile format for Apache

Apache can be configured to log data in a variety of custom formats. I recommend using the full NCSA log format in your httpd.conf file, as shown here:

```
LogFormat "%h %v %u %t "%r" %>s %b "%{Referer}i" "%{User-Agent}i"
"%{Cookie}i"" combined
```

Note the use of double quotes throughout. In addition, this statement must be a single line in your config file.

For Microsoft IIS, the format can be as follows:

```
2007-10-01 01:56:56 68.222.73.77-- GET /__utm.gif
utmn=1395285084&utmsr=1280x1024&utmsa=1280x960 &utmsc=32-
bit&utmbs=1280x809&utmul=en- us&utmje=1&utmce=1&utmtz=-
0500&utmjv=1.3&utmcn=1&utmr
=http://www.yoursite.com/s/s.dll?spage=search%2Fresultshome
1.htm&startdate=01%2F01%2F2010&
man=1&num=10&SearchType=web&string=looking+for+mysite.com&
imageField.x=12&imageField.y=6&utmp =/ 200 878 853 93 - -
Mozilla/4.0+(compatible;+MSIE+6.0;+Windows+NT+5.1;+SV1;+ .NET+CLR+1.0.3705;
+Media+Center+PC+3.1;+.NET+CLR+1.1.4322) - http://www.yoursite.com/
```

In both examples, the augmented information applied by the GATC is the addition of *utmX* name–value pairs. This is known as a *hybrid data collection method* and is discussed in Chapter 2.

Note, there are overhead considerations to keeping a local copy of visitor data, and these were discussed in Chapter 3. Because web server logfiles can get very large very quickly and swamp hard disk space, I generally do not recommend keeping a local copy of your data unless you have a specific reason for doing so. For example, maintaining a local copy of your data provides you with the option to do the following:

1. Maintain greater control over your data.
2. Troubleshoot Google Analytics implementation issues.
3. Process historical data as far back as you wish—using Urchin.
4. Reprocess data when you wish—using Urchin.

Note: The use of Urchin software is discussed in Chapter 3.

Let's take a look at these benefits in detail:

1. Maintain greater control over your data.

   Some organizations simply feel more comfortable having their data sitting physically within their premises and are prepared to invest in the IT resources to do so. You cannot run this data through an alternative web analytics vendor, as the GATC page tag information will be meaningless to anyone else. However, you do have the option of passing your data to a third-party auditing service. Third-party audit companies are used by some website owners to verify their visitor numbers—useful for content publishing sites that sell advertising and therefore wish to validate their rate cards.

Warning: Be aware that when you pass data to a third party, protecting end-user privacy (your visitors) is your responsibility, and you should be transparent about this in your privacy policy.

2. Troubleshoot Google Analytics implementation issues.

   A local copy of Google Analytics visit data is very useful for troubleshooting complex Google Analytics installations. This is possible because your logfile entries show each pageview captured in real time. Therefore, you can trace whether you have implemented tracking correctly—particularly nonstandard tracking such as PDF, EXE, other download files types, and outbound exit links.

3. Process historical data as far back as you wish—using Urchin.

   As mentioned previously, Google Analytics currently stores reports for up to 25 months. If you wanted to keep your reports longer, you could purchase Urchin software and process your local data as far back as you wish. The downloadable software version runs on a local server and processes web server logfiles, including hybrids. Although it is not as feature-rich as Google Analytics or as tightly integrated with other Google services, Urchin does enable you to view historical data over any period you have collected. It also provides some complementary information to Google Analytics, as described in Chapter 3.

Warning: Reports from Urchin will not align 100% with reports from Google Analytics, as these are two different data collection techniques. For example, a logfile solution tracks whether a download completes, whereas a page tag solution only tracks the `onclick` event—and these are not always going to be the same thing. Data alignment and accuracy issues are discussed in Chapter 2.

4. Reprocess data when you wish—using Urchin.

With data and the web analytics tool under your control, you can apply filters and process data retroactively. For example, say you wish to create a separate profile just to report on blog visitors. This is typically done by applying a page-level filter—that is, including all pageview data from the `/blog` directory. For Google Analytics, reports are populated as soon as that profile filter is applied—that is, from that point forward. For Urchin software, you can also reprocess older data to view the blog reports historically.

When and How to Use Accounts and Profiles

A Google Analytics profile is a set of configuration parameters that define a report. You need at least one profile in order to be able to view your visitor data. Figure 6.2 showed the penultimate step of creating a new Google Analytics account. The last step, following the click of the Continue button, automatically creates your first profile, and this is all you need to get started viewing reports.

However, one website may have numerous separate reports. For example, perhaps you want a profile that reports on U.S. visitors only, and a separate profile just for U.K. visitors. That would be one Google Analytics account with two profiles (configurations), which generates two reports. This is best explained using the diagram shown in Figure 6.3a.

Another scenario is when you have multiple websites, as shown in Figure 6.3b. For example, if you have two product websites, then you could have reports for each within the same Google Analytics account with the same or different filters applied to each. Creating and applying filters to profiles is described in detail in Chapter 8.

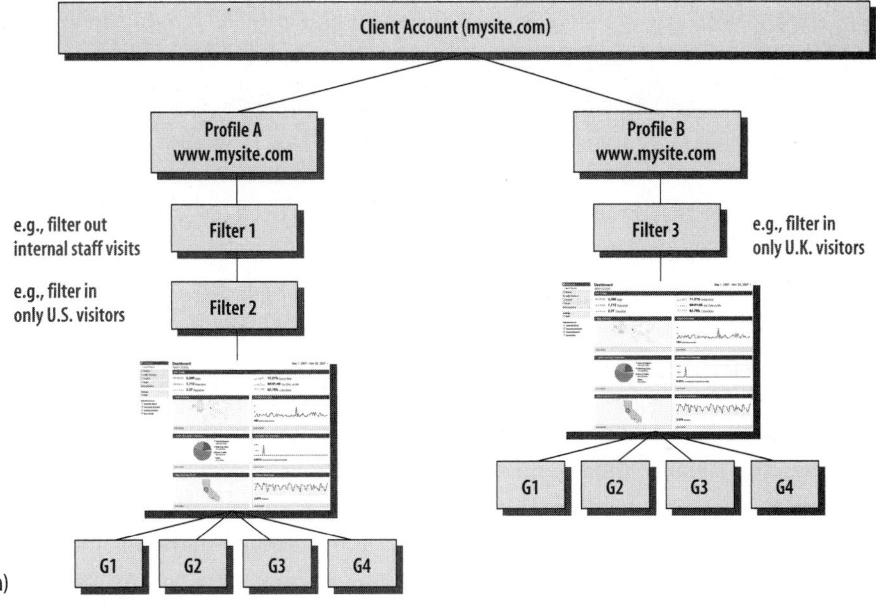

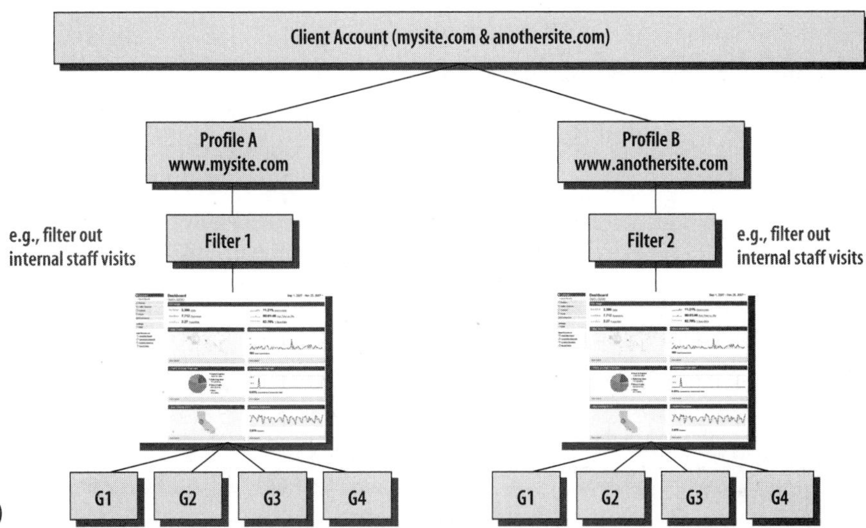

Figure 6.3 (a) Multiple profiles (reports) for the same website within one account; (b) Multiple profiles for different websites within the same account.

Note: The maximum number of profiles for a Google Analytics account is currently 50.

> **An important note on aggregation**
>
> After you have defined your profiles, you cannot produce an aggregate report at a later date—that is, you cannot roll up the individual reports. The strategy, therefore, is to produce an aggregate report first and then use filters to generate the separate reports, or add an extra page tag and collect the data into a separate Google Analytics account, as described under "Collecting Data into Multiple Google Analytics Accounts," earlier in this chapter.

Agencies and Hosting Providers: Setting Up Client Accounts

It is tempting to think that Figure 6.3b is an excellent route for agencies and hosting providers to take on behalf of their clients—that is, have all client reports in one Google Analytics account. However, in accordance with the Google Analytics Terms of Service (found on www.google.com/analytics), any party setting up Google Analytics on behalf of clients must set up a separate Google Analytics account for each separate client. This is the same way AdWords operates and should therefore be familiar to existing AdWords users.

Other limitations include the constraint of 50 profiles per Google Analytics account and the fact that if you also import AdWords data, then by default it is applied to all profiles in your account; clearly, that is undesirable.

For agencies (or hosting providers) to move efficiently between different client accounts, Google Analytics has a similar feature to the My Client Center of AdWords. As long as you use the same Google account e-mail address for each new Google Analytics account you create or manage, you will see a drop-down menu on the right side of your reports interface that lists all the accounts to which you have access. You can also create new accounts for clients from this area, as shown in Figure 6.4.

More information on the My Client Center feature of AdWords can be found here:

```
http://adwords.google.com/support/bin/answer.py?answer=7725
```

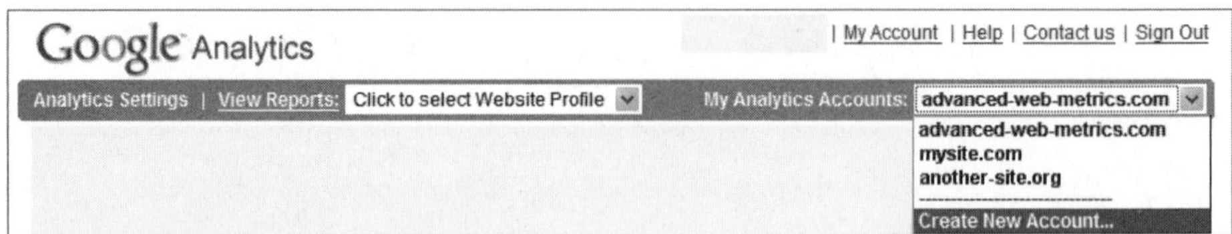

Figure 6.4 The 'My Analytics Accounts' area is the equivalent of the My Client Center for AdWords

Note: A maximum of 25 accounts can be created using the Create New Account option on the drop-down menu shown in Figure 6.4. However, no limit is set on the number of Google Analytics accounts that can be associated with your Google account. That is, any number of clients can add your Google account e-mail address as their administrator or report viewer and these will appear in your My Analytics Account drop-down menu.

Getting AdWords Data: Linking to Your AdWords Account

If you're an online advertiser, chances are good that you are using Google AdWords as part of your marketing mix. AdWords is a way of targeting text ads to visitors using the Google search engine by the keywords they use. That way, your advertisement is displayed to people who are actually looking for something related.

Note: Google AdWords are also shown in a similar way on Google partner sites such as Ask.com, AOL.com, and the AdSense network. For more information about AdSense, see: `http://adsense.google.com`.

Google AdWords is an extremely effective and efficient way of marketing online, because the auction system used is based on how many visitors click on your ad, rather than just its display. Hence, this method of advertising is referred to as pay-per-click (PPC) or cost-per-click (CPC). Yahoo! Search Marketing, Miva, and Mirago operate similar advertising networks. Google Analytics can track visits and conversions from all of these.

As you might expect, Google Analytics, being a part of Google, offers enormous benefits when it comes to integrating data from its AdWords pay-per-click network. In a manner unique for a web analytics tool, getting your AdWords data in is simply a matter of ticking two check boxes—one in your AdWords account, the other in your Google Analytics account:

- First, within your AdWords account, go to the My Account > Account Preferences area. Click the "edit" link next to Tracking. Select the box that says "Destination URL Auto-tagging" and then click Save Changes (see Figure 6.5a).
- Next, still within your AdWords account, click the Analytics tab and choose Analytics Settings > Profile Settings > Edit Profile Information. Place a check next to "Apply Cost Data," and select Save Changes (see Figure 6.5b).

That's it! All your AdWords data (impressions, clicks, cost) will automatically be imported into your account. The import takes place once per day.

a)

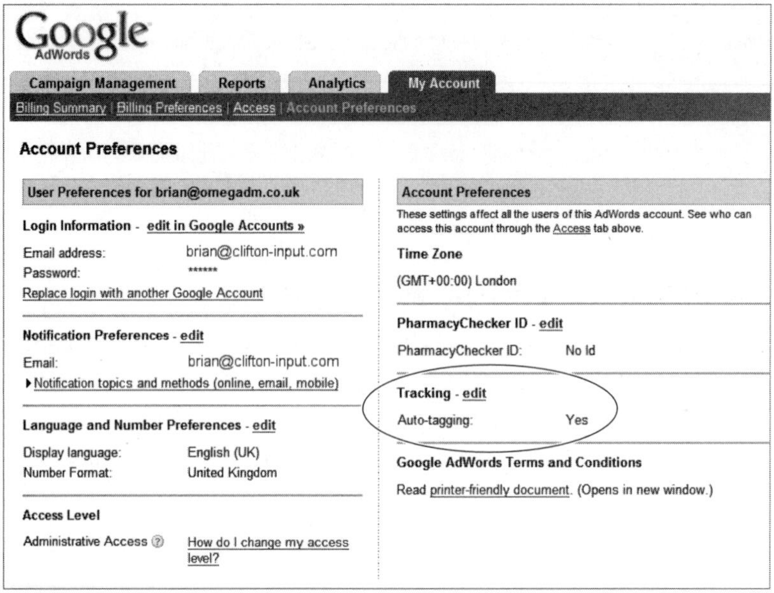

Google AdWords

Campaign Management | Reports | Analytics | My Account

Billing Summary | Billing Preferences | Access | Account Preferences

Account Preferences

User Preferences for brian@omegadm.co.uk

Login Information - edit in Google Accounts »

Email address: brian@clifton-input.com
Password: ******
Replace login with another Google Account

Notification Preferences - edit

Email: brian@clifton-input.com
▸ Notification topics and methods (online, email, mobile)

Language and Number Preferences - edit

Display language: English (UK)
Number Format: United Kingdom

Access Level

Administrative Access ? How do I change my access level?

Account Preferences

These settings affect all the users of this AdWords account. See who can access this account through the Access tab above.

Time Zone

(GMT+00:00) London

PharmacyChecker ID - edit

PharmacyChecker ID: No Id

Tracking - edit

Auto-tagging: Yes

Google AdWords Terms and Conditions

Read printer-friendly document. (Opens in new window.)

b)

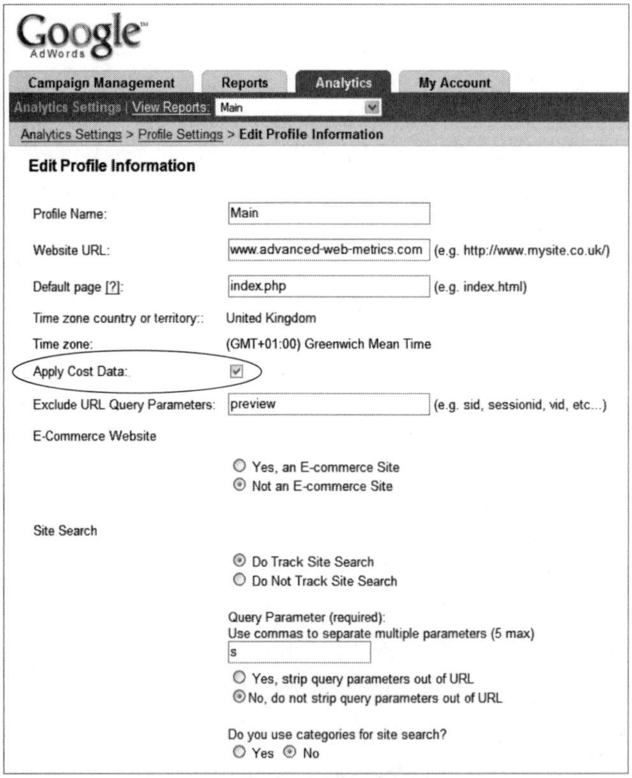

Google AdWords

Campaign Management | Reports | Analytics | My Account

Analytics Settings | View Reports: Main

Analytics Settings > Profile Settings > Edit Profile Information

Edit Profile Information

Profile Name: Main

Website URL: www.advanced-web-metrics.com (e.g. http://www.mysite.co.uk/)

Default page [?]: index.php (e.g. index.html)

Time zone country or territory:: United Kingdom

Time zone: (GMT+01:00) Greenwich Mean Time

Apply Cost Data: ☑

Exclude URL Query Parameters: preview (e.g. sid, sessionid, vid, etc...)

E-Commerce Website

○ Yes, an E-commerce Site
◉ Not an E-commerce Site

Site Search

◉ Do Track Site Search
○ Do Not Track Site Search

Query Parameter (required):
Use commas to separate multiple parameters (5 max)
s

○ Yes, strip query parameters out of URL
◉ No, do not strip query parameters out of URL

Do you use categories for site search?
○ Yes ◉ No

Figure 6.5 (a) Setting auto-tagging within your AdWords account; (b) Applying AdWords cost data

Importing cost data from multiple AdWords accounts

You may wish to import cost data from multiple AdWords accounts—for example, if you are running campaigns in the U.S. and the U.K., or you have two separate agencies managing two separate campaigns. Should you wish to do this, you need to submit a support ticket to Google from within your Google Analytics account.

Bear in mind that when importing multiple cost sources into one Google Analytics account, data needs to be aligned. That is, all time zone and currency settings will be aligned with the one AdWords account to which your Google Analytics account is linked—the one you log into via the AdWords interface. This may not be desirable. An alternative is to add multiple GATCs to your pages as described previously in the section "Collecting Data into Multiple Google Analytics Accounts."

With auto-tagging enabled you will notice an additional parameter showing in the landing page URLs of your AdWords ads, should you click through to them. For example:

```
www.mysite.com/?gclid=COvQgK7JrY8CFSUWEAodKEEyuA
```

The gclid parameter is a keyword-specific parameter unique to your account. AdWords appends this for Google Analytics tracking, and this must remain in place when visitors arrive on your website in order for them to be detected as AdWords visitors. If the gclid parameter is missing or corrupted, then the visitor will be incorrectly assigned as "google (organic)" as opposed to "google (cpc)."

Testing After Enabling Auto-tagging

As discussed in "Unparallel Results: Why PPC Vendor Numbers Do Not Match Web Analytics Reports," in Chapter 2, third-party ad tracking systems can inadvertently corrupt or remove the gclid parameter required by Google Analytics AdWords tracking. For example, systems such as Atlas Search, Blue Streak, DoubleClick, and Efficient Frontier use redirection URLs to collect visitor statistics independently of your organization. These may inadvertently break the AdWords gclid. Therefore, after enabling auto-tagging, always test a sample of your AdWords ads by clicking through from a Google search results page.

If the test fails, then contact your third-party ad tracking provider, as there may be a simple fix. For example, your AdWords auto-tagged landing page URL may look like this:

```
http://www.mysite.com/?gclid=COvQgK7JrY8CFSUWEAodKEEyuA
```

If a third-party tracking system is used for redirection, it could look end up as this:

```
http://www.redirect.com?http://www.mysite.com/?gclid=COvQgK7JrY8CFSUWEAodKEEyuA
```

This is an invalid URL—you cannot have two question marks. Some systems may allow you to replace the second ? with a # so the URL can be processed correctly. This has to be done within the third-party ad tracking system—not within AdWords. Another workaround is to append an encoded dummy variable to your landing page URL, as shown here:

```
http://www.mysite.com/%3Fdum=1
```

AdWords auto-tagging will then append the gclid as

```
http://www.mysite.com/%3Fdum=1&gclid=COvQgK7JrY8CFSUWEAodKEEyuA
```

so that when using your third-party ad tracking system the URL becomes the following:

```
http://www.redirect.com?http://www.mysite.com/%3Fdum=1&gclid=COvQgK7JrY8CFSU
WEAodKEEyuA
```

This will work. That is, the URL will retain the gclid parameter for Google Analytics tracking in the correct format. You can then exclude the tracking of the dummy variable in Google Analytics by setting the configuration (see "Initial Configuration" in Chapter 8).

Note: If you already have parameters in your landing page URLs you do *not* need to add a dummy parameter. However, you will need to change your ? to its encoded equivalent, %3F.

Answers to Common Implementation Questions

1. Can we use our existing tracking software with Google Analytics?

   Yes. Google Analytics will happily sit alongside any other page tagging, logfile, or web analytics solution. As long as there are no JavaScript errors on your web pages, Google Analytics will collect visitor information independently. Similarly, for tracking paid campaigns, Google Analytics variables are simply appended to your existing landing page URLs—regardless of whether another vendor also has tracking variables.

2. Can we track visitors across different websites?

   Yes. You can track whether a visitor traverses through many website domains owned or managed by you—for example, a visitor passing from www.mysiteA.com to www.mysiteB.com. This is achieved by ensuring that the links to the subsequent domains are modified to include a JavaScript function call to either _link (when using an href link) or _linkByPost (when using a form). This is discussed in detail in "E-Commerce Tracking" in Chapter 7.

3. Can we track transactions on a third-party payment gateway?

   Yes, provided you are able to add your GATC to your template pages hosted on the third-party site. Ensure that you use either _link (when using an href link) or _linkByPost (when using a form) when linking to the third-party payment gateway website. This is discussed in detail in "E-Commerce Tracking" in Chapter 7.

4. Do we have to modify the GATC in order to cross-segment data?

   No. Cross-segmentation is built into the Google Analytics product by drilling down into data when clicking on links within the various reports. In addition, cross-segment drop-down menus exist in most reports.

5. Does Google Analytics use first-party cookies, and what happens if the visitor disables these?

   All Google Analytics data is collected via first-party cookies only. If cookies are disabled or blocked by the visitor, their data will not be collected.

6. Is the AdWords gclid auto-tagging parameter bespoke?

   Yes, the gclid parameter is unique for each keyword in your AdWords account.

7. Can Google reprocess my historical data?

   Google cannot currently reprocess historical data, so it is important to always have a default catch-all profile with no filters applied, in case you introduce an error in your filters and lose data. Filters are discussed in Chapter 8.

8. Can we customize the reports?

   Yes, to an extent. For example, if you always want a certain report to be visible when you first log in to Google Analytics, use the Add to Dashboard link that is present at the top of all reports. Up to 12 reports can be added to the dashboard.

9. Can I schedule a report to be e-mailed to me or a colleague regularly?

   Yes, each report has an Email link. The feature includes a scheduler to automate future e-mailings.

10. Can I import cost data from Yahoo! Search Marketing or Microsoft adCenter?

    At present, this is not possible. Yahoo! Search Marketing visitors (or any other pay-per-click network) can be tracked in the same way other paid visitors can be tracked—using campaign variables appended to the landing page URLs. However, cost and impression data cannot be imported.

11. How many goals can I track?

    By default, you can track up to four goals in Google Analytics; by creating more profiles, you could also track additional goals. However, if you have numerous goals—for example, you have a PDF library you wish to track—it is better to have a pseudo e-commerce configuration. That is, trigger a virtual transaction for each goal completed. That way, each goal is considered a product, and the entire e-commerce reporting section of Google Analytics is available to you. See "Monetizing a Non-E-Commerce Website," in Chapter 11, for further details.

12. Can I monetize goals?

Yes. You can assign a goal value within the goal configuration section of the Admin area of your Google Analytics account. In fact, this is strongly encouraged, particularly for non-e-commerce sites, so that you may see the intrinsic value of your website. Also see "Monetizing a Non-E-Commerce Website," in Chapter 11.

13. Is there a relationship between the Google Analytics map overlay and the geo-targeting options available in AdWords?

Yes, the geo-ip database used for both services is the same, so you can use the map overlay information presented in Google Analytics to measure existing AdWords geo-targeted campaigns or help you target new markets.

14. Does Flash break Google Analytics?

No, flash actions can be tracked, but it requires your input—that is, you need to implement event tracking within your FLA file. Chapter 7 discusses this in detail.

15. Will tagging my pages with the GATC slow them down?

The GATC calls the ga.js file, which is approximately 18KB in size, from Google servers. The ga.js file is the same for every page you tag on your site. Therefore, once a visitor has downloaded the file from their initial pageview, it will be cached on their machine—so no further requests for the file are required. In addition, the ga.js file is the same for all users of Google Analytics. Therefore, if a visitor to your website has previously visited another website running Google Analytics tracking, then the ga.js file will already be cached on their machine or Internet service provider's caching server. The result is an industry leading minimal download time for tagging your pages.

16. Are gclid's still valid if accounts are not linked?

Yes, the gclid is added so that Google Analytics can track AdWords visitors. The linking of accounts enables you to log into Google Analytics via your AdWords account, and enables the import of cost and impression data from AdWords into Google Analytics daily. Therefore, if you don't link your accounts, you will still track visitors from AdWords, but you will have no impression or cost data imported. You will also need to log in to Google Analytics via the stand-alone interface (www.google.com/analytics).

Summary

In Chapter 6, you have learned the following:

- How to create your Google Analytics account either as part of your AdWords account or via the stand-alone version
- How to tag your pages; the help that server-side delivered tags can offer in simplifying the process; and how to get data stored into multiple accounts
- How to back up traffic data in your local web server logfiles to give you greater flexibility and options for Google Analytics troubleshooting, auditing, and reprocessing
- How to use accounts and profiles, and what to consider if you are setting up accounts on behalf of clients as an agency or hosting provider
- How to link Google Analytics with your Google AdWords account and the importance of testing the auto-tag feature, especially when using AdWords in conjunction with a third-party tracking tool that employs redirects
- Answers to common implementation questions

Advanced Implementation

Now that you understand the basics of getting your web visitor data into Google Analytics, this chapter looks at the more advanced setup considerations you may require. Capturing e-commerce transactions, tagging your marketing campaigns, and tracking events (those actions visitors make on your website that are not a standard pageview) are discussed in detail.

In addition, you'll learn how to customize the GATC for your specific needs. For example, do you want to convert dynamic URLs into something more readable? Do you use multiple domains or subdomains? Do you have nonstandard requirements such as changing timeout settings, controlling keyword preferences, or sampling rates? All these scenarios and more are covered here.

7

In Chapter 7, you will learn how to do the following:

Use the `_trackPageview()` function to create virtual pageviews
Capture e-commerce transactions
Track online campaigns in addition to AdWords
Customize the GATC for your specific needs

_trackPageview(): The Google Analytics Workhorse

As discussed in Chapter 6, the GATC contains a call to the JavaScript routine _trackPageview(). This is the main function for tracking a page within Google Analytics. _trackPageview() sets up all the required cookies for the session and submits the data to the Google servers. Table 7.1 lists the cookies that Google Analytics sets.

Note: If you are interested in viewing the values of your HTTP headers and the information transmitted from the GATC to Google servers, try the Firefox extension LiveHTTPheaders at http://livehttpheaders.mozdev.org/installation.html

▶ **Table 7.1** The Five Cookie Names and Types Used by Google Analytics

Cookie name	Time to live, Type	Purpose
__utma	24 months, first-party	Stores domain and visitor identifiers For example, unique ID, timestamp of initial visit, number of sessions to date
__utmb	Session, first-party	Stores session identifiers Changes to identify each unique session
__utmc	Session, first-party	Stores session identifiers Expires after 30 minutes of inactivity
__utmk	Session, first-party	Used for quality control Checks data integrity when using the _link() and _linkByPost() functions
__utmv	24 months, first-party	Stores custom label For example, customer, subscriber, registered user
__utmz	6 months, first-party	Stores campaign variables For example, referrer, keyword (if search engine), medium type (CPC, organic, banner, e-mail)

Note: A note on page tagging: _trackPageview() contains a self-check variable to keep it from executing twice, so if you wish to track data in multiple Google Analytics accounts, use the method described in Chapter 6.

With an understanding of how _trackPageview() works, you can leverage it to track virtual pageviews and file downloads, as discussed next.

Virtual Pageviews for Tracking Dynamic URLs

If you have a site that includes a shopping cart or has more than a few dozen pages of content, chances are good that you are using dynamic URLs. In this context, these are pages generated on-the-fly—that is, the visitor requests them by clicking on page links, as opposed to pre-built static HTML content. Dynamic URLs work by using a server-side scripting language, such as CGI-PERL, PHP, ASP, or Python, that pulls non-formatted information into a common design template. Usually, it is URL parameters that defines the content. You can tell if you are using dynamic URLs by your page names. Static URLs have page filenames ending in `.htm` or `.html`. Dynamic ones end in `.cgi`, `.pl`, `.php`, `.asp`, or `.py`, respectively. That does not mean all page names ending in `.php` are generated dynamically. However, if your website URLs also include a query (?) symbol followed by parameters such as name–value pairs, they are most likely to be dynamic URLs, as shown in the following two examples:

Example 1:

```
http://www.mysite.com/catalogue/product.php?sku=123&lang=en&sect=leather
```

Example 2:

```
http://www.mysite.com/catalogue/product.php?sku=148&lang=en&sect=suede
```

In these examples, the query parameters `sku`, `lang`, and `sect` define the content of the page within a design template.

Note: Some web servers may use an alternative to ? to define dynamic URL parameters, such as #.

For the purposes of Google Analytics, a URL structure is broken down as shown in Figure 7.1.

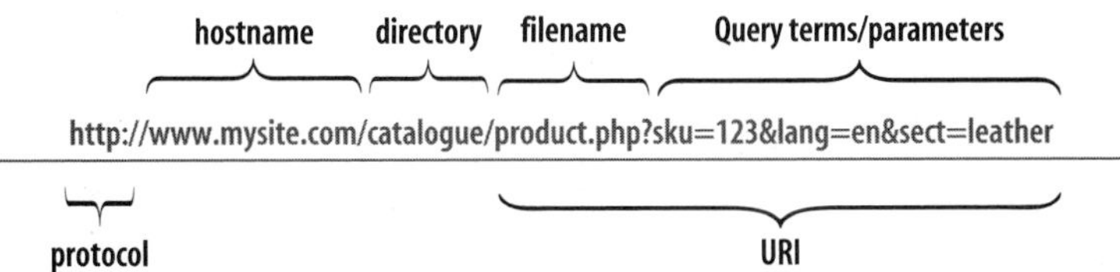

Figure 7.1 Parts of a URL

For this scenario, the query terms used in the vast majority of cases are completely meaningless to the human reader. It is therefore preferable to rewrite the query terms as product descriptions. However, this does not mean all query terms should be rewritten. Only those that are important in identifying specific pages should be rewritten, as some may be required for reporting on other information such as internal site search.

By default, Google Analytics tracks your viewed pages by calling the JavaScript routine _trackPageview() in the GATC. As described in Chapter 6, the standard GATC calls _trackPageview() without an argument (without a value in the parentheses).

With the parentheses empty, Google Analytics records the URI directly from your browser address bar and displays this in the reports as the pageview. However, you can override this behavior by modifying the _trackPageview() call to create virtual pageviews. For example:

```
pageTracker._trackPageview('/catalogue/products/english/➥
leather/blue tassel shoe');
pageTracker._trackPageview('/catalogue/products/english/➥
suede/high heeled boot');
```

The parentheses contain the virtual pageview and path. This overrides the URI value. By using virtual pageviews, reports become much easier to read and interpret. As long as the argument begins with a forward slash, virtual pageview names may be organized into any directory style structure you wish.

Of course, variables used to create the virtual pageview need to be available within your web environment such as your shopping cart or content management system, and a good webmaster or web developer will be able to set this up quite quickly. At the very least, simply using what is already available in the original URL, you could have the following:

```
pageTracker._trackPageview ('/catalogue/products/eng/leather/prod code ➥
123');
```

Clearly, this is not the finished article for you, but it does at least help with your analysis. As stated previously, you should only use this technique to rewrite dynamic URLs that are necessary to you. In addition, discuss the full consequences with your webmaster. For example, it is not necessary or desirable to rewrite the following:

```
http://www.mysite.com/search?q=shoes
```

In this example, the URI relates to an onsite search query that you will want to view in your Site Search reports. Rewriting this will break those reports. Taking this further, if your URL contains a mix of variables, some of which you want to overwrite and some that you do not, then you can achieve this by including the variables in the virtual pageview. For example, consider the following dynamic URL that contains a site search query term plus other dynamic variables:

```
http://www.mysite.com/search?q=shoes&lang=en&sect=leather
```

As a virtual URL, this could be written as:

```
pageTracker._trackPageview('/products/eng/leather/?q=shoes');
```

Here, the original q=shoes query is written back into the virtual pageview, enabling you to view Site Search reports as normal. As with all URLs, if you wish to write query variables in your virtual pageviews, then use the standard convention—a question mark (?) to begin the variable definition and an ampersand (&) to separate multiple name–value pairs.

Note: A consequence of using virtual pageviews is that they will break the Site Overlay reports as well as the Visit This Page link in the Content reports as the page doesn't exist in the real world. If these features are important to you, then don't rewrite your URLs. However, you'll likely find that the greater clarity virtual pageviews bring to the reporting of complex URLs far outweighs the loss of these features.

Virtual Pageviews for Tracking File Downloads

By default, Google Analytics will not track your file downloads (for example, PDF, EXE, DOC, XLS, ZIP), as these pages cannot be tagged with the GATC. However, it is easy to track these by modifying the download link on your web pages using the virtual pageview technique just described.

In the following example, it is the link itself within your web page that is modified, not the GATC. Here is the original HTML link that cannot be tracked:

```
<a href="mydoc.pdf">Download a PDF</a>
```

This is the new link that is tracked in the virtual `/downloads` directory:

```
<a href="mydoc.pdf"
onclick="pageTracker._trackPageview('/downloads/mydoc.pdf');">Download a
PDF</a>
```

Whether you track file downloads as virtual pageviews or as events is a matter of preference—event tracking is discussed later in this chapter. I prefer the virtual pageview method, as it seems reasonable to me that a document view should be considered in the same way as a pageview.

Virtual Pageviews for Tracking Partially Completed Forms

Virtual pageviews can also be used to track the partial completion of forms. This is particularly useful if you have a long or multi-page form, such as a registration form or a feedback survey. Using virtual pageviews in this way enables you to see where visitors bail out before getting to the Submit button. This is achieved using the Funnel Visualization report, as discussed in the section "Goals: Funnel Visualization Report," in Chapter 5.

In order to accomplish this, use the `onBlur` event handler to modify your HTML form fields as follows:

```
<form action="cgi-bin/formhandler.pl" method="post" name="theForm">
<input type="text" name="firstname"
```

```
onBlur="if(document.theForm.firstname.value != '');➡
pageTracker._trackPageview('/forms/signup/firstname')">

<input type="text" name="lastname" onBlur="
if(document.theForm.lastname.value != '');➡
pageTracker._trackPageview('/forms/signup/lastname')">

<input type="text" name="dob" onBlur=" if(document.theForm.dob.value !=
'');➡
pageTracker._trackPageview('/forms/signup/dob')">

<input type="text" name="address1" onBlur="
if(document.theForm.address1.value != '');➡
pageTracker._trackPageview('/forms/signup/address1')">

⋮
</form>
```

The if()!= '' statement is included to confirm that each form field has content before creating the event. Of course, not all form fields will be compulsory to the visitor, so use the if statement appropriately.

Note: Be warned! The virtual pageviews tracked in this example are labels that enable you to confirm whether a field has been completed—they are not personal information submitted by the visitor. It is against the Google Analytics Terms of Service to track personally identifiable information. For more information see www.google.com/analytics/tos.html.

E-Commerce Tracking

Before describing how to capture e-commerce data, consider the salient points to take into account when collecting visitor transactional data:

- Within Google Analytics, the transaction and item values are currency agnostic—that is, although you can specify the currency symbol used in your configuration (see Chapter 8), this is simply a report label. If you are running multiple websites with localized currency values, then these will not be converted into USD (or whatever currency label you configure).

  Of course, you can perform an exchange rate calculation on each of your websites to unify the currency, and then forward this to Google Analytics, but that is likely to confuse your regional marketing departments. The best practice is to use one Google Analytics account for each localized website. This makes sense when you consider that each localized website is also likely to be running in its own time zone and with its own AdWords campaigns, where the cost data is also localized.

If you want an aggregate report of all local websites, add a second GATC to your pages. Chapter 6 discusses this scenario in more detail in "Collecting Data into Multiple Google Analytics Accounts."

- Use Google Analytics E-commerce reports to measure the effectiveness of your website and its marketing campaigns at deriving revenue from online channels. As such, it should not be used as a substitute for your back office or customer relationship management system, as there will always be discrepancies between these data sources.

  For example, JavaScript-disabled browsers, cookies blocked or deleted, visitor multiple clicks, Internet connection blips, returned orders, mistakes, and so on, all add errors bars when it comes to aligning web visitor data with order fulfillment systems.
- Google Analytics does not collect any personally identifiable information and it is against the Terms of Service to attempt to collect such information.

Capturing Secure E-Commerce Transactions

Google Analytics supports a client-side data collection technique for capturing e-commerce transactions. With some simple additions to the GATC on your purchase receipt page, Google Analytics can be configured to record transaction and product information. You can use the following GATC to do this:

```
<script type="text/javascript">
   var gaJsHost = (("https:" == document.location.protocol) ? "https://ssl."
: "http://www.");
   document.write(unescape("%3Cscript src='" + gaJsHost + "google-
analytics.com/ga.js' type='text/javascript'%3E%3C/script%3E"));
</script>
<script type="text/javascript">
   var pageTracker = _gat._getTracker("UA-12345-1");
   pageTracker._initData();
   pageTracker._trackPageview();
   pageTracker._addTrans(
     "1234",                        // order ID - required
     "Mountain View Book Store",    // affiliation or store name
     "89.97",                       // total - required
     "6.30",                        // tax
```

```
      "5",                              // shipping
      "San Jose",                       // city
      "California",                     // state or province
      "USA"                             // country
    );
    pageTracker._addItem(
      "1234",                           // order ID - required
      "DD44-BJC",                       // SKU code (stock keeping unit)
      "Advanced Web Metrics",           // product name
      "Web, Technical",                 // category or variation
      "29.99",                          // unit price - required
      "3"                               // quantity - required
    );
    pageTracker._trackTrans();
</script>
```


For this example, three additional lines have been added within the GATC:

- The transaction line, as defined by _addTrans(), which is an array of comma-separated values, delimited by quotation marks
- The product item line, as defined by _addItem(), which is an array of comma-separated values, delimited by quotation marks
- A call to the JavaScript function _trackTrans() , which sends the information to Google Analytics

The order of these lines within your GATC is important, so maintain the order shown here on your receipt page.

Because _addTrans() and _addItem() are arrays, they can be written on multiple lines for clarity. They can also be written on a single line, which may be an easier format for you to use with transactions containing multiple items, for example:

```
pageTracker._addTrans("1234","Mountain View Book Store",➡
"89.97","6.30","5","San ˜CA Jose","California","USA");
pageTracker._addItem("1234","ISBN-9780470253120",➡
"Advanced Web Metrics","Web,"29.99","2");
pageTracker._addItem("1234","ISBN-9780321344755",➡
"Don't Make me Think","Web,"29.99","1");
```

For each transaction, there should be only one _addTrans() entry. This line specifies the total amount for the transaction and the purchaser's city, state, and country. For each item purchased, there needs to be an _addItem() line. That is, two purchased items requires two _addItem() lines, and so forth. Item lines contain the product names, codes, unit prices, and quantities. The variable values required are shown in Table 7.2. You obtain these from your e-commerce shopping system.

▶ **Table 7.2** E-Commerce Parameter Reference Guide

Transaction Line Variables	Description
order-id	Your internal, unique order ID number
affiliation	Optional affiliation or store name
total	Total value of the transaction
tax	Tax amount of the transaction
shipping	The shipping amount of the transaction
city	Purchaser's city address to correlate the transaction with
state	Purchaser's state or province address to correlate the transaction with
country	Purchaser's country address to correlate the transaction with
Item Line Variables	
order-id	Your internal, unique order ID (must match the transaction line)
sku-code	Product Stock Keeping Unit code
product-name	Product name or description
category	Category of the product or variation
price	Unit price of the product
quantity	Quantity ordered

If you don't have data for a certain variable, simply leave the quotation marks for the variable empty (with no spaces). For example, if you have no affiliate network, and shipping is included in the purchase price, you would use the following:

```
<script type="text/javascript">
  var pageTracker = _gat._getTracker("UA-12345-1");
  pageTracker._initData();
  pageTracker._trackPageview();
  pageTracker._addTrans(
    "1234",                        // order ID - required
    "",                            // affiliation or store name
    "89.97 ",                      // total - required
    "6.30 ",                       // tax
    "",                            // shipping
    "San Jose",                    // city
    "California",                  // state or province
    "USA"                          // country
  );
  pageTracker._addItem(
    "1234",                        // order ID - required
    "DD44-BJC",                    // SKU code (stock keeping unit)
```

```
        "Advanced Web Metrics",       // product name
        "Web, Technical",             // category or variation
        "29.99",                      // unit price - required
        "3"                           // quantity - required
    );
    pageTracker._trackTrans();
</script>
```

Note: In the preceding example there are no spaces between the double quotes (""). Also note the deliberate spaces at the end of the total transaction and tax amounts. I highlight these simply to illustrate that they do not affect the reporting because they are removed by Google Analytics during processing. Spaces between words in variable values are not trimmed. For example, "San Jose" remains as defined.

The importance of unique transactions

It is important to use unique transaction IDs (consisting of numbers or text or a mixture of both) for each transaction. Otherwise, separate transactions that have the same ID will be compounded, rendering the data meaningless. This can happen to you inadvertently if customers multiple click on the final purchase button. For best practice, prevent this behavior. Below is a JavaScript example:

```
<script>
var firsttime;
function validator(){
 if (firsttime == "Y"){
   alert("Please wait, your payment is being processed.");
   return (false);
 }
 firsttime = "Y";
 return (true);
}
</script>
```

Paste the above code into the <head> area of your HTML page that contains the final e-commerce checkout link or button. Then within your HTML of the same page, modify your submission form as follows:

```
<FORM METHOD="POST" ACTION="authorize.cgi" onSubmit="return validator()">
```

The importance of unique transactions *(Continued)*

The onSubmit event handler will prevent multiple submissions of the form thus avoiding any duplicate transaction IDs being captured by Google Analytics.

If your purchase form already has an onSubmit event handler, simply append the validator call as follows:

```
<FORM METHOD="POST" ACTION="authorize.cgi" onSubmit="return
checkEmail;return validator()">
```

Using a Third-Party Payment Gateway

If your website initiates a purchase checkout process on a separate store site (for example, if you send customers from www.mysite.com to a payment gateway, such as www.secure-site.com), you need to make additional changes to your web pages. This is because Google Analytics uses first party cookies. As discussed in Chapter 2, this means only the domain which sets the cookies can read or modify them—a security feature built into to all web browsers by default. You can overcome this, and pass your Google Analytics first party cookies to your third party domain, with the following method.

First, ensure that you have installed the GATC on both your primary site pages and all store site pages. Two of these pages require further modification of the GATC: the last page of the checkout process that occurs on www.mysite.com and the entry page visitors use to complete their checkout on www.secure-site.com. For both of these pages, modify the GATC as follows:

```
<script type="text/javascript">
   var gaJsHost = (("https:" == document.location.protocol) ? "https://ssl."
: "http://www.");
   document.write(unescape("%3Cscript src='" + gaJsHost + "google-
analytics.com/ga.js' type='text/javascript'%3E%3C/script%3E"));
</script>
<script type="text/javascript">
   var pageTracker = _gat._getTracker("UA-12345-1");
   pageTracker._setDomainName("none");
   pageTracker._setAllowLinker(true);
   pageTracker._initData();
   pageTracker._trackPageview();
</script>
```

You then need to modify the web page on www.mysite.com that calls the third party gateway site, in one of two ways:

1. If your website uses a link to pass visitors to the third-party site, then modify it to look like this:

```
<a href=https://www.secure-site.com/?store=parameters
onclick="javascript:pageTracker._link(this.href); return
false;">Continue to Purchase</a>
```

   With this method, the Google Analytics cookies are passed to the receiving domain by appending them to the URL string. If you see *__utma*, *__utmb* and *__utmc* parameters in your third-party landing page URL, then this has worked.

Note: Note the use of `return false;` here. This ensures that for visitor browsers that have JavaScript disabled, the href link will be followed without error. Of course, if JavaScript is disabled, Google Analytics tracking won't occur. However, the modified link will still work.

2. If your website uses a form to pass visitors to the third-party site, then modify the form as follows:

```
<form method="post" action="http://www.secure-site.com/process.cgi"
onSubmit="pageTracker._linkByPost(this)">
```

   With this method, the Google Analytics cookies are passed to the receiving domain via the HTTP headers. This will work even for forms where method="GET". You can verify if this has worked by viewing the HTTP headers sent in Firefox using the add-on LiveHTTPheaders (http://livehttpheaders.mozdev.org).

What to do when a third-party gateway does not allow tracking

If your third-party payment gateway does not allow you to modify their payment pages—that is, add your GATC—you cannot directly capture completed transactions. However, there is a workaround: use the onClick or onSubmit event handlers. By placing one of these on your website at the point where visitors are just about to click through onto the payment gateway, you can call the _trackTrans() function and capture the transaction details. The addTrans and addItem arrays also must be configured on the same page.

An example call via a link would be as follows:

```
<a href=https://www.secure-site.com/?store=parameters
onclick="javascript:pageTracker._trackTrans(); return false;">Continue to
Purchase</a>
```

What to do when a third-party gateway does not allow tracking *(Continued)*

For a form it would be like this:

```
<form action="http://www.secure-site.com/process.cgi"
onSubmit="pageTracker._trackTrans()">
```

The caveat with this method is that you are not tracking completed transactions, merely the intent to complete. Perhaps the visitor's credit card details are declined or they change their mind at the last minute before completing payment. Whatever the reason, your Google Analytics E-commerce reports will be a good indicator of transaction activity, but they are unlikely to exactly align with reports provided by your third-party gateway company.

Negative Transactions

All e-commerce organizations have to deal with product returns at some point, whether because of damaged or faulty goods, order mistakes, or other reasons. It is possible to account for these within your Google Analytics reports by processing a *negative transaction*. However, I don't recommend this for two reasons:

- Aligning web visitor data with internal systems does not yield perfect results. A negative transaction usually takes place well after the original purchase—therefore, in a different reporting period. This is generally more confusing than simply leaving the returned transaction in your reports.
- Consider carefully the purpose of including a negative transaction. If I search for "running shoes" and then make a purchase from your website, that is a perfectly good transaction—one which reflects the effectiveness of your website and your marketing campaigns.

  If subsequently I decide I don't like the shoes and return them, this would be because of the product, perhaps a quality issue. That is a separate issue from the effectiveness of your marketing or your website; just because I return my running shoes does not mean that no further marketing investment should be made.

For completeness, how to process a negative transaction is included here. First, create an internal-only version of your completed purchase form that can be edited for the negative details. The form should be edited in a text editor and *not* loaded in a browser at this stage. To remove an order, edit as follows:

For the `_addTrans` line:

- Use the same *order-id* for the transaction as the one used for the original purchase.
- Ensure that the *total* variable is negative.
- Ensure that the *tax* and *shipping* variables are negative.

For the _addItem line:

- Use the same *order-id* for the transaction as the one used for the original purchase.
- Ensure that the *price* is positive.
- Ensure that the *quantity* is negative.

Process the form details by loading the modified copy of your order receipt page into your browser. This will call the *pagetracker._trackTrans* function as if it were a regular purchase.

Note: You will still be able to see the actual transaction and the duplicate negative transaction when you select the day on which these transactions were recorded. However, when you select a date range that includes both the original and the negative transaction, the transaction will not be included in the total revenue reported.

Online Campaign Tracking

Being able to track online campaigns depends on the use of landing page URLs. A *landing page* is the destination page on which you want visitors to enter your website, following a click-through on a referring website. In most cases, you can control what destination page your visitors arrive at (land on) by specifying the URL. For example, if you have a link on a product portal directory that specializes in all things widget, then you may decide to point your link URL to a specific product landing page such as www.mysite.com/widgets.htm, as opposed to your generic home page. That way, you improve the experience for visitors who click through, by showing them a specific page relevant to their interests.

For the product portal directory example, nothing more is required. You will see how many visitors and conversions are received from that website in your Google Analytics Traffic Sources > Referring Sites report. However, if the referrer has a mixture of paid and non-paid links to your website, you will need to differentiate these links; otherwise, they appear as one single source. The way to differentiate them is to tag your landing page URLs.

Tagging Your Landing Page URLs

Tagging your landing page URLs to differentiate paid versus non-paid links from the same referrer is the most common use of this technique. The principle and process are straightforward—you simply append additional Google Analytics parameters to the end of your URLs.

The following are two examples (that will be discussed in more detail) of tagging landing pages for use in paid campaigns on the Yahoo! Search Marketing network:

- Tagging a static landing page

  Original landing page URL: http://www.mysite.com/widgets.htm

  Tagged landing page URL: http://www.mysite.com/widgets.htm?utm_source=yahoo&utm_medium=ppc&utm_term=widgets
- Tagging a dynamic landing page

  Original landing page URL: http://www.mysite.com/widgets.htm?prod=101

  Tagged landing page URL: http://www.mysite.com/widgets.htm?prod=101&utm_source=yahoo&utm_medium=ppc&utm_term=widgets

Note: Manually tagging your landing page URLs is not required for AdWords campaigns. This is done for you automatically (see "Getting AdWords Data: Linking to Your AdWords Account," in Chapter 6).

In addition to pay-per-click networks, banners, links within documents (PDF, DOC, XLS, etc.), and e-mail marketing campaigns can all be tracked in this straightforward way. It allows for the complete differentiation of your visitors. Here is a two-step process to get you started:

Step 1: Tag Only What You Need

Generally speaking (AdWords being the exception), you need to tag all of your paid keyword links, such as Microsoft adCenter, Yahoo! Search Marketing, banners, and any other form of online advertising. You should also tag the links inside promotional e-mail messages and embedded links within digital collateral such as DOC, XLS, and PDF files.

If you don't tag these non-paid and paid links, then visitor click-throughs will still be tracked, but the referrer information becomes aggregated. For example, a non-tagged paid link from Yahoo! Search Marketing will show as the same referrer as an organic link from Yahoo! Search—that is, it will show as "yahoo (organic)" for both visits. Similarly, non-tagged links in email messages and digital collateral will show as "direct" visits—that is, grouped with those visitors that either type in your web address directly into their browser, or click on a previously saved bookmark or favorite. Clearly marketers wish to differentiate these visit referrals.

There are certain links that you don't need to tag. For instance, you cannot tag organic (non-paid) keyword links from search engines, and it isn't necessary to tag links that come from referral sites where your link listing is free, such as web portals. In addition, you should not attempt URL tagging for internal links (links within your website). Doing so will overwrite existing referrer campaign variables, which will result in a data misalignment.

Step 2: Use the URL Builder

Campaign links consist of a URL address followed by a ? (or & if you have existing parameters), followed by two or more of your campaign variables, as described in Table 7.3:

► **Table 7.3** Landing Page Campaign Variables

Tag variables	Description
utm_source	**Required.** Used to identify a particular search engine, newsletter, or other referral source.
utm_medium	**Required.** Used to identify a medium, for example CPC, PPC, banner, e-mail, PDF, DOC, XLS, etc.
utm_term	**Optional.** Used for paid search to note the keywords being targeted for a particular ad.
utm_content	**Optional.** Used for ad version testing to distinguish different ads that link to the same landing page.
utm_campaign	**Recommended.** Used to identify different strategic campaigns from the same source–medium combination. For example, for an e-mail newsletter, using "spring promotion" or "summer promotion."

It is these additional variables appended to your landing page URLs that enables Google Analytics to differentiate visitors—for example, between an organic visitor from Yahoo! and a pay-per-click visitor from Yahoo!, or a direct visitor from one who clicked on one of your marketing e-mails.

Because up to five variables are allowed, the URLs can appear complicated. To avoid worrying about syntax, use the URL Builder tool at www.google.com/support/googleanalytics/bin/answer.py?answer=55578&topic=10998.

The URL Builder tool creates the tagged links for you—simply copy and paste the resultant URL as your ad landing page URL. Once you understand the structure of the tagged URLs, you may want to switch to using a spreadsheet of these for bulk upload into your pay-per-click account or other management system.

Note: If you are also using a third-party ad tracking system to track click-throughs to your website, your visitors will be passed through redirection URLs. If this describes your scenario, be sure to test your tagged landing page URLs, as redirection may break them. You can test by clicking the resultant combined link (third-party link plus campaign tagged link). See "Testing After Enabling Auto-tagging," in Chapter 6.

The following examples demonstrate the best ways to tag the four most common kinds of online campaigns: banner ads, e-mail campaigns, paid keywords (pay-per-click campaigns), and digital collateral. Note that a landing page URL is specific to the campaign you create it for—do not use it anywhere else!

Tagging Banner Ad URLs

Consider the following hypothetical marketing scenario on the AOL.com website: You have a graphical banner for branding purposes and an organic listing from the non-paid organic listings. AOL has informed you that the banner will only display when a visitor searches for the term *shoes*; and in this case the banner campaign is about *Sprint shoes*. That's two different campaigns, from the same domain name (reported as aol.com), that can refer a visitor to your website.

Using the URL Builder tool, shown in Figure 7.2, you can differentiate visitors from banner click-throughs by supplying the resultant tagged landing page URL to the person or agency setting up your AOL banner. It is not necessary (or possible) to tag your AOL organic listing, as this will be detected automatically.

Google Analytics URL Builder

Fill in the form information and click the **Generate URL** button below. If you're new to tagging links or this is your first time using this tool, read How do I tag my links?

If your Google Analytics account has been linked to an active AdWords account, there's no need to tag your AdWords links - auto-tagging will do it for you automatically.

Step 1: Enter the URL of your website.

Website URL: * http://www.mysite.com/products/shoes.htm
(e.g. *http://www.urchin.com/download.html*)

Step 2: Fill in the fields below. **Campaign Source** and **Campaign Medium** are required values.

Campaign Source: *	AOL US	(referrer: google, citysearch, newsletter4)
Campaign Medium: *	Banners	(marketing medium: cpc, banner, email)
Campaign Term:	Shoes	(identify the paid keywords)
Campaign Content:		(use to differentiate ads)
Campaign Name*:	Sprint sales	(product, promo code, or slogan)

Step 3

Generate URL | Clear

http://www.mysite.com/products/shoes.htm?utm_source=AOL%2BUS&utm_

Figure 7.2
Tagging banner ad URLs

Tagging E-mail Marketing Campaigns

Continuing with the previous example, suppose you also plan to run a monthly e-mail newsletter that begins in July 2008. The newsletter is for the shoe department and concerns a summer promo. You want to ensure that all click-throughs from the e-mail campaign are tracked in your Google Analytics reports.

In addition to sending the e-mails, your marketing department wants to compare the effectiveness of sending plain-text e-mails versus HTML format, which includes rich-text formatting and images. They would like to know whether visits and conversions vary depending on the format of the sent e-mail (this is the basis of A/B split testing).

Tracking these two e-mail campaigns can be achieved using the example landing page URLs shown in Figure 7.3. In both cases, the Campaign Content field is used to differentiate the formatting.

You then supply the resultant tagged landing page URL to the person setting up your e-mail marketing. Of course, links in your e-mail message may point to different landing pages on your website, so the URL should be adjusted accordingly. For example, shoes.htm may become boots.htm. However, the tracking parameters will remain the same.

a)

Google Analytics URL Builder

Fill in the form information and click the **Generate URL** button below. If you're new to tagging links or this is your first time using this tool, read How do I tag my links?

If your Google Analytics account has been linked to an active AdWords account, there's no need to tag your AdWords links - auto-tagging will do it for you automatically.

Step 1: Enter the URL of your website.

Website URL: * http://www.mysite.com/products/shoes.htm
(e.g. *http://www.urchin.com/download.html*)

Step 2: Fill in the fields below. **Campaign Source** and **Campaign Medium** are required values.

Campaign Source: *	July-08 Newsletter	(referrer: google, citysearch, newsletter4)
Campaign Medium: *	Email	(marketing medium: cpc, banner, email)
Campaign Term:	Shoes	(identify the paid keywords)
Campaign Content:	text	(use to differentiate ads)
Campaign Name*:	Summer promo	(product, promo code, or slogan)

Step 3
Generate URL | Clear

http://www.mysite.com/products/shoes.htm?utm_source=July-08%2BNewsl

b)

Google Analytics URL Builder

Fill in the form information and click the **Generate URL** button below. If you're new to tagging links or this is your first time using this tool, read How do I tag my links?

If your Google Analytics account has been linked to an active AdWords account, there's no need to tag your AdWords links - auto-tagging will do it for you automatically.

Step 1: Enter the URL of your website.

Website URL: * http://www.mysite.com/products/shoes.htm
(e.g. *http://www.urchin.com/download.html*)

Step 2: Fill in the fields below. **Campaign Source** and **Campaign Medium** are required values.

Campaign Source: *	July-08 Newsletter	(referrer: google, citysearch, newsletter4)
Campaign Medium: *	Email	(marketing medium: cpc, banner, email)
Campaign Term:	Shoes	(identify the paid keywords)
Campaign Content:	html	(use to differentiate ads)
Campaign Name*:	Summer promo	(product, promo code, or slogan)

Step 3
Generate URL | Clear

http://www.mysite.com/products/shoes.htm?utm_source=July-08%2BNewsl

Figure 7.3
Tagging e-mail campaigns as (a) text format; (b) HTML format

Plain text versus HTML email

Studies show that recipients are more likely to click on links in html emails than plain text—see for example MailerMailer Email Metrics Report, Jan-Jun 2006.

According to E-consultancy's Online Marketing Benchmarks 2004 for the UK, HTML generally generates 20-40 percent more response than an equivalent plain text version. The caveat is that this is very dependent on the target market and products/services in question.

Tagging Paid Keywords

As discussed earlier in this section, Google automatically tags your paid keywords (Google AdWords campaigns). However, campaigns running on other paid networks do require tagging. Otherwise, a paid visitor will be reported as an organic (non-paid) visitor. Figure 7.4 shows an example URL Builder to differentiate Yahoo! organic (non-paid) visitors from pay-per-click Yahoo! Search Marketing visitors.

You supply the resultant tagged landing page URL to the person setting up your pay-per-click campaigns. A similar approach should be used for other pay-per-click accounts that you run—for example, Microsoft adCenter. The only difference is that the Campaign Source would be set as “adCenter” (or any phrase you wish to use to identify such visitors).

Google Analytics URL Builder

Fill in the form information and click the **Generate URL** button below. If you're new to tagging links or this is your first time using this tool, read How do I tag my links?

If your Google Analytics account has been linked to an active AdWords account, there's no need to tag your AdWords links - auto-tagging will do it for you automatically.

Step 1: Enter the URL of your website.

Website URL: * http://www.mysite.com/products/shoes.htm
(e.g. *http://www.urchin.com/download.html*)

Step 2: Fill in the fields below. **Campaign Source** and **Campaign Medium** are required values.

Campaign Source: *	YSM	(referrer: google, citysearch, newsletter4)
Campaign Medium: *	ppc	(marketing medium: cpc, banner, email)
Campaign Term:	Red Shoes	(identify the paid keywords)
Campaign Content:		(use to differentiate ads)
Campaign Name*:	Shoes	(product, promo code, or slogan)

Step 3
Generate URL | Clear

http://www.mysite.com/products/shoes.htm?utm_source=YSM&utm_mediur

Figure 7.4
Tagging paid keywords

Note: Google AdWords auto-tagging always labels AdWords visitors as `medium = cpc` (cost-per-click). You may wish to continue this labeling convention for Yahoo! Search Marketing, Microsoft adCenter, and other pay-per-click networks so they are reported together when viewing medium reports. However, as AdWords is currently so prevalent for online advertising, I have found it useful to group all other pay-per-click networks as `medium = ppc` and treat them as if they were a separate medium. This enables them to be compared against AdWords as a whole.

Tagging Embedded Links within Digital Collateral

If you host non-HTML content on your website, such as `catalogue.pdf`, `spec-sheet.doc`, `price-matrix.xls`, you probably have links within those documents that point back to your website. By tagging these links, you can track visits that result from those documents, which in turn will enable you to monetize your digital collateral. Without tagging, visitors from your digital collateral are labeled as direct—that is, they are grouped together with visitors who typed the URL directly into their browser or bookmarked your site from a previous visit.

Using the method shown in Figure 7.5 ensures that links from your digital collateral are given credit for referring visitors to your website. Supply the resultant tagged landing page URL to the people that create such documents. Alternatively, coach your content creators to use the URL Builder tool themselves. That way, they will be tracking links as an integral part of their content creation and design process.

Google Analytics URL Builder

Fill in the form information and click the **Generate URL** button below. If you're new to tagging links or this is your first time using this tool, read How do I tag my links?

If your Google Analytics account has been linked to an active AdWords account, there's no need to tag your AdWords links - auto-tagging will do it for you automatically.

Step 1: Enter the URL of your website.

Website URL: * http://www.mysite.com/products/shoes.htm
(e.g. *http://www.urchin.com/download.html*)

Step 2: Fill in the fields below. **Campaign Source** and **Campaign Medium** are required values.

Campaign Source: *	Catalogue	(referrer: google, citysearch, newsletter4)
Campaign Medium: *	pdf	(marketing medium: cpc, banner, email)
Campaign Term:		(identify the paid keywords)
Campaign Content:		(use to differentiate ads)
Campaign Name*:		(product, promo code, or slogan)

Step 3

Generate URL | Clear

http://www.mysite.com/products/shoes.htm?utm_source=Catalogue&utm_m

Figure 7.5 Tagging embedded links within digital collateral

Custom Campaign Fields

If you have been using another tracking methodology or tool, you have probably already manually tagged your landing page URLs for paid campaigns, banners, e-mail, and digital collateral. Rather than disregard these, or append the additional Google Analytics variables, it is possible to configure Google Analytics to recognize your existing tags.

Note: This technique is only applicable for landing page URLs that have been previously manually tagged for other tracking purposes. It is not applicable, or required, for AdWords tracking—assuming you are auto-tagging your AdWords campaigns as described in "Getting AdWords Data: Linking to Your AdWords Account," in Chapter 6.

Add the following highlighted code to your GATC, replacing *orig-name* with the variable name that you are currently using. If no original value exists, then omit that line from your GATC.

```
<script type="text/javascript">
   var gaJsHost = (("https:" == document.location.protocol) ? "https://ssl."
: "http://www.");
   document.write(unescape("%3Cscript src='" + gaJsHost + "google-
analytics.com/ga.js' type='text/javascript'%3E%3C/script%3E"));
</script>
<script type="text/javascript">
   var pageTracker = _gat._getTracker("UA-12345-1");
   pageTracker._setCampNameKey("orig_campaign");     // default: utm_medium
   pageTracker._setCampMediumKey("orig_medium");     // default: utm_medium
   pageTracker._setCampSourceKey("orig_source");     // default: utm_source
   pageTracker._setCampTermKey("orig_term");         // default: utm_term
   pageTracker._setCampContentKey("orig_content");   // default: utm_content
   pageTracker._initData();
   pageTracker._trackPageview();
</script>
```

At a minimum, *orig-source* and *orig-medium* are required. If these are not present in your current landing page URLs, you need to include the Google Analytics equivalents.

Event Tracking

Google Analytics is capable of tracking any browser-based event, including Flash and JavaScript events. Event activity is reported separately from your pageview activity and can be used to track the following:

- Any Flash-driven element, such as a Flash website or a Flash movie player
- Embedded Ajax page elements such as, onClick, onSubmit, onReset, onMouseOver, onMouseOut, onMouseMove, onSelect, onFocus, onBlur, onKeyPress, onChange, onLoad, onUnload etc.
- Page gadgets
- File downloads
- Load times for data

Event tracking uses standard JavaScript method calls and provides a hierarchy of objects and actions. The data model includes objects, actions, labels, and values.

Note: A word of caution: At the time of publishing, Event Tracking is a beta feature of Google Analytics. It is therefore likely that the final implementation syntax will vary. Keep up to date with any changes by visiting: www.advanced-web-metrics.com.

Setting Up Event Tracking

Follow these four steps to set up event tracking:

1. Define the set of events you want to track.
2. Enable Event Tracking in your reporting profile.
3. For each set of events, create an event tracker instance.
4. Call the _trackEvent() method in your web page source code.

Note: Event tracking reporting is not enabled by default in Google Analytics. Enable this in your Analytics Settings page. Within the desired profile, click edit link by the "Main Website Profile Information" section. Select the Event Tracking Enabled option and save your changes. This makes the event reports visible in the Content section of Google Analytics.

Assuming you have completed steps 1 and 2, the next step is to create an event tracker instance.

Creating an Event Tracker Instance

In this example, a video tracker object is created with the name "Video" within your GATC. This can be added specifically to the page for which you wish to track events, or to every page with the GATC:

```
<script type="text/javascript">
   var gaJsHost = (("https:" == document.location.protocol) ? "https://ssl."
: "http://www.");
   document.write(unescape("%3Cscript src='" + gaJsHost + "google-
analytics.com/ga.js' type='text/javascript'%3E%3C/script%3E"));
</script>
<script type="text/javascript">
   var pageTracker = _gat._getTracker("UA-12345-1");
   pageTracker._initData();
   pageTracker._trackPageview();

   //creates an event tracker object with the name "video"
   var videoTracker = pageTracker._createEventTracker("Video");
</script>
```

Note: The _createEventTracker() declaration is order-dependent and has to be called after the page tracker code _trackPageview has loaded and initialized.

Calling the _trackEvent() Method in Your Source Code

Insert this method in the source code for your video, gadget, or other web element. The syntax for the _trackEvent() method is as follows:

```
_trackEvent(action, optional_label, optional_value)
```

- *action* (required)—A string you pass to the class instance to track event behavior or elements
- *optional_label*—An optional string you pass to the class instance to provide additional classification for the object. Note that any spaces used in the label parameter must be encoded as %20
- *optional_value*—An integer that you can use to provide numerical data about the user event, such as time or a dollar amount

Flash Events

This example illustrates how to track a visitor interaction with the play button on a Flash video player. To begin, define the object videoTracker as the last entry in your GATC within your site's HTML, with the name "Video," as follows:

```
//creates an event tracker object with the name "Video"
var videoTracker = pageTracker._createEventTracker('Video');
```

Then, within your Flash application, call the *videoTracker* object and pass the term "Play" to use as the action associated with the user event and a label to identify the video name.

```
onRelease (button) {
   getURL ("javascript:videoTracker._trackEvent('Play','Ratatouille');")
}
```

Note: onRelease() and getURL() are supported under ActionScript 1.0 and 2.0.

Here, the action name and label for the movie are supplied in the Flash code for the play button. Used in this way, an example event summary could be as per Table 7.4.

► **Table 7.4** Event Reporting Example

Object	Action	Label
Video	Play	Ratatouille, The Incredibles, Ice Age 2

Other Flash buttons can have their events defined in a similar way, such as Stop and Pause. Multiple videos can be tracked by passing different labels, assuming they are hosted on pages that have the same *videoTracker* defined on their GATC. Thus, to track three movies, your video object might be reported as per Table 7.5. An example of an Event Tracking Labels report is shown in Figure 7.6.

Table 7.5 Event Reporting Example

Object	Action	Label
Video	Play	Ratatouille, The Incredibles, Ice Age 2
	Pause	Ratatouille, The Incredibles
	Stop	Ratatouille

Figure 7.6 Event Tracking Labels report

Extending the Flash example further, when the video is placed on the web page, you can use the FlashVars parameter to provide individual *label* and *value* input values. FlashVars is the Flash counterpart to a URL query string. That is, it's a way to pass data or variables from HTML to a Flash movie. Variables passed via FlashVars are placed into the _root level of the Flash movie, as shown in the following example:

```
<object classid="clsid:D27CDB6E-AE6D-11cf-96B8-444553540000" ➥
codebase="http://download.macromedia.com/pub/shockwav➥
e/cabs/flash/swflash.cab#version=7,0,19,0" width="300" height="400">
```

```
    <param name="FlashVars" value="label=The%20Incredibles&value=9" />
    <param name="movie" value="movie1.swf" />
    <param name="quality" value="high" />
    <embed src="movie1.swf" ➡
    FlashVars="label=The%20Incredibles&value=9" quality="high" ➡
    pluginspage="http://www.macromedia.com/go/getflashpla➡
    yer" type="application/x-shockwave-flash" width="300"➡
    height="400"></embed>
</object>
```

This makes your Flash code within the player more generic and therefore easier to maintain—you reuse the same code for each movie. For example, within your Flash application, call the *videoTracker* object as follows:

```
onRelease (button) {
   getURL ("javascript:videoTracker._trackEvent('Pause'" + label + "," +
value + ");")
}
```

Note: The FlashVars parameter works with Flash Player 6 (Flash MX or newer).

Page Load Time

This example demonstrates how page load time can be measured, in milliseconds, by passing a value for an event. The example shown creates a timestamp at the top and bottom of an HTML page using the JavaScript Date() method. The difference between the two timestamps is passed to the _trackEvent() call:

```
<body>
<script type="text/javascript">
   var Begin = new Date();
   var Start = Begin.getTime();
</script>

   [ ... PAGE BODY CONTENT ... ]

<script type="text/javascript">
   var gaJsHost = (("https:" == document.location.protocol) ? "https://ssl."
: "http://www.");
   document.write(unescape("%3Cscript src='" + gaJsHost + "google-
analytics.com/ga.js' type='text/javascript'%3E%3C/script%3E"));
</script>
```

```
<script type="text/javascript">
   var pageTracker = _gat._getTracker("UA-12345-1");
   pageTracker._initData();
   pageTracker._trackPageview();

   //creates an event tracker object with the name "Page Load"
   var loadTracker = pageTracker._createEventTracker('Page Load');

   var End = new Date();
   var Stop = End.getTime();
   var timeElapse = Stop - Start; // stored as milliseconds
   loadTracker._trackEvent('Load –
   Time','products/pageX.htm',timeElapse);
</script>
</body>
```


The event summary is shown in Table 7.6.

► **Table 7.6** Load Time Reporting Example

Object	Action	Label	Value
Page Load	Load Time	products/pageX.htm	3724
	Load Time	demo/pageY.htm	4842
	Load Time	products/pageZ.htm	7703

From your event tracking reports, you can also determine the average time for page loads across your entire site, as well as the average page load for individually tracked pages.

Note: Note that the values in Table 7.6 are shown in milliseconds. This is because only integers can be stored in the value field for an event. By default, computer operating systems report time in milliseconds.

Banners and Other Outgoing Links

If you publish advertising banners on your site or refer visitors to other websites, there is an easy way for you to track which banners and links visitors click to leave your site. You can also monetize these individually. First define the object exitTracker as the last entry in your GATC, with the name “Exit Points”:

```
//creates an event tracker object with the name "Exit Points"
var exitTracker = pageTracker._createEventTracker('Exit Points');
```

For an animated GIF or other non-Flash banner ad, modify the outgoing link as follows:

```
<a href="http://www.advertiser-site.com"
onClick="exitTracker._trackEvent('Click','advertisername - Ad version A',
4)"><img src="bannerA.gif"></a>
```

Note that a value of *4* has been assigned to this event (a click-through). The equivalent code used within a Flash banner, assigned with a higher monetary value, is as follows:

```
onRelease (button) {
   getURL ("javascript:exitTracker._trackEvent('Click','advertisername - Ad
version B', 5);")
}
```

I prefer to use action names to distinguish object elements. For example, rather than aggregate clicks on all banner types together, you could differentiate between Flash and GIF banner click-throughs as follows:

GIF banner event tracking

```
<a href="http://www.advertiser-site.com"
onClick="exitTracker._trackEvent('Click - GIF banner','advertisername - Ad
version A', 4)"><img src="bannerA.gif"></a>
```

Flash banner event tracking

```
getURL ("javascript:exitTracker._trackEvent('Click - FLASH
banner','advertisername - Ad version A', 5);")
```

To wrap up this series of outbound click tracking, for an outbound link, use the following example:

Link event tracking

```
<a href="http://www.advertiser-site.com"
onClick="exitTracker._trackEvent('Click - link','linkURL', 1)">View our
Partner</a>
```

Mailto: Clicks

The mailto: link is another outgoing link that can be tracked in exactly the same way as described above. I discuss it here separately simply to emphasize the importance of tracking mailto: clicks—particularly for non-e-commerce websites, where any action that can bring a visitor closer to lead generation for you has an intrinsic value. As your sales department follows these contacts up, you will be able to assess the conversion rate and average order value of such leads and therefore monetize the mailto: onClick event.

Use the same exitTracker object defined in the preceding section as the last entry of your GATC. Then modify your mailto: links as follows:

```
<a href="mailto:mail@mysite.co.uk" onClick="exitTracker._trackEvent('Click -
email','mail@mysite.co.uk')">mail@mysite.co.uk</a>
```

Add a monetary value to this event as desired. The tracking of mailto: click-throughs is shown in the report of Figure 7.6 (row 6).

Customizing the GATC

For the majority of websites, you won't need to make any customizations to your GATC. However, should the need arise, the following sections describe some available options you can use.

Subdomain Tracking

Google Analytics uses first-party cookies, which means collected information is associated with your fully qualified host name—for example, www.mysite.com. Only your fully qualified host name can read or set its first party cookies. This is a built-in security feature of all web browsers.

A *subdomain* is one that is a part of the parent domain. In this example, the parent domain is mysite.com, so www is actually a subdomain of mysite.com. Other example subdomains include secure.mysite.com, ww2.mysite.com, en.mysite.com, and so on.

Note: Any name can be used as a subdomain as long as it contains only alphanumeric characters and the hyphen (-). Of course, you can only use a subdomain if your DNS has been configured for it.

Subdomains have their own fully qualified hostnames. That means by default Google Analytics cannot track visitors that traverse different subdomains on your website because it uses first-party cookies. Fortunately, modifying this behavior for your own domains is straightforward. This is achieved by combing all your subdomain data under the one parent domain. To accomplish this, set your parent domain in the GATC so that the Google Analytics first-party cookies can be shared across your subdomains, as highlighted below:

```
<script type="text/javascript">
   var gaJsHost = (("https:" == document.location.protocol) ? "https://ssl."
: "http://www.");
   document.write(unescape("%3Cscript src='" + gaJsHost + "google-
analytics.com/ga.js' type='text/javascript'%3E%3C/script%3E"));
</script>
```

```
<script type="text/javascript">
   var pageTracker = _gat._getTracker("UA-12345-1");
   pageTracker._setDomainName("mysite.com");
   pageTracker._initData();
   pageTracker._trackPageview();
</script>
```

No further modifications are required. However, bear in mind when doing this that you cannot distinguish to which subdomain the visit occurred. For example, visits to sub.mysite.com/index.html and www.mysite.com/index.html will be shown in your Google Analytics reports as the same page—that is, both /index.html. You can differentiate these two pages by applying the filter as shown in Figure 7.7.

Add Filter to Profile
Choose method to apply filter to Website Profile
Please decide if you would like to create a new filter or apply an existing filter to the Profile.
Add **new** Filter for Profile OR Apply **existing** Filter to Profile
Add new Filter for Profile
Filter Name: Insert subdomain
Filter Type: Custom filter
Exclude
Include
Lowercase
Uppercase
Search and Replace
Lookup Table
Advanced
Field A -> Extract A: Hostname (.*)
Field B -> Extract B: Request URI (.*)
Output To -> Constructor: Request URI /$A1$B1
Field A Required: Yes No
Field B Required: Yes No
Override Output Field: Yes No
Case Sensitive: Yes No
Cancel Finish »

Figure 7.7 Filter to differentiate identical subdomain page names

Note: The filter shown in Figure 7.7 will make site overlay inoperable and may require you to modify your goal settings accordingly. However, I find the loss of the site overlay report is more than compensated by the greater insight that applying this filter provides.

By using this filter, page names will include your subdomains. For example, in the Content > Top Content reports will be sub.mysite.com/index.html and www.mysite.com/index.html respectively.

The use of filters is discussed in detail in Chapter 8.

How to track subdomains into separate profiles

This is the default behavior when you setup your Google Analytics profiles. That is, enter the correct subdomain name in the URL field of the "Add a Profile for a new domain" section. This will generate a unique tracking code for each subdomain, so ensure that you add the correct code to your pages.

Creating profiles is discussed in Chapter 8 in "Creating Additional Profiles."

Multiple Domain Tracking

As discussed in the previous section, web browsers have built-in security features that prevent the sharing of first-party cookies with other domains. If your website passes a visitor around to different parent domains, then this needs special consideration.

Consider the following example: Your main website is www.mysite.com and you host regional variations (language, currency, etc.) on different parent domains such as www.mysite.co.uk. Both sites are tagged with your GATC. A visitor arrives on www.mysite.com by clicking a link from a search results page on www.google.com for example. Next, they click the option to select your regional version at www.mysite.co.uk. A conversion is then made on this site.

Note: Google Analytics cannot track visitors traversing the Web to unrelated domains. It can only track visitors across domains that you own or control and to which you can add your GATC.

By default, the visitor converting at www.mysite.co.uk will be reported as a referral visitor from www.mysite.com. The original referral information (search at www.google.com) is lost because the cookie information cannot follow the visitor to the second (third party) domain.

If you maintain separate Google Analytics profiles for these two websites, then all page metrics (time on site, page depth, bounce rate, etc.) will be counted separately—in this example, a one-page visit for www.mysite.com and $x + 1$ page visits for www.mysite.co.uk. On the other hand, if you have configured data for both websites to be collected into a single profile, then your page metrics will be skewed with overinflated numbers of single-page visits. Clearly, this is not the outcome you want.

The solution for tracking visitors across multiple sites is to maintain the session by transferring cookies across the multiple domains. There are two methods of achieving this, depending on how you forward visitors to your other domains. These are similar to those discussed earlier (see "E-Commerce Tracking—Using a Third-Party Gateway"), as in both cases first-party cookies need to be handed over to a third-party domain.

Method 1: Track a visitor across domains when using a link.

This is achieved by sending cookie information via URL parameters (HTTP GET) to the receiving domain. First, modify your GATC for all pages on all your domains as shown in the following highlighted code:

```
<script type="text/javascript">
   var gaJsHost = (("https:" == document.location.protocol) ? "https://ssl."
: "http://www.");
   document.write(unescape("%3Cscript src='" + gaJsHost + "google-
analytics.com/ga.js' type='text/javascript'%3E%3C/script%3E"));
</script>
<script type="text/javascript">
   var pageTracker = _gat._getTracker("UA-12345-1");
   pageTracker._setDomainName("none");
   pageTracker._setAllowLinker(true);
   pageTracker._initData();
   pageTracker._trackPageview();
</script>
```

Then, within your web page HTML documents, modify all links to your other domains as follows:

```
<a href="http://www.mysite.co.uk"
onclick="pageTracker._link('http://www.mysite.co.uk/');
return false;">Go to our UK web site</a>
```

With this method, the Google Analytics cookies are passed to the receiving domain by appending them to the URL string. If you see *__utma*, *__utmb* and *__utmc* parameters in the URL of the landing page, then this has worked.

Note: Note the use of `return false;` here. This ensures that for visitor browsers that have JavaScript disabled, the href link will be followed without error. Of course, if JavaScript is disabled, Google Analytics tracking won't occur. However, the modified link will still work.

Method 2: Track a visitor across domains when using a form.

Use this second method when you are passing visitors to another domain using a form. In this case, sending cookie information is achieved via HTTP POST to the receiving domain. Exactly as you did for Method 1, modify your GATC on all the pages of all your domains.

Then, within your web page HTML documents, modify all form references to your other domains as follows:

```
<form method="post" onsubmit="pageTracker._linkByPost(this)">
...
</form>
```

If you already have an onSubmit validation routine, you append the cross domain modification to your existing function call as follows:

```
<form method="post"
onsubmit="validate_routine(this);pageTracker._linkByPost(this)">
...
</form>
```

With this method, the Google Analytics cookies are passed to the receiving domain via the HTTP headers. This will work even for forms where method="GET". You can verify if this has worked by viewing the HTTP headers sent in Firefox using the add-on LiveHTTPheaders (http://livehttpheaders.mozdev.org).

Note: It is possible to only modify the GATCs of the pages where your cross-domain linking occurs. In the example given, this would be the home pages of www.mysite.com and www.mysite.co.uk respectively. However, in such scenarios it is very common to have multiple cross-domain points throughout the website, and so it is better to make these changes site-wide to ensure good data alignment.

Restricting Cookie Data to a Subdirectory

By default, Google Analytics first-party cookies can be viewed by any page on your domain. If you want to restrict the use of cookies to a subdirectory—for example, in cases where you only own a subdirectory of the parent domain—you can set the preferred cookie path in your GATC using the _setCookiePath() function:

```
<script type="text/javascript">
   var gaJsHost = (("https:" == document.location.protocol) ? "https://ssl."
: "http://www.");
   document.write(unescape("%3Cscript src='" + gaJsHost + "google-
analytics.com/ga.js' type='text/javascript'%3E%3C/script%3E"));
</script>
```

```
<script type="text/javascript">
   var pageTracker = _gat._getTracker("UA-12345-1");
   pageTracker._setCookiePath("/path/of/cookie/");
   pageTracker._initData();
   pageTracker._trackPageview();
</script>
```

To copy existing cookies from other subdirectories on your domain, use the function _cookiePathCopy() as follows:

```
<script type="text/javascript">
   var gaJsHost = (("https:" == document.location.protocol) ? "https://ssl."
: "http://www.");
   document.write(unescape("%3Cscript src='" + gaJsHost + "google-
analytics.com/ga.js' type='text/javascript'%3E%3C/script%3E"));
</script>
<script type="text/javascript">
   var pageTracker = _gat._getTracker("UA-12345-1");
   pageTracker._initData();
   pageTracker._trackPageview();
   pageTracker._cookiePathCopy("/new/path/for/cookies/");
</script>
```


Controlling Timeouts

Two cookie timeouts can be controlled from within your GATC: the session timeout and the campaign conversion timeout.

By default, a visitor's session (visit) times out after 30 minutes of inactivity, so if a visitor continues browsing your website after 31 minutes of inactivity, that visitor is counted as a returning visitor. The original referral information is maintained as long as a new referral source was not used to continue their session.

The 30-minute rule is the unwritten standard across the web analytics industry. However, their may be instances when you wish to change this. Typical examples include when your visitors are engaging with music or video or reading lengthy documents during their visit. The latter is a less likely scenario, as large documents are usually printed and read offline by visitors. However, music and video sites are common examples in which visitors set and forget their actions, only to return and complete another action on your site when the content has finished playing.

If inactivity is likely to last longer than 30 minutes for a continuous visit, then consider increasing the default session timeout as follows:

```
<script type="text/javascript">
   var gaJsHost = (("https:" == document.location.protocol) ? "https://ssl."
: "http://www.");
```

```
    document.write(unescape("%3Cscript src='" + gaJsHost + "google-
analytics.com/ga.js' type='text/javascript'%3E%3C/script%3E"));
</script>
<script type="text/javascript">
    var pageTracker = _gat._getTracker("UA-12345-1");
    pageTracker._setSessionTimeout("3600");
    // increased to 1 hour
    pageTracker._initData();
    pageTracker._trackPageview();
</script>
```

Note: In Google Analytics, time is measured in seconds. Therefore, 30 minutes = 1,800 seconds, one hour = 3,600 seconds, and so forth.

Another timeout that can be adjusted is the length of time for which Google Analytics credits a conversion referral. By default, the campaign conversion timeout is six months (15,768,000 seconds), after which the referral cookie (__utmz) expires. An example where you may wish to reduce this value is when you are paying a commission to affiliates. This can be achieved as follows:

```
<script type="text/javascript">
    var gaJsHost = (("https:" == document.location.protocol) ? "https://ssl."
: "http://www.");
    document.write(unescape("%3Cscript src='" + gaJsHost + "google-
analytics.com/ga.js' type='text/javascript'%3E%3C/script%3E"));
</script>
<script type="text/javascript">
    var pageTracker = _gat._getTracker("UA-12345-1");
    pageTracker._setCookieTimeout("2592000");
    // decreased to 30 days
    pageTracker._initData();
    pageTracker._trackPageview();
</script>
```

The value of the campaign conversion timeout can also be increased. However, it doesn't make much sense to go beyond six months, due to the increased risk that the original cookie information is likely to be lost—making your conversion referral data less reliable. See "Getting Comfortable with Your Data and Its Accuracy," in Chapter 2.

Setting Keyword Ignore Preferences

You can configure Google Analytics to treat certain keywords as direct traffic (i.e., not as a referral)—for example, visitors who type your domain (www.mysite.com) into a search engine.

Use _addIgnoredOrganic() to treat a keyword as a referral, or _addIgnoredRef() to treat a referral as direct, as shown here:

```
<script type="text/javascript">
   var gaJsHost = (("https:" == document.location.protocol) ? "https://ssl."
: "http://www.");
   document.write(unescape("%3Cscript src='" + gaJsHost + "google-
analytics.com/ga.js' type='text/javascript'%3E%3C/script%3E"));
</script>
<script type="text/javascript">
   var pageTracker = _gat._getTracker("UA-12345-1");
   pageTracker._addIgnoredOrganic("mysite.com");
   pageTracker._addIgnoredRef("sistersite.com");
   pageTracker._initData();
   pageTracker._trackPageview();
</script>
```

Although these variables are available for you to adjust, I recommend that you do not use them. Discovering that your brand is being used in the search engines as a keyword is an important piece of information that can be used to evaluate your brand effectiveness.

In terms of treating a particular referral as direct, if you have multiple domain names, then you probably want to see the interaction between them. If not, then consider using 301 redirect codes on your web server (or .htaccess file) to ensure that all visitors and search engine robots are forwarded to your main domain.

Note: Further information on redirection for the Apache web server can be found at http://httpd.apache.org/docs/1.3/mod/mod_alias.html#redirect

Controlling the Collection Sampling Rate

By default, Google Analytics collects pageview data for every visitor. For very high traffic sites, the amount of data can be overwhelming, leading to large parts of the "long tail" of information to be missing from your reports, simply because they are too far down in the report tables. You can diminish this issue by creating separate profiles of visitor segments—for example, /blog, /forum, /support, etc. However, another option is to *sample* your visitors.

Sampling occurs at the visitor level and is specified as a percentage of the total to sample using the _setSampleRate() function, as shown here:

```
<script type="text/javascript">
   var gaJsHost = (("https:" == document.location.protocol) ? "https://ssl."
: "http://www.");
   document.write(unescape("%3Cscript src='" + gaJsHost + "google-
analytics.com/ga.js' type='text/javascript'%3E%3C/script%3E"));
</script>
<script type="text/javascript">
   var pageTracker = _gat._getTracker("UA-12345-1");
   pageTracker._setSampleRate(25);
   // set sample rate to 25%
   pageTracker._initData();
   pageTracker._trackPageview();
</script>
```

A sample rate of 25 percent means that every fourth visitor is counted for Google Analytics tracking. Unless you receive more than one million visitors per day, it is unlikely you will need to use the _setSampleRate() function.

Summary

Having read this far, you will have now tagged all of your website pages with the GATC, tagged your landing page URLs, adjusted your setup for tracking file downloads and event tracking, and modified your checkout completion page for the capture of e-commerce transactions, if you have such a facility on your site.

With all that in place, your installation is complete. Take an initial look at some of your reports and get comfortable with using them, as described in Part II.

In Chapter 7, you have learned how to do the following:

- Use the _trackPageview() function to create virtual pageviews
- Capture e-commerce transactions
- Track online campaigns in addition to AdWords
- Customize the GATC for your specific needs

Best Practices Configuration Guide

Having read so far, you should now have your Google Analytics account set up and collecting good quality data. To gain a better understanding of visitor behavior, this chapter will help you with your configuration. No modifications of the Google Analytics Tracking Code (GATC) or your pages are required here; all configuration is managed within the Google Analytics administration interface.

For this chapter, it is important that the marketer and webmaster work together to understand each other's needs. The marketer will be building the marketing strategy, and that requires working in conjunction with the webmaster to implement the necessary configuration changes. If you are a part of a large organization, then it is the analyst who manages and oversees this part of the project.

In Chapter 8, you will learn about the following:

Best practices for configuring Google Analytics
The importance of defining goals and funnels
The importance of visitor segmentation
How to use filters for segmentation

Initial Configuration

Once you have established your first Google Analytics profile—created as part of your initial account creation process—there are a couple of options you should configure, as shown in Figure 8.1. To access this area within your account you need to have administrator access. From the initial login area, click the Edit link next to your profile name.

Figure 8.1 Initial profile setup options

Note: "Receiving data" will show a green tick (as shown in Figure 8.1) once you have added your GATC to your home page. Allow 24 hours for this. Note, Google Analytics will only check your home page for the presence of a correctly formatted GATC—not other web pages on your site. If you include the GATC as part of another loaded JavaScript file this will not work.

Apart from the time zone and localization of currency options, you should enter your default page and any URL query parameters for which reports are not required. Click the Edit link in the top, right corner to do this, which takes you to the screen as shown in Figure 8.2.

Setting the Default Page

Your Google Analytics settings, shown in Figure 8.2, contain a field to specify your default page. The default page is the web page your server defaults to when no page is specified—that is, the filename of your home page. This is usually `index.html`, `index.htm`, `index.php`, or `default.asp`, but it can be anything your web hosting company or webmaster has specified. By entering your default page, Google Analytics is able to combine visits to `www.mysite.com` and `www.mysite.com/index.html`, which are in fact the same page. If the default page is not specified, then these are reported as two separate pages, which is not desirable.

Excluding Unnecessary Parameters

If your site uses unique session IDs or displays other query parameters in your URLs that are of no interest to you, then exclude these parameters by entering them in the Exclude

URL Query Parameters field. In fact, it is best practice to do this, as it reduces the amount of superfluous data collected, making reports load faster and easier to read. Enter the variable name that you wish to exclude as it appears in your URLs. Variable name–value pairs follow a query symbol (?) in your URL and are separated by ampersands (&). Enter the name part you wish to exclude here—the part before the equals sign (=).

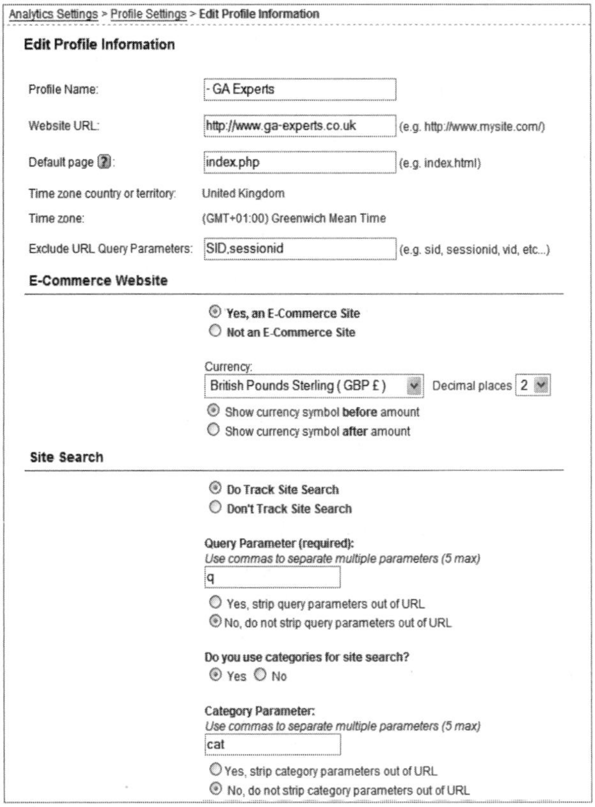

Figure 8.2
Editing profile information

Enabling E-Commerce Reporting

If your site has an e-commerce facility, you will want to see this data in your reports so that you can follow the complete visitor journey from referral source and pages viewed, through to check out. Selecting "Yes, an E-Commerce Site," as shown in Figure 8.2, enables this reporting; you will see it as a separate menu item on the left side of the reports and an additional tab within most report tables. If you have an e-commerce website, select your currency label and its placement, as well as the number of decimal places. Otherwise, keep the default selection of "Not an E-Commerce Site."

Enabling e-commerce provides additional reports within your account. To actually collect e-commerce data, you need to apply additional tags to the receipt page of your checkout system—see "E-Commerce Tracking," in Chapter 7.

Enabling Site Search

If your site has an internal search engine to help visitors locate content, you will want to see how this facility affects your visitor's experience. To do this, first select "Do Track Site Search," as shown in Figure 8.2. This enables an additional Google Analytics report menu that can be found in the Content > Site Search section.

With this feature enabled you need to define which query parameter in your URLs contains the visitor's site search term. You can usually discover this quickly by performing a site search yourself and looking for your search term in the result page URL. This is typically of the form `?q=mykeyword` or `&search=mykeyword`. For these examples, the query parameter names are "q" and "search," respectively.

Notice also that there is an option to strip your defined site search query parameters from the URL after site search processing has been completed. This can be helpful if those query parameters are of no further use to you for the purpose of Google Analytics reporting. However, those parameters may be important for defining your goals, your funnels steps, or your filters (see the next two sections). Site search query parameters could also be important if you are using virtual pageviews to aid in the reading of your reports (discussed under "trackPageview(): The Google Analytics Workhorse," in Chapter 7). Therefore, only strip query parameters if absolutely necessary.

Google Analytics Site Search also provides the option to define categories. Use this if your site search facility allows visitors to select a category for their search. For example, a retail site may have categories such as "menswear," "ladies wear," and so on. A real estate website may have categories such "apartments," "condos," "houses," and so on. Categories help users find information easier by focusing the search.

As with defining the Site Search query parameter, category parameters are obtained from the result page URL—for example, `?cat=menswear` or `§=condo`. For these examples, the category parameter names are "cat" and "sect," respectively. As with your defined query parameter, you can also strip your defined category parameters from the URL after site search processing has been completed. However, for the same reasons, only strip query parameters if absolutely necessary.

Note: Site Search processing takes place before filter processing. Although it is possible to apply filters that modify the site search query or category parameters (perhaps making them more reader friendly), these will not show in your Site Search reports.

What if my URLs don't contain Site Search parameters?

For this situation you can employ virtual pageviews to insert the parameters for you. For example, if your Site Search result page contains the visitor's query term as an environment variable, for example *%searchterm*, then you can use this as a virtual pageview. The following example is a modified GATC to achieve this:

```
<script type="text/javascript">
   var gaJsHost = (("https:" == document.location.protocol) ?
"https://ssl." : "http://www.");
   document.write(unescape("%3Cscript src='" + gaJsHost + "google-
analytics.com/ga.js' type='text/javascript'%3E%3C/script%3E"));
</script>
<script type="text/javascript">
   var pageTracker = _gat._getTracker("UA-12345-1");
   pageTracker._initData();
   pageTracker._trackPageview('/site search/?q=%searchterm');
</script>
```

In this example I have created a virtual pageview with a query parameter of "q" and its value set as the environment variable *%searchterm*. You can then use "q" as your Site Search query parameter as if this were the physical URL. The use of virtual pageviews is discussed in the section "trackPageview(): The Google Analytics Workhorse," in Chapter 7.

Goals and Funnels

As emphasized throughout this book, collecting data is only the first step in understanding the visitor performance of your website. Google Analytics has more than 80 built-in reports by default; that's impressive for fine-grain analysis, but it can be quite daunting to attempt to absorb all of this information, even for experienced users. In fact, I recommend you don't event attempt this.

To help you distill visitor information, configure Google Analytics to report on goal conversions. Think of goal conversions as specific measurable actions that can be applied to every visit. The path a visitor takes to reach a goal is known as the *funnel*; this is shown schematically in Figure 8.3. As you can see, the number of visitors entering the funnel process decreases at each step.

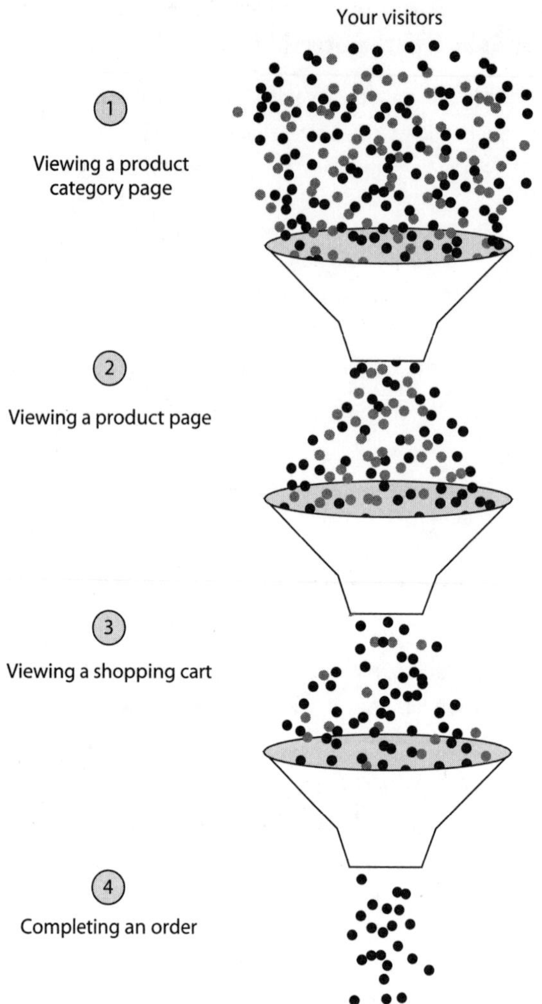

Figure 8.3 Schematic funnel and goal process

The Importance of Defining Goals

Defining your website goals is probably the single most important step of your configuration process, as it enables you to define success. An obvious goal for an e-commerce website is the completion of a transaction—that is, a purchase. However, not all visitors complete a transaction on their first visit; so another useful e-commerce goal is adding an item to the shopping cart, whether they complete or not—in other words, beginning the shopping process.

Whether you have an e-commerce website or not, your website has goals. A goal is typically the reason or reasons why you put up a website in the first place: Was it to sell directly, to gain more sales leads, to keep your clients informed and up to date, to provide a centralized product support forum, or to attract visitors to your stores? As you begin this exercise, you will realize that you actually have many website goals.

Also consider that goals don't have to include the full conversion of a visitor into a customer—that is obviously very important, but only part of the picture. If your only goal is to gain customers, then how will you know just how close non-customers came to converting? You can gain insight into this by using additional goals to measure the building of relationships with your visitors. For example, for most visitors arriving on your website, it is unlikely they will instantly convert, so the page needs to *persuade* them to go deeper—that is, get them one step closer to your goal. Table 8.1 lists some example goals.

Further reading on designing goal-driven websites

Bryan Eisenberg, his brother Jeffrey, and Lisa T. Davis have written extensively on the persuasion process technique and coined the phrase "persuasion architecture." Their books include *Call to Action* and *Waiting for Your Cat to Bark*.

Another worthwhile read when considering website goals and funnels is the excellent book *Don't Make Me Think* by Steve Krug (www.sensible.com/about.html). It's a commonsense approach to web usability written in an easy-to-read and humorous way.

Table 8.1 Sample Website Goals

Non-e-commerce goals	Examples
Visitors downloading a document	Brochure, manual, whitepaper, price list (file types include PDF, XLS, DOC, PPT, etc.)
Visitors looking at specific pages or sections of pages	Price list, special offers, login page, admin page, location and contact details, terms and conditions, help desk or support area
Visitors completing a form	Login, registration, feedback form, subscription
Visitor engagement	Adding a blog comment, submitting a forum post, adding or editing a profile, uploading content
e-commerce Goals	
Transaction completed	Credit card thank-you page
Transaction failed	Credit card rejection page
Visitors entering shopping system	Add to cart page

Note: Currently, it is not possible to configure an event as a goal in Google Analytics.

Apart from the goals shown in Table 8.1, your website may process negative goals—that is, goals for which you would like to decrease or minimize the conversion rate. For example, if onsite search is an important aspect of your website navigation structure, then minimizing the number of zero search results returned for a query is a valid ambition. Perhaps minimizing the number of searches per visitor is also an indication of an efficient onsite search tool; the theory could be that fewer searches conducted means visitors are finding what they are looking for more quickly. Negative goals are common for product support websites—that is, when the best visitor experience is for the least amount of engagement, such as time on site or page depth.

Defining and measuring goals is the basis for building your key performance indicators (KPIs). Chapter 10 defines and discusses KPIs in more detail, but essentially they enable you to incorporate web data into your overall business model.

Your Google Analytics profile can be configured for up to four goals

This limit may appear small, but your website goals should be focused enough that four goals covers your requirements. If they don't, then consider looking at the number of goals you wish to measure again. An obvious efficiency is to use wildcards—for example, `*.pdf` rather than individual PDF files. You can also create multiple carbon copy profiles with additional goals defined.

If you truly need more than four goals to measure your website effectiveness, read Chapter 11, "Monetizing Non-E-Commerce Websites," which is applicable for all non-e-commerce goals.

What Funnel Shapes Can Tell You

Many website owners and marketers want to see a 100 percent goal conversion rate. In the real world that just isn't feasible; in fact, it is not as desirable as you might think. Consider your funnel as acting like a sieve, qualifying visitors along the way. As with the offline world, it is important to qualify your web visitors so that your support or returns department is not swamped with calls from disappointed customers.

Losing visitors via your funnel is not necessarily a bad thing. Conversely, if you have verified all the qualifications before the visitor enters the funnel, then you would expect a high conversion rate. The outcome is highly dependent on how good your funnel pages are at doing their job—that is, persuading visitors to continue to the next step. Figure 8.4 shows example schematic funnel shapes.

Note: A detailed funnel analysis for a website is performed in Chapter 11.

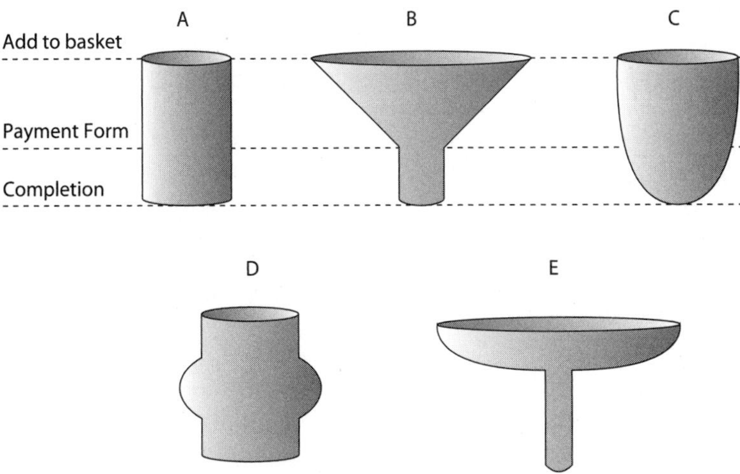

Figure 8.4 Schematic conversion funnel shapes

Figure 8.4 explained:

Shape A The near-impossible 100 percent conversion rate

Shape B The most common funnel shape, showing a sharp decrease in visitors until the payment form step. Assuming there are no hidden surprises to the visitor at this point, the vast majority of visitors who reach this point should convert.

Shape C A well-optimized conversion funnel process, with only a gradual decrease in visitors. This is the optimum shape you will wish to obtain for all your funnels.

Shape D An ill-defined funnel—visitors are entering the funnel midway through the process.

Shape E A poorly converting funnel with a serious barrier to progress

The most common shapes I have come across are B, D, and E. Shape A only occurs for a small section of an overall funnel process (if at all). Shape C is very rare and is where your greatest opportunity lies.

The Goal Setup Process

To set up your goals, log in to your Google Analytics account and click Edit in the Settings column, next to the profile to which you want to add a goal (or funnel). In the Conversion Goals and Funnel section, click Edit again. You will see the page shown in Figure 8.5.

Analytics Settings > Profile Settings > Goal Settings

Goal Settings: G1

Enter Goal Information

1 Goal URL: /club/submit_order\.asp (e.g. http://www.mysite.com/thankyou.html)
When the user navigates to this page, they have reached the conversion goal (Checkout Complete, Registration Confirmation, etc.).

Goal name: Order Complete
Goal name will appear in Conversion reports.

Active Goal: ◉ On ○ Off

Define Funnel (optional)

2 A funnel is a series of pages leading up to the Goal URL. For example, you might define the checkout steps that lead up to a completed purchase as a funnel. In this example, the funnel generally would not include individual product pages -- rather, it would consist only of those final pages that are common to all transactions.

The Defined Funnel Navigation report will show you how effectively you retain visitors throughout the conversion process.

	URL	Name	
Step 1	/club/search_text\.asp	Search Properties	☐ Required step
Step 2	/club/search_results\.asp	Search Results	
Step 3	/club/search_availability\.asp	Availability	
Step 4	/club/search_item_details\.asp	Booking Details	
Step 5	/club/search_extras\.asp	Extras?	
Step 6	/club/search_book_item\.asp	Enter Details	
Step 7	/club/order_payment\.asp	Order Payment	
Step 8			
Step 9			
Step 10			
Goal (see above)	/club/submit_order\.asp	Order Complete	

Additional settings

3 Case sensitive ☐
URLs entered above must exactly match the capitalization of visited URLs.

Match Type: Regular Expression Match

Goal value: 850

Figure 8.5 Example goal and funnel configuration

Define your goals using the three sections shown in Figure 8.5:

Section 1: Enter Goal Information First, specify a page URL that can only be reached by achieving a goal. Clearly, if your goal page can be reached by visitors who have not completed the goal, then your conversion rates will be inflated and not representative. An example goal for a visitor sign-up registration process would be the final Thank-You page URL.

Second, specify the name you will recognize when viewing reports. Examples of names you might use include "E-mail sign-up," "Article AB123 download," "Enquiry form sent," and "Purchase complete." Ensure that Active Goal is set to On.

Section 2: Define Funnel (optional) You may specify up to 10 page URLs in a defined funnel. These pages represent the path that you expect visitors to take on their way to converting to the goal. Defining these pages enables you to see which pages lead to goal abandonment and where they go next. For an e-commerce goal, these pages might be the Begin Checkout page, Shipping Address Information page, and Credit Card Information page—a three-step funnel.

By using wildcards in the configuration, you could extend this with a View Product Category page and a View Product Description page. This then provides a five-step funnel for analysis.

Use Funnels where appropriate

Not all goals have funnels. That is, not all conversions are achieved by visitors following a clearly defined linear path. An obvious linear path to conversion is an e-commerce shopping cart. You should certainly configure a defined funnel to analyze such a process.

However, for non e-commerce conversions, consider carefully whether a funnel is necessary. For example, if there are many paths to achieve a PDF download, then analyzing this with a funnel would be pointless at best and misleading at worse. If such downloads are a key element of measuring your website's success, then you could consider the addition of a registration form, to provide a funnel process for analysis.

What Is a Required Step? As you can see in Figure 8.5, there is a check box labeled Required Step next to the first funnel step. If this check box is selected, users reaching your goal page without traveling through this funnel page will not be counted as conversions.

The required step can be an important differentiator for you. For example, consider visitors accessing a password-protected area of your website. You want to define two goals: new auto-signups for access to this area and the log in of an existing user.

Figure 8.6 illustrates the scenario. In this case, completion of the registration process for new visitors to gain access leads to the same page that existing users visit when they log in—that is, the same goal URL page.

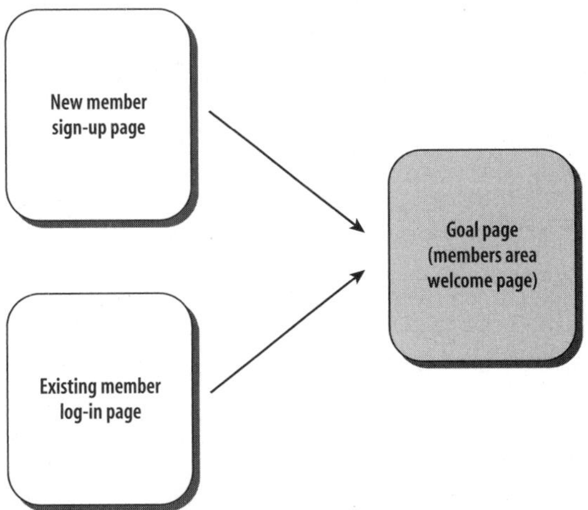

Figure 8.6 Differentiating two goals that have the same defined URL

Although the goal URL for this example is the same, the initial page is different—two different entry points leading to the same goal page. To differentiate these two, use the Required Step check box.

Note: Using this method to differentiate goals with the same URL will only show in reports that have funnel visualization in them. Other goal reports will show the same conversion rate for both examples, as it is only the funnel path that differentiates them.

Section 3: Additional settings

- Case sensitive

  If you want to differentiate URLs that are identical except for the fact that one uses uppercase characters and the other uses lowercase characters—for example `productx.html` and `productX.html`—then you should tick the "Case sensitive" check box. Most people do not change this, but it is there if needed.

- Match Type

  This determines how your defined URLs are matched. There are three ways to achieve this: Exact Match, Head Match, and Regular Expression Match.

  1. Exact Match

     This means exactly what it says—the exact URL of the page you want to define. No dynamic session identifiers and no wildcards can be used here, so it is best to cut and paste your URL from the address bar of your browser to define your goal.

  2. Head Match

     If your URL is always the same for this step of your funnel but is followed by a unique session identifier or other parameters, use the Head Match filter and omit the unique values. For example, if the URL for a particular page is

     ```
     www.mysite.com/checkout.cgi?page=1&id=9982251615
     ```

     but the `id` varies for every user, enter

     ```
     www.mysite.com/checkout.cgi?page=1
     ```

  3. Regular Expression Match

     Uses regular expressions to match your URLs—for example, using wildcards and meta-characters. This is useful when the URL, query parameters, or both vary between users:

     ```
     http://sports.mysite.com/checkout.cgi?page=1&id=002
     http://news.mysite.com/checkout.cgi?page=1&language=fr&id=119
     ```

To match against a single goal for this example, you would use the regular expression *.*page=1*.* to define the constant element.

Head Match and Exact Match are by far the most common ways to define simple goal and funnel steps, but e-commerce systems often require the use of regular expressions.

Note: Figure 8.5 utilizes regular expressions to match page URLs. E-commerce tracking is not used in this example; therefore, a goal value, corresponding to the average order value, was applied to monetize the process. Head Match could also have been used here with the same result.

- Goal value

  For non-e-commerce goals, Google Analytics uses your assigned goal value to calculate ROI, $Index, and other metrics. A good way to value a goal is to evaluate how often the visitors who reach the goal become customers. If, for example, your sales team can close 10 percent of people who request to be contacted, and your average transaction is $500, then you might assign $50 (10 percent of $500) to your "Enquiry form sent" goal. Conversely, if only 1 percent of mailing list signups result in a sale, then you might only assign $5 to your "E-mail sign-up" goal. Monetizing goals is discussed in detail in "Monetizing a Non-E-Commerce Website" in Chapter 11.

Note: To define an e-commerce goal, set your receipt page as the goal URL and leave the Goal Value (Revenue) field blank. Then, set up your receipt page as described in "E-Commerce Tracking," in Chapter 7.

Tracking Funnels for Which Every Step Has the Same URL

You may encounter a situation where you need to track a visitor's progress through a funnel that has the same URL for each step. For example, your sign-up funnel might look like this:

- Step 1 (Sign Up)

  `www.mysite.com/sign_up.cgi`

- Step 2 (Accept Agreement)

  `www.mysite.com/sign_up.cgi`

- Step 3 (Finish)

  `www.mysite.com/sign_up.cgi`

To get around this, call the _trackPageview() function to track virtual pageviews within each step, as discussed in "Virtual Pageviews for Tracking Dynamic URLs," in Chapter 7. For example, within the GATC of the pages in question, create virtual pageviews to be logged in Google Analytics as follows:

```
pageTracker._trackPageview("/funnel_G1/step1.html")
pageTracker._trackPageview("/funnel_G1/step2.html")
pageTracker._trackPageview("/funnel_G1/step3.html")
```

With these virtual pageviews now being logged instead of sign_up.cgi, you would configure each step of your funnel as follows:

- Step 1 (Sign Up)

  `http://www.mysite.com/funnel_G1/step1.html`

- Step 2 (Accept Agreement)

  `http://www.mysite.com/funnel_G1/step2.html`

- Step 3 (Finish)

  `http://www.mysite.com/funnel_G1/step3.html`

Why Segmentation Is Important

To understand the importance of segmentation, we first need to examine how averages are used in web analytics. When discussing averages, we are generally referring to the arithmetic mean that is computed by adding a group of values together and dividing by the total number of values in the group. It's used in mathematics to approximate the statistical norm or expected value.

The arithmetic mean works well when the distribution under consideration is close to normal, that is, Gaussian or bell-shaped. For normal distributions the average value is also the most common (modal) value. For example, assuming a normal distribution for visitor time on site, if the average time is calculated at 95 seconds, then it is also true that the average visitor spends 95 seconds on your website. However, this is not true when the distribution is not normal. See Figure 8.7.

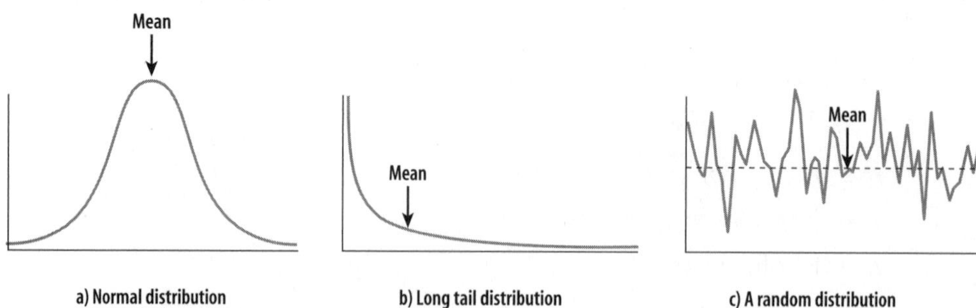

Figure 8.7 Sample visitor distributions for time spent on site

Note: For the vast majority of web metrics, the distribution of values is not Gaussian. In many cases, when considering the whole data set, the distribution appears random. The whole data set can include new visitors, returning visitors, existing customers, people researching products, people purchasing products, job seekers, spammers, mistaken visitors (wrong address), employees, competitors, and so on.

Figure 8.7 shows that for non-normal distributions, a typical visitor will not exhibit the average (mean) behavior—staying on the site for the mean length of time, in other words.

> *"Plans based on average assumptions are wrong on average."*
>
> —from "The Flaw of Averages" by Sam Savage, www.stanford.edu/%7Esavage/faculty/savage/Flaw%20of%20averages.pdf

For the random distribution in Figure 8.7c, the mean value for the time spent on site is misleading, as the distribution indicates many types of behavior are being exhibited. Perhaps the difference is indicating a mix of personas on your website—visitors, customers, blog readers, demographic differences, geographic difference. Whatever the reason, simply reporting an average is a blunt metric, and is precisely the reason why you rarely see averages reported in Google Analytics. When averages are reported, they are segmented—for example, shown for a specific page URL.

To illustrate this, Figure 8.8a shows a significant number of one-page visits that are probably not representative of an interested website visitor. Quoting an average depth would hide the fuller picture.

In Figure 8.8b, there are two maxima—indicating two types of visitor. If only an average is quoted without looking at the distribution, then you lose the clue that your site needs to cater to different visitor needs (personas).

The vast majority of web analytics vendors (including Google Analytics) only use the arithmetic mean when referring to the average. That is perfectly acceptable if the mean is calculated for segmented visitors, as this improves the statistical distribution (so long as the sample size is not too small).

As described in the section "Cross-Segmentation" in Chapter 4, segmentation in Google Analytics takes place as you drill down through your reports (clicking on data links) or by using the Segment drop down menu located at the top of report tables (see Figure 4.11 for example). However, there are circumstances when you will wish to have a dedicated report on a particular visitor segment. In Google Analytics, the method to do this is to use filters, which is described next.

a)

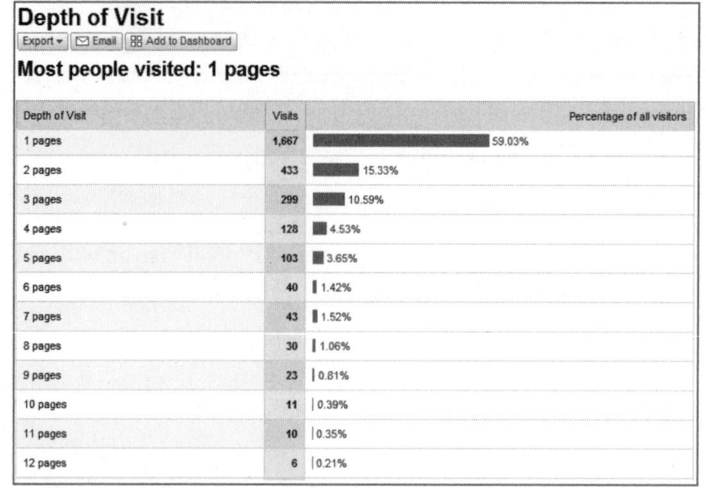

Depth of Visit

Export ▾ | Email | Add to Dashboard

Most people visited: 1 pages

Depth of Visit	Visits	Percentage of all visitors
1 pages	1,667	59.03%
2 pages	433	15.33%
3 pages	299	10.59%
4 pages	128	4.53%
5 pages	103	3.65%
6 pages	40	1.42%
7 pages	43	1.52%
8 pages	30	1.06%
9 pages	23	0.81%
10 pages	11	0.39%
11 pages	10	0.35%
12 pages	6	0.21%

b)

Length of Visit

Export ▾ | Email | Add to Dashboard

Most visits lasted: 0-10 seconds

Length of Visit	Visits	Percentage of all visitors
0-10 seconds	846	55.01%
11-30 seconds	119	7.74%
31-60 seconds	146	9.49%
61-180 seconds	201	13.07%
181-600 seconds	154	10.01%
601-1,800 seconds	63	4.10%
1,801+ seconds	9	0.59%

Figure 8.8 Typical non-normal distributions of visitors to a website

Filtering: Segmenting Visitors Using Filters

Everything discussed so far in this book has been concerned with the collection of good-quality data—ensuring that the report numbers are as comprehensive, accurate, and informative as possible. In this section, we consider the removal of data by using filters—that is, segmenting your visitors.

Segmentation helps you gain a better understanding of visitor types, to avoid interpreting an average of averages, which, as we have just seen, can be meaningless. Often, rather than simply remove data, you will wish to collect the excluded data in a separate profile—that is, view it in a separate, siloed report.

To segment the visitors on your website, you apply filters to the data. For example, you may want to remove visits to your website from your own employees, as these visits can be significant, especially if your website is set as the default home page in their browsers. Or you may want to report on visitors only within the same country you market or deliver to—for example, excluding all visitors outside the U.S. That way, your website conversion rates and ROI metrics will more accurately reflect their true value—assuming you are not actively acquiring visitors from markets you cannot supply.

When a filter is created within your profile, it's immediately applied to new data coming into your account. New filters will not affect historical data, and it is not possible to reprocess your old data through the new filter.

Best practice tip: Keep a profile without filters

Always keep raw data intact. That is, keep your original profile and apply new filters to a duplicate profile in your account. That way, if a mistake is made in applying a new filter, you always have the original profile to fall back on.

To create a duplicate profile, log in to your account as administrator and click the Add Website Profile link. From the next page, ensure that you select the option to Add Profile for an Existing Domain. Select your existing domain and provide a new profile name.

Using this method, data will be imported simultaneously into both the original and the new report profiles. Note that any existing filters applied to the first profile will not be copied, so you need to reapply these using the Filter Manager.

Creating a Filter

To create a filter, within your Google Analytics account click the Add Filter link. If you already have a filter defined in your account, the link will be labeled Filter Manager. You will then be presented with the Add Filter to Profile page, shown in Figure 8.9.

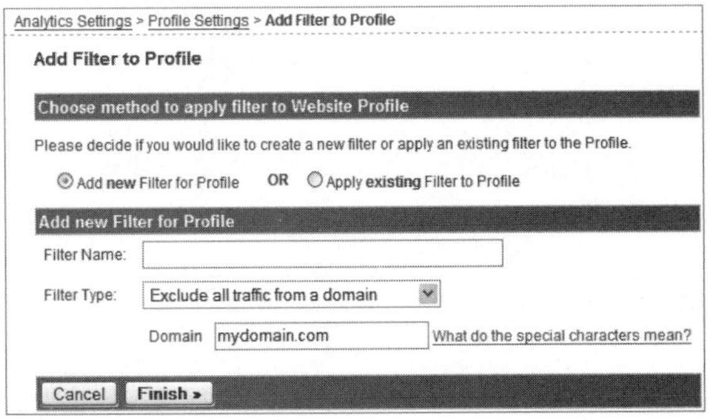

Analytics Settings > Profile Settings > Add Filter to Profile

Add Filter to Profile

Choose method to apply filter to Website Profile

Please decide if you would like to create a new filter or apply an existing filter to the Profile.

Add new Filter for Profile OR Apply existing Filter to Profile

Add new Filter for Profile

Filter Name:

Filter Type: Exclude all traffic from a domain

Domain mydomain.com What do the special characters mean?

Cancel Finish »

Figure 8.9 Adding a new filter

Google Analytics provides you with three predefined filter types, as well as numerous custom filter options:

- Predefined filters are a quick and easy way to accomplish some of the most common filtering tasks, as shown in Table 8.2. Creating a predefined filter is covered online in "How do I create a predefined filter?"

  `www.google.com/support/googleanalytics/bin/answer.py?answer=55496`

- Custom filters allow for more advanced manipulation of data, and these are listed in Table 8.3. Creating a custom filter is covered online in "How do I create a custom filter?"

  `www.google.com/support/googleanalytics/bin/answer.py?answer=55492`

► **Table 8.2** Predefined Filters

Filter name	Description
Exclude all traffic from a domain	Excludes traffic from a specific domain, such as an ISP or company network
Exclude all traffic from an IP address	Excludes clicks from certain sources. You can enter a single IP address or a range of addresses.
Include only traffic to a subdirectory	Includes visitors only viewing a particular subdirectory on your website, such as `www.mysite.com/helpdesk`

► **Table 8.3** Custom Filters

Custom Filter	Description
Exclude Pattern	This type of filter excludes log file lines (hits) that match the filter pattern. Matching lines are ignored in their entirety; for example, a filter that excludes Netscape will also exclude all other information in that log line, such as visitor, path, referral, and domain information.
Include Pattern	This type of filter includes log file lines (hits) that match the filter pattern. All nonmatching hits are ignored, and any data in nonmatching hits is unavailable.
Uppercase/Lowercase	Converts the contents of the field into all uppercase or all lowercase characters. These filters only affect letters, and will not affect special characters or numbers.
Search & Replace	This is a simple filter that can be used to search for a pattern within a field and replace it with an alternate form.
Lookup Table	Not yet implemented
Advanced	This type of filter enables you to build a field from one or two other fields. The filtering engine will apply the expressions defined in the two Extract fields and then construct a field using the Constructor expression. See the "Advanced Filters" section below for more information.

If the filter being applied is an exclude filter and the pattern matches, then the pageview data entry is thrown away and Google Analytics continues with the next entry. If the pattern does not match, then the next filter is applied to that hit. This means that you can create either a single exclude filter with multiple patterns separated by pipe characters (|) or you can create multiple exclude filters with a single pattern for each.

Include filters are applied with the reverse logic. When an include filter is applied, the data entry is thrown away if the pattern does *not* match the data. If multiple include filters are applied, then the data entry must match *every applied include filter* in order for the data entry to be saved. To include multiple patterns for a specific field, create a single include filter that contains all of the individual expressions separated by pipe characters (|).

Note: Filter patterns must not be longer than 255 characters.

Using multiple filters

It is important to understand how filter logic works, as adding more than one include filter to a profile can cause data to not appear in your reports. For example, if you include visitors from a certain IP address range, then all other IP ranges will be excluded from the data set.

For example, if you applied an include filter for your internal (employee) visitors using your network IP address, then it would not make sense to then add an additional include filter for, say, all Google visitors. The combination will *not* result in reports of internal visitors plus Google visitors. The report will only be for internal visitors, assuming this filter is applied first.

Best practice advice is to assign a maximum of one include filter to each of your profiles unless you have a specific need to do otherwise and understand the logic.

What Information Do Filter Fields Represent?

Tables 8.4 and 8.5 list all available fields and their purposes. Table 8.4 lists the regular fields—those automatically captured by Google Analytics. Table 8.5 lists the user-defined variables whose values are determined by your implementation of Google Analytics, such as the landing page URL tags, e-commerce fields, and so on.

▶ **Table 8.4** Regular Field List

Filter Name	Description
Request URI	Includes the relative URL (the piece of the URL after the hostname). For example: for `http://www.mysite.com/requestURL/index.html?sample=text`, the Request URI is `/requestURL/index.html?sample=text`.
Hostname	The full domain name of the page requested. For example: for `http://www.mysite.com/requestURL/index.html?sample=text`, the hostname is `www.mysite.com`.
Referral	The external referrer, if any. This field is only populated for the initial external referral at the beginning of a session.
Page Title	The contents of the `<title>` tags in the HTML of the delivered page
Visitor Browser Program	The name of the browser program used by the visitor
Visitor Browser Version	The version of the browser program used by the visitor
Visitor Operating System Platform	The visitor's operating system platform
Visitor Operating System Version	The visitor's operating system version
Visitor Language Settings	The language setting in the visitor's browser preferences
Visitor Screen Resolution	The resolution of the visitor's screen, as determined from the browser program
Visitor Screen Colors	The color capabilities of the visitor's screen, as determined from the browser program
Visitor Java Enabled?	Whether Java is enabled in the visitor's browser program
Visitor Flash version	The version of Flash installed in the visitor's browser program
Visitor IP Address	The visitor's IP address
Visitor Geographic Domain	The visitor's ISP; for example, `aol.com` or `aol.co.uk` for AOL users, derived from the geographic database
Visitor ISP Organization	The ISP organization registered to the IP address of the user. This is the ISP the visitor is using to access the Internet.
Visitor Country	The visitor's geographic country location obtained by information registered with the IP address
Visitor Region	The visitor's geographic region or state location obtained by information registered with the IP address
Visitor City	The visitor's geographic city location obtained by information registered with the IP address
Visitor Connection Speed	The visitor's connection speed, obtained by information registered with the IP address
Visitor Type	Either "New Visitor" or "Returning Visitor," based on Google Analytics identifiers

► **Table 8.4** Regular Field List *(Continued)*

Filter Name	Description
Custom Field 1	An empty, custom field for storage of values during filter computation. Data is not stored permanently in this field, but can be used by subsequent filters.
Custom Field 2	An empty, custom field for storage of values during filter computation. Data is not stored permanently in this field, but can be used by a subsequent filters.

► **Table 8.5** User-Defined Variables

Filter Name	Description
Campaign Source	The resource that provided the click, e.g., "google." This variable is automatically generated for AdWords hits when auto-tagging is turned on through the AdWords interface.
Campaign Medium	The medium used to generate the request, e.g., "organic," "cpc," or "ppc." This variable is automatically generated for AdWords hits when auto-tagging is turned on through the AdWords interface.
Campaign Name	The name given to the marketing campaign or used to differentiate the campaign source, e.g., "October Campaign." This variable is automatically generated for AdWords hits when auto-tagging is turned on through the AdWords interface.
Campaign Term	The term used to generate the ad from the referring source or campaign source, such as a keyword. This variable is automatically generated for AdWords hits when auto-tagging is turned on through the AdWords interface.
Campaign Content	Typically defines multivariate or split testing, or is used to disseminate campaign target variables in an advertising campaign. This variable is automatically generated for AdWords hits when auto-tagging is turned on through the AdWords interface.
Campaign Code	Can be used to refer to a campaign lookup table (not yet implemented in Google Analytics)
User Defined	A custom variable name, for use by the end user
E-Commerce Transaction Id	An unique ID variable correlated with a designated transaction
E-Commerce Transaction Country	Used to designate the country defined by the transaction process, obtained by information registered with the IP address
E-Commerce Transaction Region	Used to designate the region defined by the transaction process, obtained by information registered with the IP address
E-Commerce Transaction City	Represents the city where the commerce transaction occurred, obtained by information registered with the IP address
E-Commerce Store or Order Location	Describes the store or affiliated site processing the transaction, e.g., U.S. store, U.K. store, Affiliate123 etc.
E-Commerce Item Name	The item name purchased
E-Commerce Item Code	The identifier or code number corresponding to the item purchased. Commonly referred to as the stock keeping unit (SKU) code

The Six Most Common Filters

The following list highlights the six most common filters applied by most users of Google Analytics:

- Include only your website's traffic—at the very least you should apply this filter to all your profiles.
- Exclude certain known visitors—this might include, for example, your employees, your web agency, and so on.
- Separate new versus returning visitors—often these visitors display different behavior.
- Segment by geographical location—Make it easy on your country managers by creating profiles of visitors only relevant to them.
- Segment by visitor campaign, medium, or source referrer—Visitors from different referrers may have different objectives.
- Segment by content—Visitors viewing particular sections of your website may display different behavior, e.g., purchase versus support sections.

These filters are discussed in more detail in the following sections. Before studying these, you should be familiar with regular expressions, as discussed in the section "Regular Expression Overview," in Chapter 6.

Include Only Your Website's Traffic

This filter ensures that your data, and only your data, is collected into your Google Analytics profile. For example, it is possible for another website owner to copy your GATC onto their own pages, causing their traffic data to become part of your profile and contaminating your results. The include filter shown in Figure 8.10 will only report on traffic to the mysite.com domain. Note the backslash character ("\") used to escape the delimiter character ("."). This is an example of using regular expression syntax. Simply substitute mysite.com for your domain using the escape character for each "." in your domain.

Of course, it may be desirable to collect data from multiple websites into one profile. In that case, add the multiple domains in the filter pattern separated with pipe characters—for example, mysite\.com|yoursite\.com.

Exclude Certain Known Visitors

Excluding visits from employees, your search marketing agency, or any known third party, such as your web developers, is an important step when first creating your profiles. These visitors generate a relatively high number of pageviews in areas that will greatly impact key metrics, such as your conversion rates.

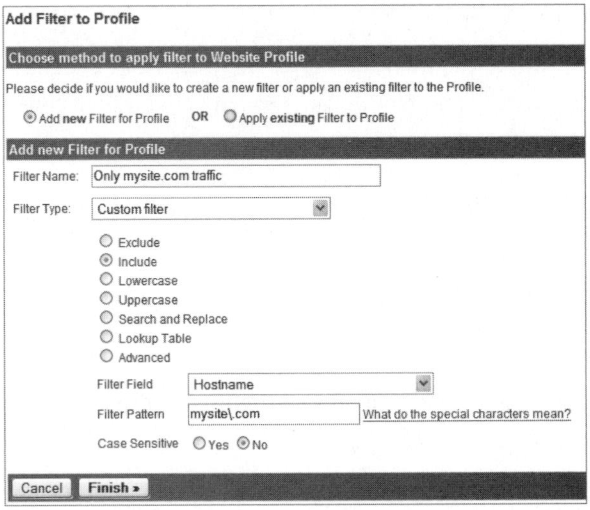

Add Filter to Profile

Choose method to apply filter to Website Profile

Please decide if you would like to create a new filter or apply an existing filter to the Profile.

Add **new** Filter for Profile OR Apply **existing** Filter to Profile

Add new Filter for Profile

Filter Name: Only mysite.com traffic

Filter Type: Custom filter

Exclude
Include
Lowercase
Uppercase
Search and Replace
Lookup Table
Advanced

Filter Field: Hostname

Filter Pattern: mysite\.com What do the special characters mean?

Case Sensitive: Yes No

Cancel Finish »

Figure 8.10 Filter to include only your website's traffic

For example, employees who have their browser home page set to the company website will show in your reports as retuning visitors every time they open their browser—and most likely as one-page visitors. Remember that the GATC deliberately breaks through any caching, so it's important to exclude employee visits from those of potential customers. Similarly, web developers heavily test checkout systems for troubleshooting purposes. These will also trigger GATC page requests, and most likely these will be for your goal conversion pages. You should therefore remove all such visits from your reports.

Excluding known visitors is straightforward. If the visitor connects to the Internet via a fixed IP address, select the predefined filter Exclude All Traffic from an IP Address from the Filter Manager, as shown in Figure 8.11.

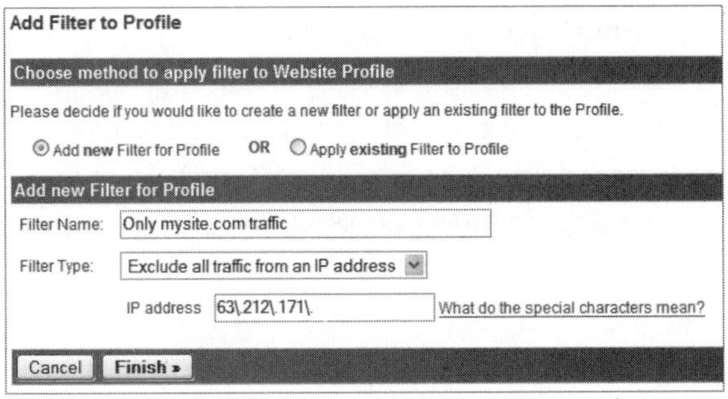

Add Filter to Profile

Choose method to apply filter to Website Profile

Please decide if you would like to create a new filter or apply an existing filter to the Profile.

Add **new** Filter for Profile OR Apply **existing** Filter to Profile

Add new Filter for Profile

Filter Name: Only mysite.com traffic

Filter Type: Exclude all traffic from an IP address

IP address: 63\.212\.171\. What do the special characters mean?

Cancel Finish »

Figure 8.11 Excluding visitors from a known IP address

What If Visitors Do Not Have a Fixed IP Address?

This is often the case for home users, where the Internet service provider (ISP) assigns a different IP address each time the home user connects; this can also happen during a connected session. The solution is to use the function _setVar() in conjunction with a custom exclude filter. The use of _setVar() to label visitors is discussed in more detail later in this section, but essentially the principle is that you direct known visitors you want to exclude to a hidden page (not used by regular visitors) that contains a JavaScript label within the GATC. The label is stored as a persistent cookie on that visitor's computer and forms part of their pageview data. An exclude filter is then used to remove any pageview data that contains this label.

To assign a custom label to visitors, call the function _setVar() within the GATC on your hidden page as follows:

```
<script type="text/javascript">
   var gaJsHost = (("https:" == document.location.protocol) ? "https://ssl."
: "http://www.");
   document.write("\<script src='" + gaJsHost + "google-analytics.com/ga.js'
type='text/javascript'>\<\/script>" );
</script>

<script type="text/javascript">
   var pageTracker = _gat._getTracker("UA-12345-1");
   pageTracker._initData();
   pageTracker._trackPageview();
   pageTracker._setVar("dynamic");
</script>
```

In this way, only one visit to, for example, www.mysite.com/hiddenpage.htm is required to label the visitor until the cookie expires (24 months)—assuming the label cookie (stored by the name *__utmv* by the visitor's browser) is not overwritten or deleted. Note that in this example _setVar() is called and set to the label "dynamic." However, any value can be used in the brackets. With each pageview from your dynamic IP visitor now labeled, Figure 8.12 shows the filter required to exclude those visits from your profile. The value of _setVar() is stored in the Google Analytics field labeled User Defined.

Separate New Versus Returning Visitors

It is straightforward within Google Analytics to select a metric and then cross-segment by visitor type, such as new versus returning visitor. However, quite often the behavior of a new visitor to a website is markedly different from that of a returning visitor. For example, a new visitor to a retail site is likely to be researching products—comparing prices, features, delivery details, and so on. The same visitor returning is more likely to become a customer and therefore has different requirements, such as wanting to know about product availability, confidence in your handling of personal information, and the speed and efficiency of the checkout process.

When optimizing a website for conversions, it can be beneficial to segregate these two types of visitors into separate profiles so that they can be studied in greater detail. Figure 8.13 shows the filter required to do this. An additional profile with a filter pattern set to Returning Visitor completes the process.

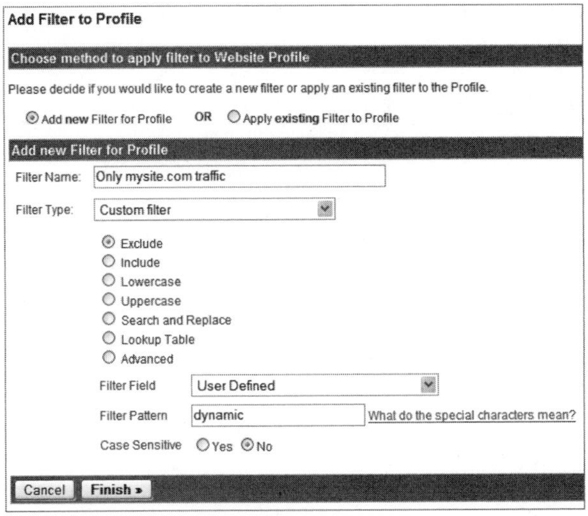

Add Filter to Profile

Choose method to apply filter to Website Profile

Please decide if you would like to create a new filter or apply an existing filter to the Profile.

(•) Add **new** Filter for Profile OR ( ) Apply **existing** Filter to Profile

Add new Filter for Profile

Filter Name: Only mysite.com traffic

Filter Type: Custom filter

(•) Exclude
( ) Include
( ) Lowercase
( ) Uppercase
( ) Search and Replace
( ) Lookup Table
( ) Advanced

Filter Field: User Defined

Filter Pattern: dynamic — What do the special characters mean?

Case Sensitive: ( ) Yes (•) No

Cancel | Finish »

Figure 8.12 Excluding labeled visitors

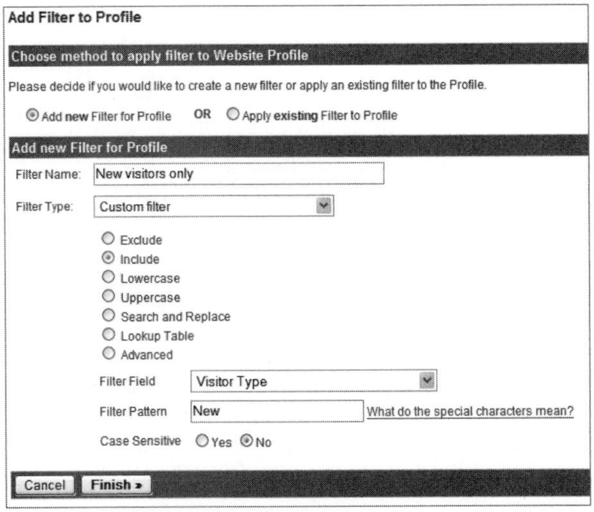

Add Filter to Profile

Choose method to apply filter to Website Profile

Please decide if you would like to create a new filter or apply an existing filter to the Profile.

(•) Add **new** Filter for Profile OR ( ) Apply **existing** Filter to Profile

Add new Filter for Profile

Filter Name: New visitors only

Filter Type: Custom filter

( ) Exclude
(•) Include
( ) Lowercase
( ) Uppercase
( ) Search and Replace
( ) Lookup Table
( ) Advanced

Filter Field: Visitor Type

Filter Pattern: New — What do the special characters mean?

Case Sensitive: ( ) Yes (•) No

Cancel | Finish »

Figure 8.13 Filter to only include new visitors

Segment by Geographical Location

Google Analytics performs an excellent job of showing you the countries from which your visitors are accessing your website. It even groups these into regions (continents: Americas, Europe, Asia, Oceania, Africa) and subregions (Northern Europe, Central Europe, Eastern Europe, Southern Europe, for example). However, if your organization operates specifically in certain markets, you may want to create a profile that focuses on reporting visitors just from those countries. For example, France, Germany, and Spain can be included in a separate profile, as shown by the filter in Figure 8.14.

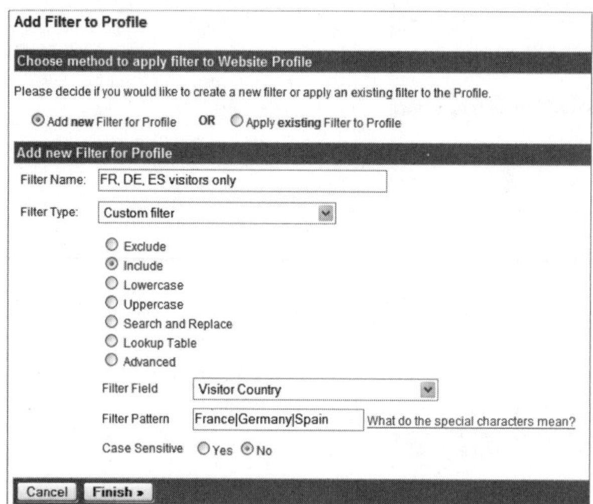

Figure 8.14
Segmenting visitors by country

Segment by Campaign, Medium, or Source Referrer

As with the use of other filters in this section, Google Analytics already does an excellent job of displaying different campaigns, mediums, or source referrers. However, in some scenarios it can be helpful to have a profile with dedicated reports to a particular campaign, medium, or source, in order to help you optimize those better. For example, if you are conducting e-mail marketing, it can be beneficial to have a report for visitors that come only from `medium = email`. Likewise, many organizations spend significant sums optimizing their pages for high rankings on Google (both paid and non-paid). Isolating visitors from these allows for quicker and easier analysis, providing a more efficient route to understanding the engagement of those visitors.

How you construct this filter depends on how you have tagged your landing page URLs (see "Online Campaign Tracking," in Chapter 7). The values you set for `utm_source`, `utm_medium`, and `utm_campaign` need to match the following filter fields:

- Campaign Name
- Campaign Source
- Campaign Medium

Google AdWords visitors are automatically tracked (assuming you have auto-tagging enabled in your AdWords account), but to isolate only these requires the application of two filters, in order, as shown in Figure 8.15a and b.

If you want to include all Google visitors, both paid and non-paid, then apply only the filter shown in Figure 8.15a.

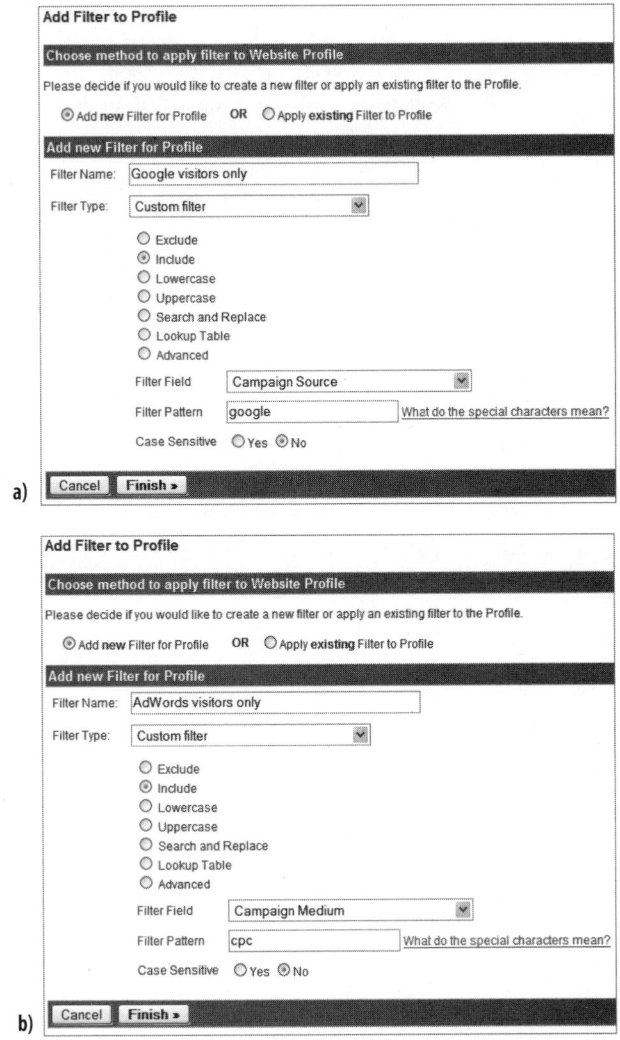

a)

Add Filter to Profile

Choose method to apply filter to Website Profile

Please decide if you would like to create a new filter or apply an existing filter to the Profile.

(•) Add **new** Filter for Profile **OR** ( ) Apply **existing** Filter to Profile

Add new Filter for Profile

Filter Name: Google visitors only

Filter Type: Custom filter

( ) Exclude
(•) Include
( ) Lowercase
( ) Uppercase
( ) Search and Replace
( ) Lookup Table
( ) Advanced

Filter Field: Campaign Source

Filter Pattern: google — What do the special characters mean?

Case Sensitive: ( ) Yes (•) No

Cancel | Finish »

b)

Add Filter to Profile

Choose method to apply filter to Website Profile

Please decide if you would like to create a new filter or apply an existing filter to the Profile.

(•) Add **new** Filter for Profile **OR** ( ) Apply **existing** Filter to Profile

Add new Filter for Profile

Filter Name: AdWords visitors only

Filter Type: Custom filter

( ) Exclude
(•) Include
( ) Lowercase
( ) Uppercase
( ) Search and Replace
( ) Lookup Table
( ) Advanced

Filter Field: Campaign Medium

Filter Pattern: cpc — What do the special characters mean?

Case Sensitive: ( ) Yes (•) No

Cancel | Finish »

Figure 8.15
Filter to include only Google visitors (a); AdWords visitors (b)

Note: If you tag all other pay-per-click campaigns, such as Yahoo Search Marketing, Microsoft adCenter, Miva, and so on, with `utm_medium = ppc`, then the filter shown in Figure 8.15b on its own would be sufficient to segment Google AdWords visitors. I use this technique as Google AdWords is currently so prevalent for online marketing. Being able to compare AdWords visitors against all other pay-per-click networks as a whole can be very useful.

Filter pattern tip

When deciding what value to place in the Filter Pattern field, always consult your reports. For example, when cross-segmenting a page by visitor type, there are two possible values:

- New Visitor
- Returning Visitor

These are the only values that can be used in the Filter Pattern field (partial matches are also allowed). Similarly, when cross-segmenting a page by country, the available values are displayed. Note that these are all in English. For example, it is Spain, Netherlands, Germany, and so on, not España, Nederland, Deutschland. Only use the values from your reports in the Filter Pattern field.

Figure 8.16 shows how to segment only e-mail visitors—that is, those visitors who have clicked on a link to your website within an e-mail message, assuming you tagged such links in the following way:

```
http://www.mysite.com/products/shoes?utm_source=July-08%20Newsletter& ➥
utm_medium=Email&utm_content=text&utm_term=Shoes& ➥
utm_campain=Summer%20Promo
```

As you can see, segmenting by campaign, source, or medium is as simple as knowing what these values are in your corresponding landing page URLs, and then applying them as field values to your include and exclude filters.

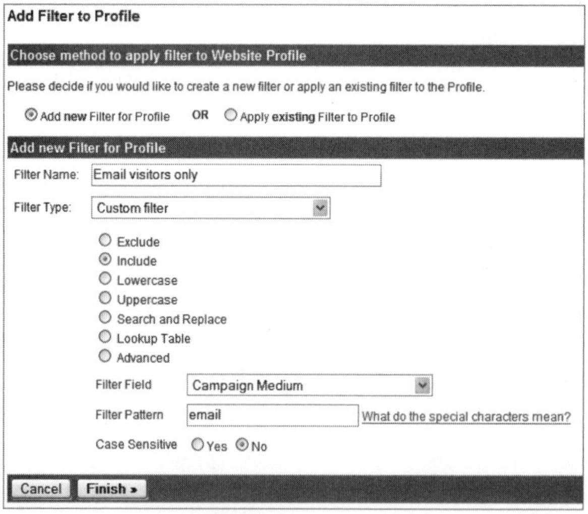

Add Filter to Profile

Choose method to apply filter to Website Profile

Please decide if you would like to create a new filter or apply an existing filter to the Profile.

(◉) Add new Filter for Profile OR ( ) Apply existing Filter to Profile

Add new Filter for Profile

Filter Name: Email visitors only

Filter Type: Custom filter

( ) Exclude
(◉) Include
( ) Lowercase
( ) Uppercase
( ) Search and Replace
( ) Lookup Table
( ) Advanced

Filter Field: Campaign Medium

Filter Pattern: email — What do the special characters mean?

Case Sensitive: ( ) Yes (◉) No

Cancel | Finish »

Figure 8.16 Filter to only include e-mail visitors

Segment by Content

Often within one website, you will be trying to satisfy the needs of very different visitors—for example, product purchase versus product support or corporate information versus customer information. Effectively measuring such different needs requires the setting of very different goals for each section—hence the creation of separate profiles using filters. Figure 8.17 is an example filter that segments by content—in this case, a support blog.

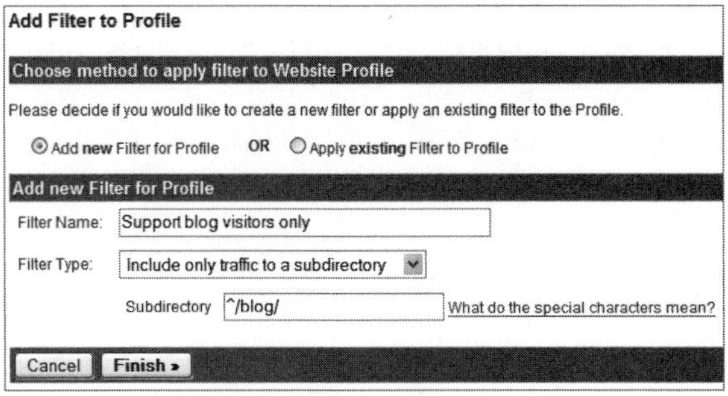

Figure 8.17 Filter to include only blog visitors

Of course, the success of this filter depends on you having a well-ordered website directory structure on which to filter content. If you do not, it is possible to achieve a virtual structure by using virtual pageviews, as described in "trackPageview(): The Google Analytics Workhorse," in Chapter 7.

Assigning a Filter Order

By default, a profile's filters are applied to the incoming data in the order in which the filters were added. However, you can easily modify the order from your Profile Settings page, using the Assign Filter Order link from within your profile settings. Filter order is important for the filters described in Figure 8.15a and b, where a must come before b.

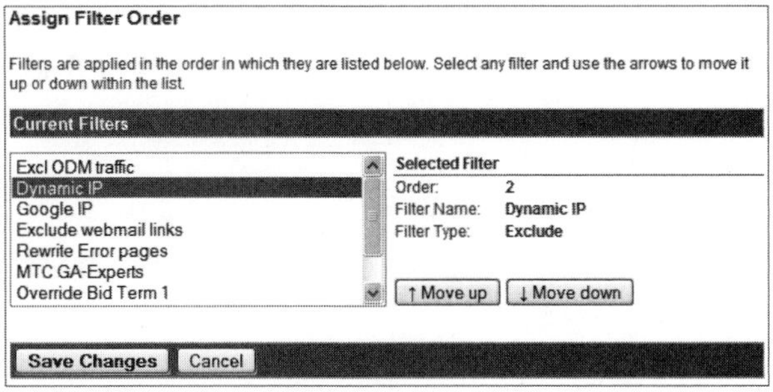

Figure 8.18 Assigning filter order

Summary

This chapter has been all about getting the most out of your data. Configuring goals provides you with conversion and engagement rates; funnels enable you to see what barriers exist on the path to achieving a goal; filtering keeps your data clean and is the method of segmenting visitors into separate profiles.

If you have followed all these steps, congratulations! You now have a best practice implementation of Google Analytics that will enable you to gain real insight into the performance of your online presence.

If you didn't follow all the steps, go back and reread this chapter. Seriously, this chapter is too important to skip over it without implementing the suggested configurations—particularly goals and funnels. Often, beyond a transaction, people get stuck on identifying goals, but time spent considering this can reap huge rewards later, so pay particular attention to this before proceeding.

In Chapter 8, you have learned how to do the following:

- Set the initial configuration of your account, including localization, e-commerce and site search settings
- Identify and set goals in order to benchmark yourself
- Understand how to configure funnels, and the significance of their shapes
- Set up filters to maintain the integrity of your data
- Segment data to gain a deeper understanding of visitor behavior
- Use filters as the method for segmentation

Google Analytics Hacks

Out of the box, Google Analytics is a powerful tool to add to your armory of search marketing, customer relationship, and other business management tools. With only a single page tag required to collect data, it is straightforward to set up; and with the addition of some filters, you can really gain an insight into your website performance.

If at this stage the reports answer all of your questions, that's great. However, you may find yourself asking further questions that by default are not answered in your reports. Fear not, you can still achieve a great deal more insight with a little bit of lateral thought; Google Analytics is incredibly flexible in that respect.

In this chapter I assume you have a strong understanding of JavaScript.

In Chapter 9, you will learn about the following:

Customizing the list of recognized search engines
Labeling and sessionizing visitors for better segmentation
Tracking error pages and broken links
Gaining a greater insight into your pay-per-click tracking
Improving site overlay, conversion and e-commerce reports

Positioning of GATC hacks

When modifying the GATC, the placement of the code edits is important. In the vast majority of cases, any edits to the GATC must take place before the `_initData()` call.

Customizing the List of Recognized Search Engines

Google Analytics currently identifies organic referrals from the following search engines in your reports:

- AOL
- About
- Alice
- Alltheweb
- AltaVista
- Ask
- Baidu
- CNN
- Clubinternet
- Gigablast
- Google
- Google.interia
- Live
- LookSmart
- Lycos
- MSN
- Mama
- Mamma
- Najdi
- Netscape
- Netsprint
- Onet
- Pchome
- Search
- Seznam
- Szukacz
- Virgilio
- Voila
- Wp
- Yahoo!
- Yam
- Yandex

Although Google Analytics adds new recognized search engines to this list regularly, there are of course a great many more search engines in the world—language- and region-specific as well as niche search engines such as price comparison and vertical portals. It is therefore possible to modify and append to the list of recognized search engines.

For example, suppose you wanted the BBC search engine to be listed as such in your reports, along with the search terms used by those visitors. First, conduct a search on the BBC website and view the resultant URL. For example, searching for motorcycle produces the following search result URL:

```
http://search.bbc.co.uk/cgi-bin/search/results.pl?q=motorcycle
```

To capture this URL and recognize it as a search engine, add the following code to your page GATC:

```
<script type="text/javascript">
   var gaJsHost = (("https:" == document.location.protocol) ? "https://ssl."
: "http://www.");
   document.write(unescape("%3Cscript src='" + gaJsHost + "google-
```

```
analytics.com/ga.js' type='text/javascript'%3E%3C/script%3E"));
</script>
<script type="text/javascript">
   var pageTracker = _gat._getTracker("UA-12345-1");
   pageTracker._addOrganic("bbc.co.uk", "q");
   pageTracker._initData();
   pageTracker._trackPageview();
</script>
```

The line pageTracker._addOrganic("bbc.co.uk", "q") simply appends this search engine to the default list of search engines contained in the GATC. As you can see, the format is:

```
pageTracker._addOrganic("search_engine_domain", "query_parameter_name");
```

You can continue to add other search engines as needed by creating additional _addOrganic lines. For example, to add the price comparison engine Kelkoo as a regular search engine, add the following:

```
<script type="text/javascript">
   var gaJsHost = (("https:" == document.location.protocol) ? "https://ssl."
: "http://www.");
   document.write(unescape("%3Cscript src='" + gaJsHost + "google-
analytics.com/ga.js' type='text/javascript'%3E%3C/script%3E"));
</script>
<script type="text/javascript">
   var pageTracker = _gat._getTracker("UA-12345-1");
   pageTracker._addOrganic("bbc.co.uk", "q");
   pageTracker._addOrganic("Kelkoo", "siteSearchQuery");
   pageTracker._initData();
   pageTracker._trackPageview();
</script>
```

Using this method, Kelkoo would be listed in the Search Engine report along with other search engines. That is useful in itself, but what provides more insight is that the corresponding Kelkoo search terms used by visitors would be listed in the Keywords report. Without this little hack, Kelkoo would simply be listed as a standard referrer with no search terms logged.

Note: The use of Kelkoo is purely for illustration purposes. For this to work in the merchant's Google Analytics reports, the price comparison site would have to transparently send traffic to its merchants so that the referrer can be detected. That may not be the case if the price comparison engine is using redirects. Redirect issues are discussed in the section "Testing After Enabling Auto-tagging," in Chapter 6.

Differentiating Regional Search Engines

Apart from adding additional search engines to the existing list provided by Google Analytics, you could also use this method to create more regional lists of the main players. For example, if you are based in the U.K., being able to differentiate google.co.uk from google.com may be of importance. You might think adding the following to the GATC of your pages would provide this:

```
pageTracker._addOrganic("google.co.uk","q");
```

However, this won't work, because when adding regional variations to the search engine list, the order becomes important. Defining the custom *addOrganic* variable in your GATC appends google.co.uk (or any other variation) to the end of the default search engine array list, but the default list is already assigning any google.* domain as "google"; therefore, appending is too late to change this.

The answer is to first clear the default search engine list from the GATC and then redefine all search engines using your custom list, as shown in this example:

```
pageTracker._clearOrganic() // clears the default list of search engines
// Define new search domains
pageTracker._addOrganic("google.com","q");
pageTracker._addOrganic("google.co.uk","q");
pageTracker._addOrganic("google.es","q");
pageTracker._addOrganic("google.pt","q");
pageTracker._addOrganic("google.it","q");


etc.
```

Rather than define a long list of search engines in your GATC, it is better to place these in a separate JavaScript file—named, for example, custom_se.js. Place this file in the root of your web hosting account. Then call the file in all your web pages by adding the following line to your GATC:

```
<script type="text/javascript">
   var gaJsHost = (("https:" == document.location.protocol) ? "https://ssl."
: "http://www.");
   document.write(unescape("%3Cscript src='" + gaJsHost + "google-
analytics.com/ga.js' type='text/javascript'%3E%3C/script%3E"));
</script>
<script src="custom_se.js" type="text/javascript"></script>
<script type="text/javascript">
   var pageTracker = _gat._getTracker("UA-12345-1");
   pageTracker._initData();
   pageTracker._trackPageview();
</script>
```

A comprehensive list of over 100 search engines is available at www.advanced-web-metrics.com/scripts/. Use this as the starting point for your own custom list.

Note: Using the _clearOrganic() function will completely remove the entire list of organic search engines that Google Analytics can identify. Therefore, use it wisely, and be sure to rebuild the entire list of search engines; otherwise, you will lose organic details of any search engine not specifically included in your customizations.

Capturing Google Image Search

At present, Google Analytics shows all traffic from Google Image search as referrals—a standard click-through from a link like any other. That means any keyword information associated with the visitor's image search is not reported on. However, perhaps this information is important to your business model. If that describes your situation, consider the following.

Conduct a search at http://images.google.com and click on an image. The resultant referrer URL will look similar to:

```
http://images.google.co.uk/imgres?imgurl=http://www.ru.is/lisalib/getfile.aspx
%3Fitemid%3D6207&imgrefurl=http://www.ru.is/%3FPageID%3D836&h=299&w=448&sz=10
&hl=en&start=1&sig2=a54dUs9R8ntHco0dMOo__Q&um=1&tbnid=jwwvkKWJlfyT6M:&tbnh=85
&tbnw=127&ei=tF2_RsyAKo2MxAGT_-yQCA&prev=/images%3Fq%3D%2522brian%2Bclifton
%2522%26ndsp%3D18%26um%3D1%26hl%3Den%26newwindow%3D1%26rls%3DGGGL,GGGL:2006-17,
GGGL:en%26sa%3DN
```

Pretty it isn't! However, the referrer URL for a Google Image search contains the search keyword in the parameter named prev, along with other surplus parameters that are not relevant to you. Because of this, viewing the Google Image search term in your reports requires a two-step process:

1. Add images.google to the search engine list of Google Analytics by modifying your GATC on all pages (or add to your custom_se.js) as follows:

```
<script type="text/javascript">
   var gaJsHost = (("https:" == document.location.protocol) ? "https://ssl."
: "http://www.");
   document.write(unescape("%3Cscript src='" + gaJsHost + "google-
analytics.com/ga.js' type='text/javascript'%3E%3C/script%3E"));
    </script>
<script type="text/javascript">
   var pageTracker = _gat._getTracker("UA-12345-1");
   pageTracker._initData();
```

```
    pageTracker._addOrganic("images.google", "prev");
    pageTracker._trackPageview();
</script>
```

2. Use an advanced filter to extract the keyword from the prev parameter, as shown in Figure 9.1.

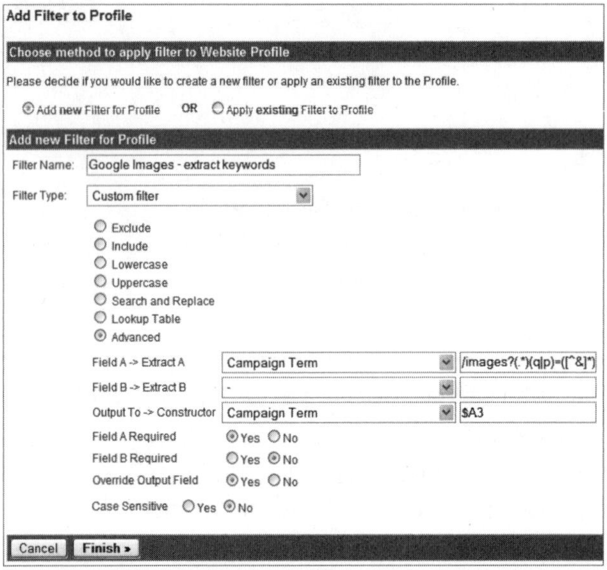

Figure 9.1 Advanced filter to extract the keyword from the prev parameter

In English, the advanced filter of Figure 9.1 reads:

a. From the referring site URL, extract from the campaign term (defined as the previous parameter in the URL) the string that contains */images?* followed by zero or more of any character, followed by a *p* or a *q*, followed by = and anything up to the next *&* character.

b. Overwrite the campaign term with the extracted contents from = to *&*.

Once implemented, you will see images.google (organic) show up in your search engine reports for visitors who use the Google Image search. Clicking on its link will display the keywords used.

Labeling Visitors

Labeling was first described in Chapter 8, where it was used in conjunction with a filter to remove visitors with dynamic IP addresses. The important step is calling the function _setVar(), which enables you to define a custom variable that labels a visitor depending on the criteria you set. In addition to labeling dynamic-IP visitors, it can be used in many other circumstances. For example, when a visitor completes a conversion and becomes a customer, you may want to label that visitor as such so that you can differentiate or filter them in your reports.

Adding a custom label to visitors is achieved by using the function _setVar() within the GATC of the page where the label is applied, as shown in the following example:

```
<script type="text/javascript">
   var gaJsHost = (("https:" == document.location.protocol) ? "https://ssl."
: "http://www.");
   document.write(unescape("%3Cscript src='" + gaJsHost + "google-
analytics.com/ga.js' type='text/javascript'%3E%3C/script%3E"));
</script>
<script type="text/javascript">
   var pageTracker = _gat._getTracker("UA-12345-1");
   pageTracker._initData();
   pageTracker._trackPageview();
   pageTracker._setVar("customer");
</script>
```


The value in parentheses can be any label you wish, though you should only use alpha-numeric characters (as well as the space character), to avoid any potential encoding issues. The value you set will be displayed in the Visitors > User Defined report and can be cross-segmented. Figure 9.2 shows several labels that have been defined with _setVar() on an example website.

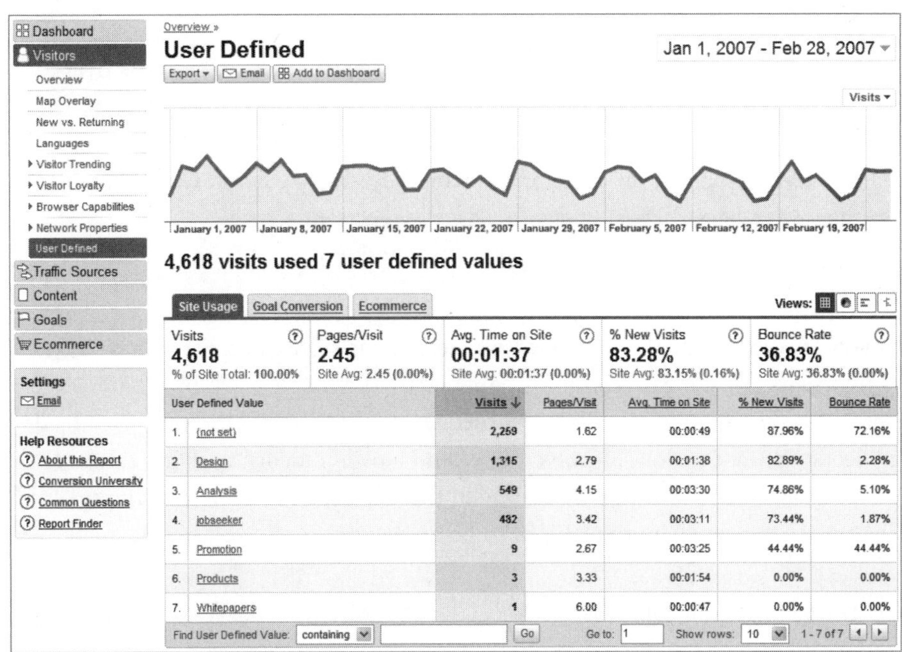

	User Defined Value	Visits	Pages/Visit	Avg. Time on Site	% New Visits	Bounce Rate
1.	(not set)	2,259	1.62	00:00:49	87.96%	72.16%
2.	Design	1,315	2.79	00:01:38	82.89%	2.28%
3.	Analysis	549	4.15	00:03:30	74.86%	5.10%
4.	jobseeker	432	3.42	00:03:11	73.44%	1.87%
5.	Promotion	9	2.67	00:03:25	44.44%	44.44%
6.	Products	3	3.33	00:01:54	0.00%	0.00%
7.	Whitepapers	1	6.00	00:00:47	0.00%	0.00%

Figure 9.2 Example user-defined report

Note that _setVar() is a visitor-centric value, as opposed to a pageview-centric value. That means, once set for a particular visit, the label is applied during data processing for the entire visit. This means the label *customer* is assigned to all pageviews during the same visit—including those that occurred before that visitor became a customer. Similarly, any filter applied that uses the label will be applied to all pageviews for that entire visit.

The value of _setVar()is stored in a persistent cookie (__utmv), so if the same visitor returns to your website at a later date, whether they purchase from you again or not, they still remain labeled as a *customer*.

As _setVar() is not pageview-centric, it should not be used to label sections of your website because the first value set will be applied to the entire visit. Even if the label changes during the same visit, it will show no effect in the reports for that visit. If _setVar() is called multiple times during a visit, the last value set remains in the cookie. The next time that visitor returns, the last value associated with the cookie is used to label the returning visit. This is obviously confusing to interpret (that is, the last value of the previous visit being used to label the current visit), so only use this labeling method for visitors. A workaround for this is discussed next.

Note: To emphasize the last point, use _setVar() to label visitors only—not pageviews or visits.

Sessionizing Visitor Labels

Despite the previous caveat of not using _setVar() to measure visits, this can be overcome with some JavaScript. Consider the following example: A publisher and content provider—for example, a newspaper website—wishes to know which section is read first when registered users log in. This can be achieved using a custom visitor label applied to the visitor once they enter the first website section.

As just described, a custom visitor label is set within the GATC as follows:

```
pageTracker._setVar("Automotive");
```

As you now know, the label applied by _setVar() is stored in a persistent cookie that lasts two years. Therefore, this simple label method will not work by default—the same label will be applied on the visitor's subsequent visits, even if they only ever read the sports section from then on. However, this can be overcome by applying the label on a per-visit basis via sessionizing the cookie in the GATC. For example, using the following code on the page where the label is to be applied:

```
labelVal = 'Automotive';
pageTracker._setVar(labelVal);
```

```
date = new Date();
date.setTime(date.getTime() + 0.5*60*60*1000);
document.cookie = "__utmv="+_udh+"."+_uES(labelVal)+";➥
path="+_utcp+"; expires="+date.toGMTString()+";"+_udo;
```

This sets the cookie value, as specified by labelVal, to expire after 30 minutes, though you can change this by adjusting the value added to date.getTime(). I choose 30 minutes to match the default session time out used by Google Analytics. Visitors that return later than this specified time period will receive a new visit label. The advantage of this technique is that you only need to apply this code on the pages where you use _setVar().

Note: When modifying the GATC, the line beginning document.cookie must be written on one continuous line.

A full GATC for such a page would look as follows:

```
<script type="text/javascript">
   var gaJsHost = (("https:" == document.location.protocol) ? "https://ssl."
: "http://www.");
   document.write(unescape("%3Cscript src='" + gaJsHost + "google-
analytics.com/ga.js' type='text/javascript'%3E%3C/script%3E"));
</script>
<script type="text/javascript">
   var pageTracker = _gat._getTracker("UA-12345-1");
   pageTracker._initData();
   pageTracker._trackPageview();

   labelVal = 'Automotive';
   pageTracker._setVar(labelVal);
   date = new Date();
   date.setTime(date.getTime() + 1*60*60*1000);
   document.cookie = "__utmv="+_udh+"."+_uES(labelVal)+";
   path="+_utcp+"; expires="+date.toGMTString()+";"+_udo;
</script>
```

This technique was used in the report displayed in Figure 9.2. That is, the labels are per visit based and therefore refer to how popular the respective sections of the website are.

Tracking Error Pages and Broken Links

With an out-of-the-box install of Google Analytics, you will not be tracking error pages or broken links on your website. This is because by default you probably have not added the GATC to your error pages. After all, how can you track a page that does not exist? To enable this, you need to add the GATC to your error page templates, which are delivered by your web server. A webmaster will typically do this. The GATC will then track your error page URLs as if they were any other pageview request. That is the caveat: Without modification, error pages are reported as regular pages, not as errors. However, you can highlight and separate error pages using a simple filter.

For this example it is assumed that your GATC is loaded with the error page template and that your web server displays its error code in the HTML <title> tag; most Apache configurations do this by default, as shown in Figure 9.3.

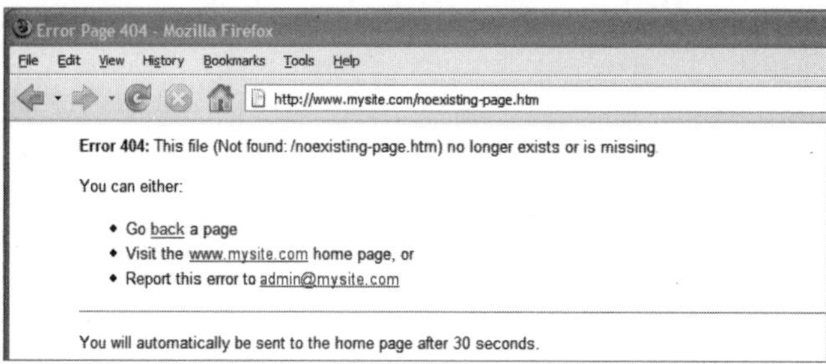

Figure 9.3 Typical 404 "not found" error page returned from the Apache web server

Web server status codes

These are the status codes, defined in the HTTP 1.0 specification, returned by your web server in its headers (see www.w3.org/Protocols/Overview.html).

2xx Success

The requested action was successfully received and understood.

- 200 OK
- 201 Created
- 202 Accepted
- 203 Provisional Information
- 204 No Response
- 205 Deleted
- 206 Modified

Web server status codes *(Continued)*

3xx Redirection

Further action must be taken in order to complete the request.

- 301 Moved Permanently
- 302 Moved Temporarily
- 303 Method
- 304 Not Modified

4xx Client Error

The request contains bad syntax or is inherently impossible to fulfill.

- 400 Bad Request
- 401 Unauthorized
- 402 Payment Required
- 403 Forbidden
- 404 Not Found
- 405 Method Not Allowed
- 406 None Acceptable
- 407 Proxy Authentication Required
- 408 Request Timeout

5xx Server Error

The server could not fulfill the request.

- 500 Internal Server Error
- 501 Not Implemented
- 502 Bad Gateway
- 503 Service Unavailable
- 504 Gateway Timeout

Using the filter shown in Figure 9.4 enables you to differentiate error pages from other pageviews within your Google Analytics reports. In English, the filter is described as follows:

- Check whether the page title contains the phrase "Error Page." If so, extract the page title and the page URI entries.
- Combine the page title and page URI entries and overwrite the page URI field.

Add Filter to Profile

Choose method to apply filter to Website Profile

Please decide if you would like to create a new filter or apply an existing filter to the Profile.

◉ Add **new** Filter for Profile **OR** ○ Apply **existing** Filter to Profile

Add new Filter for Profile

Filter Name: Rewrite Error pages

Filter Type: Custom filter

○ Exclude
○ Include
○ Lowercase
○ Uppercase
○ Search and Replace
○ Lookup Table
◉ Advanced

Field A -> Extract A	Page Title	(Error page.+)
Field B -> Extract B	Request URI	(.*)
Output To -> Constructor	Request URI	$A1$B1
Field A Required	◉ Yes ○ No	
Field B Required	◉ Yes ○ No	
Override Output Field	◉ Yes ○ No	
Case Sensitive	○ Yes ◉ No	

Cancel | Finish »

Figure 9.4 Filter to highlight error pages

For the example shown in Figure 9.4, the resultant entries for error pages would show in the Top Content report as `/Error Page 404/noexisting-page.htm`. This provides you with two very important pieces of information: the type of error (error code) and the URL of the page that produced this. Figure 9.5 shows an example Top Content report for the error pages. Note that the report uses the inline filter to highlight these—that is, bring them to the top. This is important, as without this, error pages are usually buried at the bottom of your pageview listings—assuming they are a small fraction of the total!

Tip: Knowing your error page URLs is clearly important, yet they typically appear at the bottom of your Top Content report—possibly hundreds of pages deep. To ensure that your web design and development team follows these up, set the inline filter to *error* (as shown in Figure 9.4) and schedule this report to be e-mailed to them on a daily or weekly basis (click the e-mail button at the top of the report and follow the instructions). E-mailing reports is discussed in the section "Scheduled Export of Data," in Chapter 4.

Of course, once you have identified error pages, you will want to know which links within your website point to these pages—that is, have broken links. From the report shown in Figure 9.5, click any of the listed error pages to get the detail for that specific page (see Figure 9.6), and then select Navigational Summary. The result is a list of pages that your visitors were on just prior to clicking through and receiving the error page, as shown in Figure 9.7.

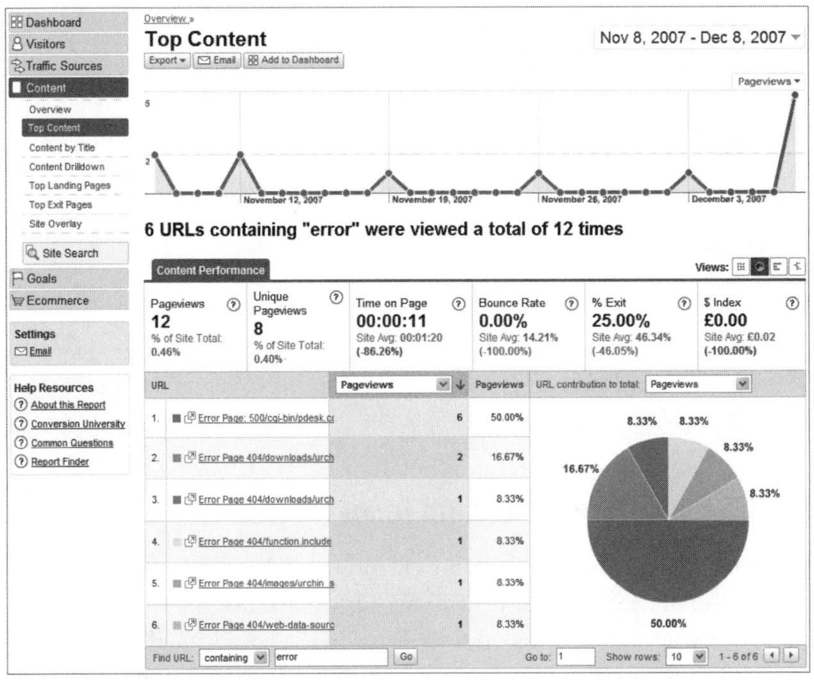

Figure 9.5 Viewing error pages

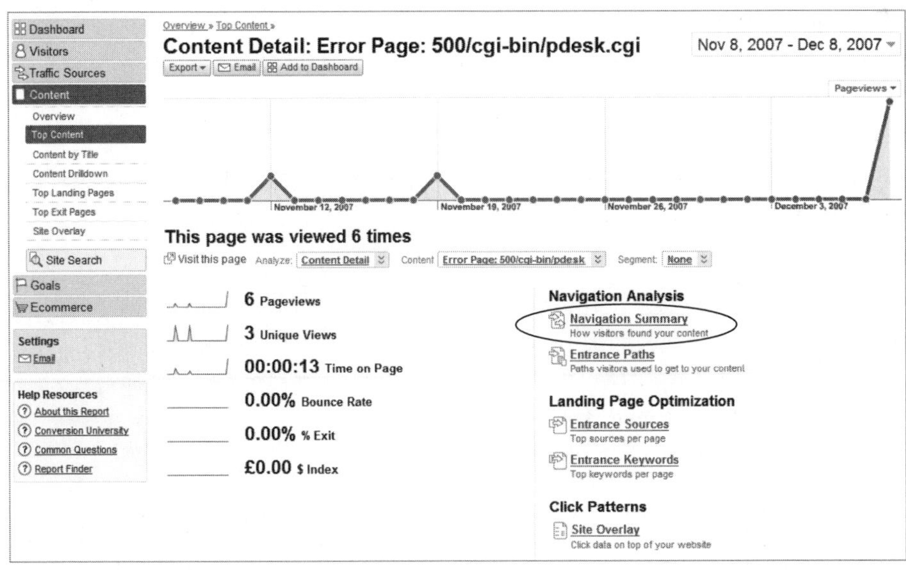

Figure 9.6 Specific page detail from the Top Content report

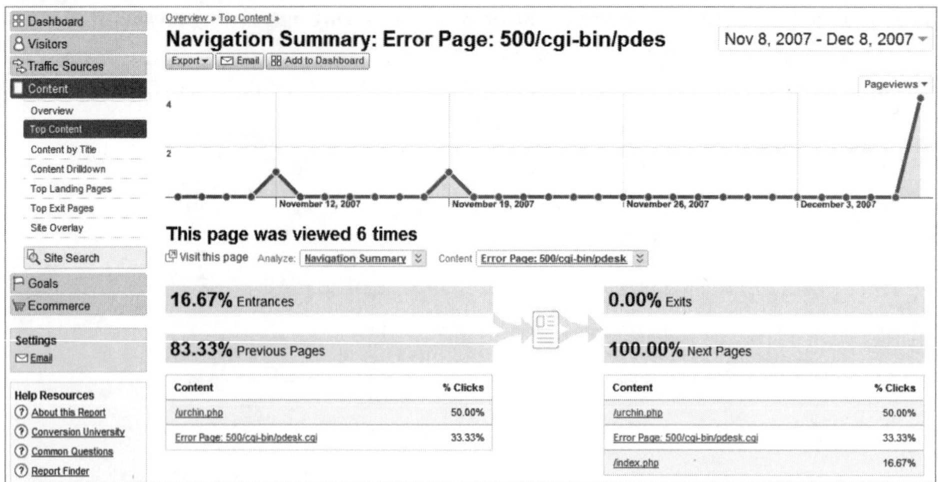

Figure 9.7 Pages leading to the error page

Tracking Pay-Per-Click Search Terms and Bid Terms

By default, Google Analytics tracks the bid term used in your AdWords account. Therefore, if you have Broad Match set in your AdWords campaigns, the Broad Match term will be displayed in your Google Analytics reports for all click-throughs for that campaign. For example, if you set up a campaign and bid on "shoes" with Broad Match set, then the reported term is "shoes" even if visitors actually searched for "blue shoes," "leather shoes," "gym shoes," and so on. This is the same when tracking other pay-per-click networks.

Of course, reporting on the bid term provides limited information. Ideally, you should have separate Ad Groups with Standard or Exact Match set so that you can view the precise keyword used by each visitor. However, when creating new pay-per-click campaigns, this may not be possible—that is, you won't yet know which terms are the best to target.

Tip: Best practice tip: Using Broad Match is a quick and easy way to get up and running with your AdWords account, but it is a blunt instrument and should only be used in the early stages of your online marketing. As you determine which keywords convert best, set up campaign groups with unique ad creatives and landing page URLs that target those specific search terms.

In the initial stages of online marketing, Broad Match can help you identify keyword themes that are successful at driving visitor click-throughs and conversions on your website, with a minimal amount of effort on your part. To move to the next level of a more targeted advertising campaign, you need to know the exact search terms used by visitors clicking on your ads. To show both the bid term and the search term, use the two-step filter shown in Figure 9.8a and 9.8b.

In English, Step 1 (refer to Figure 9.8a) reads:

a. For every pageview whose *Referral* has the pattern '(\?|&)(q|p)=([^&*])', AND

b. whose visit session has a Campaign Medium of cpc or ppc

c. Copy the third matching element from *Referral* to a variable called *$A3*, and then

d. Copy the contents of *$A3* to *Custom Field 1*.

For Step 1a, the query parameter (search term) for AdWords, Microsoft adCenter, and Yahoo! Search Marketing pay-per-click campaigns are named as either a q or p parameter in the referral URL.

For Step 1b, note that cpc or ppc is used here to match the Campaign Medium. This is because for this example, the website owner manually tagged Yahoo! Search Marketing and Microsoft adCenter campaigns with utm_medium=ppc, while AdWords auto-tagging assigns the medium as cpc. This enables the owner to differentiate AdWords pay-per-click advertising from other pay-per-click networks combined. For further information on tagging your online campaigns, refer to "Online Campaign Tracking," in Chapter 7.

Step 1c copies the search term used from the referral, and then Step 1d stores this in a temporary field.

In Step 2 (refer to Figure 9.8b), all contents of the search term (held in Custom Field 1) are combined with all contents of the Campaign Term (bid term), and the result is written back to the Campaign Term, overwriting its original value. With this method, both the bid term and the search term are visible in the Traffic Sources > AdWords > AdWords Campaigns Search Engine Marketing report, as shown in Figure 9.9. The highlighted example shows that the search terms "companies in crawley" and "companies in horsham" both match the bid term "businesses in sussex" (table rows 5 and 6, respectively).

Note: The filter order is important for this to work. Figure 9.8a must be applied before Figure 9.8b.

a)

Add Filter to Profile

Choose method to apply filter to Website Profile

Please decide if you would like to create a new filter or apply an existing filter to the Profile.

◉ Add **new** Filter for Profile OR ○ Apply **existing** Filter to Profile

Add new Filter for Profile

Filter Name: Override bid term

Filter Type: Custom filter

○ Exclude
○ Include
○ Lowercase
○ Uppercase
○ Search and Replace
○ Lookup Table
◉ Advanced

Field A -> Extract A: Referral — (\?|&)(q|p)=([^&]*)
Field B -> Extract B: Campaign Medium — cpc|ppc
Output To -> Constructor: Custom Field 1 — $A3
Field A Required: ◉ Yes ○ No
Field B Required: ◉ Yes ○ No
Override Output Field: ◉ Yes ○ No
Case Sensitive: ○ Yes ◉ No

Cancel Finish »

b)

Add Filter to Profile

Choose method to apply filter to Website Profile

Please decide if you would like to create a new filter or apply an existing filter to the Profile.

◉ Add **new** Filter for Profile OR ○ Apply **existing** Filter to Profile

Add new Filter for Profile

Filter Name: Override bid term 2

Filter Type: Custom filter

○ Exclude
○ Include
○ Lowercase
○ Uppercase
○ Search and Replace
○ Lookup Table
◉ Advanced

Field A -> Extract A: Custom Field 1 — (.*)
Field B -> Extract B: Campaign Term — (.*)
Output To -> Constructor: Campaign Term — $B1,($A1)
Field A Required: ◉ Yes ○ No
Field B Required: ◉ Yes ○ No
Override Output Field: ◉ Yes ○ No
Case Sensitive: ○ Yes ◉ No

Cancel Finish »

Figure 9.8 (a) Step 1: Obtain the search term from your pay-per-click referrals; (b) Step 2: Overwrite the current bid term with the bid term + the search term.

Combining pageview fields with session fields

Note, there is a slight caveat when working with the filter described in Figure 9.8a: It combines a per-pageview field (Referral) with a per-session field (Campaign Source). A pageview field is a field that is populated with every pageview recorded by Google Analytics, whereas a session field is set and maintained throughout a visitor's time on the site.

For example, each time a pageview is made, the page title, URL, and referral are updated to match the current page; but the session fields (returning visitor versus new visitor indicator, or campaign name, for example) are the same regardless of the page currently being viewed. Referral is a pageview field, in that each pageview will have its own unique referral, whereas Campaign Medium will always have the same value across the entire session.

Because visitors can remove cookies during a session, it is possible that applying a different filter may alter a session field *within* a visitor's session. This can cause a data misalignment, potentially resulting in an unpredicted data value showing in the reports. This is rare but it occasionally happens (see row 7 of Figure 9.12, for example).

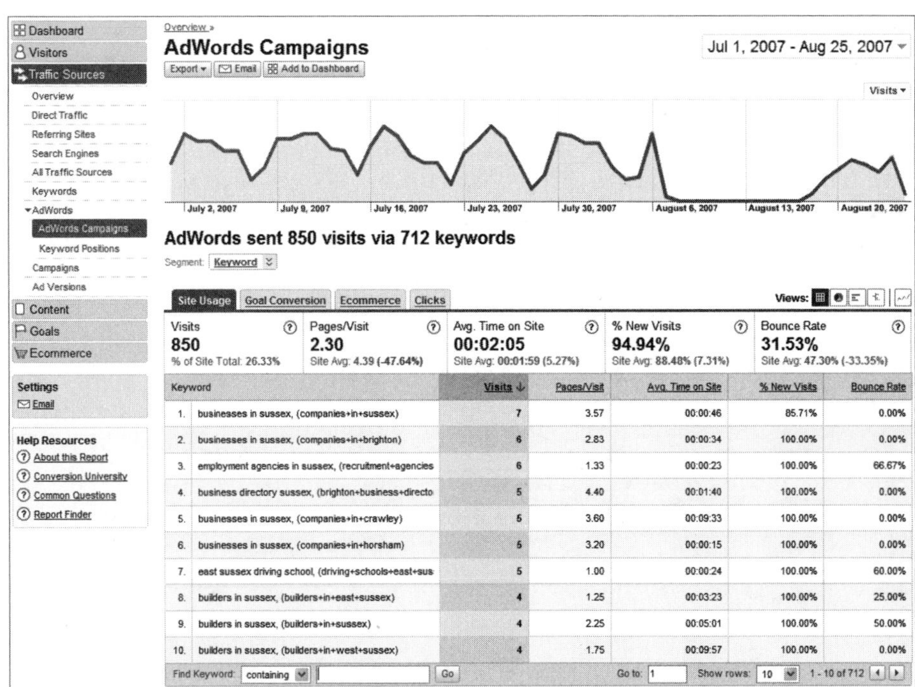

	Keyword	Visits	Pages/Visit	Avg. Time on Site	% New Visits	Bounce Rate
1.	businesses in sussex, (companies+in+sussex)	7	3.57	00:00:46	85.71%	0.00%
2.	businesses in sussex, (companies+in+brighton)	6	2.83	00:00:34	100.00%	0.00%
3.	employment agencies in sussex, (recruitment+agencies	6	1.33	00:00:23	100.00%	66.67%
4.	business directory sussex, (brighton+business+directo	5	4.40	00:01:40	100.00%	0.00%
5.	businesses in sussex, (companies+in+crawley)	5	3.60	00:09:33	100.00%	0.00%
6.	businesses in sussex, (companies+in+horsham)	5	3.20	00:00:15	100.00%	0.00%
7.	east sussex driving school, (driving+schools+east+sus	5	1.00	00:00:24	100.00%	60.00%
8.	builders in sussex, (builders+in+east+sussex)	4	1.25	00:03:23	100.00%	25.00%
9.	builders in sussex, (builders+in+sussex)	4	2.25	00:05:01	100.00%	50.00%
10.	builders in sussex, (builders+in+west+sussex)	4	1.75	00:09:57	100.00%	0.00%

Figure 9.9 Sample report showing the "bid term, (search term)" combination

Tracking Referral URLs from Pay-Per-Click Networks

As well as displaying ads on their own search properties, pay-per-click networks often partner with other websites to display their advertisements, sharing revenue from resultant ad click-throughs with the partner. An example is the relationship between Google and Ask.com. Ask is an independent search engine with its own search technology for displaying organic search results. However, for paid search, Ask partners with Google AdWords. If you advertise on AdWords, then your advertisement will also appear on the Ask.com website. In this way, pay-per-click partner networks are a great additional distribution channel for your advertisement, enabling you to reach a wider audience.

Note: AdWords has a search network opt-out feature that enables you to advertise only on Google web properties if desired.

By default, reports in Google Analytics group all pay-per-click partner click-throughs for AdWords as "google/cpc." For example, you will not see pay-per-click visitors that originate from Ask labeled as such—just google/cpc, as shown in Figure 9.10. The same is true for other pay-per-click networks such as the Yahoo! partners—Alta Vista, Lycos, and Excite. However, a simple filter can be applied to show more fully where your pay-per-click visitors are originating from, as shown in Figure 9.11.

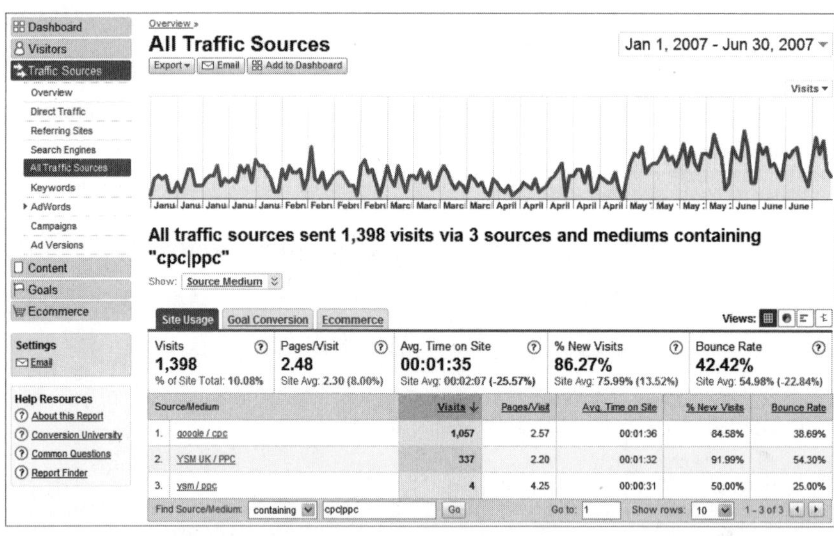

Figure 9.10 Different paid networks

Add Filter to Profile

Choose method to apply filter to Website Profile

Please decide if you would like to create a new filter or apply an existing filter to the Profile.

(•) Add **new** Filter for Profile OR ( ) Apply **existing** Filter to Profile

Add new Filter for Profile

Filter Name: Show referrer network

Filter Type: Custom filter

( ) Exclude
( ) Include
( ) Lowercase
( ) Uppercase
( ) Search and Replace
( ) Lookup Table
(•) Advanced

Field A -> Extract A	Referral	^http://([^/]*)
Field B -> Extract B	Campaign Source	(.*)
Output To -> Constructor	Campaign Source	$B1.$A1

Field A Required (•) Yes ( ) No
Field B Required (•) Yes ( ) No
Override Output Field (•) Yes ( ) No
Case Sensitive ( ) Yes (•) No

Cancel | Finish »

Figure 9.11 Filter to include the original referrer from different pay-per-click networks

In English, Figure 9.11 reads as follows:

a. For every pageview, extract the referring domain, omitting the "http://" text and anything after the next slash (/).

b. Copy the contents of the Campaign Source field.

c. Append the referring domain to the Campaign Source variable and overwrite it.

Notice that both steps a and b must be executed in order for the filter to proceed to step c. The result is a report that lists both the original referral and the Google Analytics–defined campaign source, as shown in Figure 9.12.

Search engine relationships

The relationships among search engines (paid and non-paid), directories, and portals are quite complex. For example, the relationship chart shown here is from a U.K. perspective. To understand this chart, view Google's relationships only; it shares its organic search results with AOL and Netscape. AdWords results are shared with AOL, Netscape, Ask, LookSmart, and Teoma; and Google receives directory results from DMOZ. The other search engines have similar multiple relationships.

Continues

Search engine relationships *(Continued)*

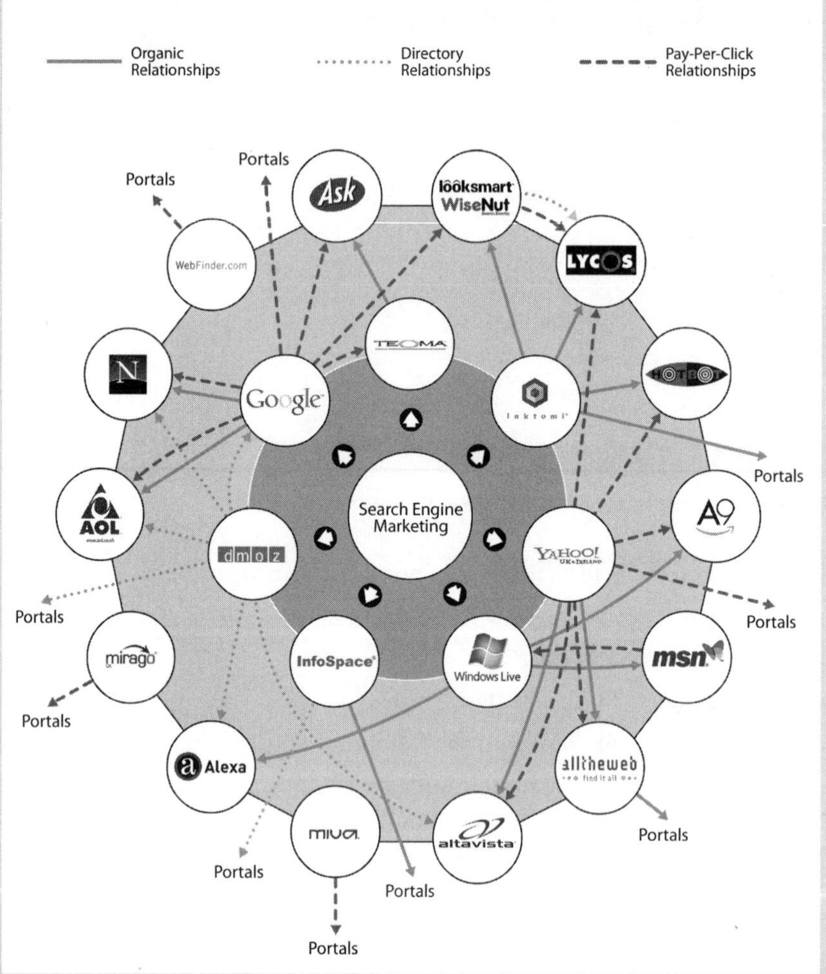

This printed version is limited; a color-coded, interactive version is available at www.omegadm.co.uk/relationship-chart.htm

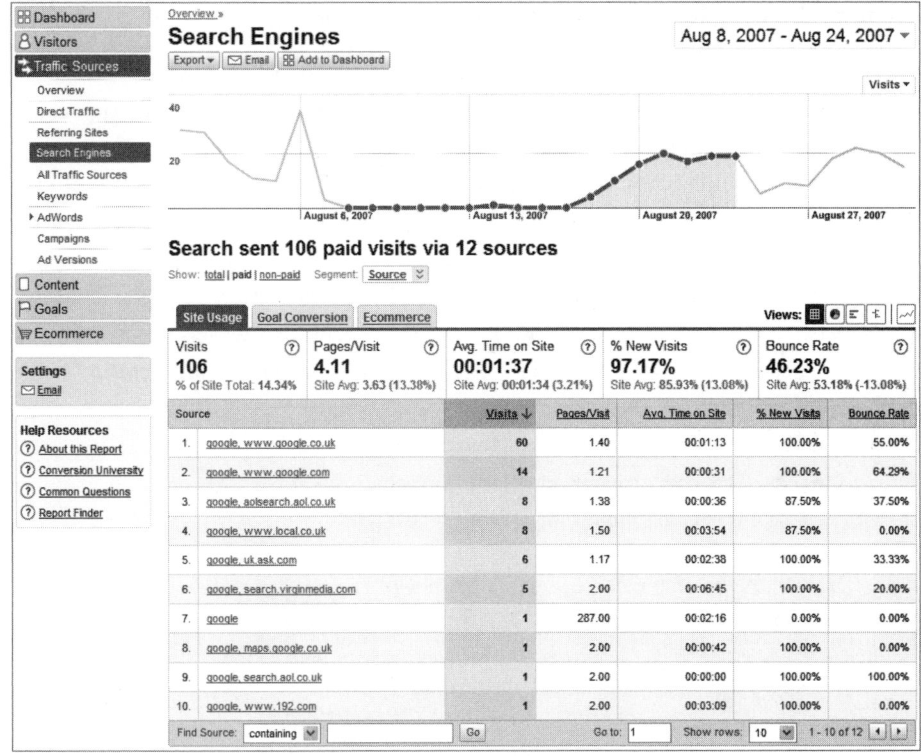

	Source	Visits	Pages/Visit	Avg. Time on Site	% New Visits	Bounce Rate
1.	google, www.google.co.uk	60	1.40	00:01:13	100.00%	55.00%
2.	google, www.google.com	14	1.21	00:00:31	100.00%	64.29%
3.	google, aolsearch.aol.co.uk	8	1.38	00:00:36	87.50%	37.50%
4.	google, www.local.co.uk	8	1.50	00:03:54	87.50%	0.00%
5.	google, uk.ask.com	6	1.17	00:02:38	100.00%	33.33%
6.	google, search.virginmedia.com	5	2.00	00:06:45	100.00%	20.00%
7.	google	1	287.00	00:02:16	0.00%	0.00%
8.	google, maps.google.co.uk	1	2.00	00:00:42	100.00%	0.00%
9.	google, search.aol.co.uk	1	2.00	00:00:00	100.00%	100.00%
10.	google, www.192.com	1	2.00	00:03:09	100.00%	0.00%

Figure 9.12 Showing the referral URLs from pay-per-click networks

As you can see, the structure of the report in Figure 9.12 is a referral source list of the form *ppc network source, referring website.*

The report shows visitors from the Google AdWords partner network, including: aol.co.uk, local.co.uk, uk.ask.co.uk, virginmedia.com, maps.google.com, and 192.com. Without this Show Referrer filter in place, the level of detail is limited to a single entry from "google."

Note: The same caveat applies here as described for Figure 9.8a. That is, the filter combines a per-pageview field (Referral) with a per-session field (Campaign Source); and sometimes, such as when visitors delete their cookies during a session, the data may not align. This can result in the occasional odd value showing in the reports. Figure 9.12 is a case in point—row 7 is clearly incorrect.

Site Overlay: Differentiating Links to the Same Page

Site overlay, first discussed in Chapter 5, is an excellent way to visualize what links your visitors are clicking and which links have the most value—that is, drive conversions. What happens if on your category page you have numerous links pointing to the same product page—for example, an image link, a menu link, and a headline link? It would be useful to know which of these is best at driving visitors through to conversions, so you know where to focus improvements.

By default, for the same URL link on a page, Google Analytics reports will show identical statistics for each one of those links (see Figure 9.13). However, it is possible to differentiate these in Google Analytics by adding a different query parameter to the identical links on the page, as shown here:

```
http://www.mysite.com/product.htm?linkid=sideMenu
http://www.mysite.com/product.htm?linkid=image
http://www.mysite.com/product.htm?linkid=textbox
```


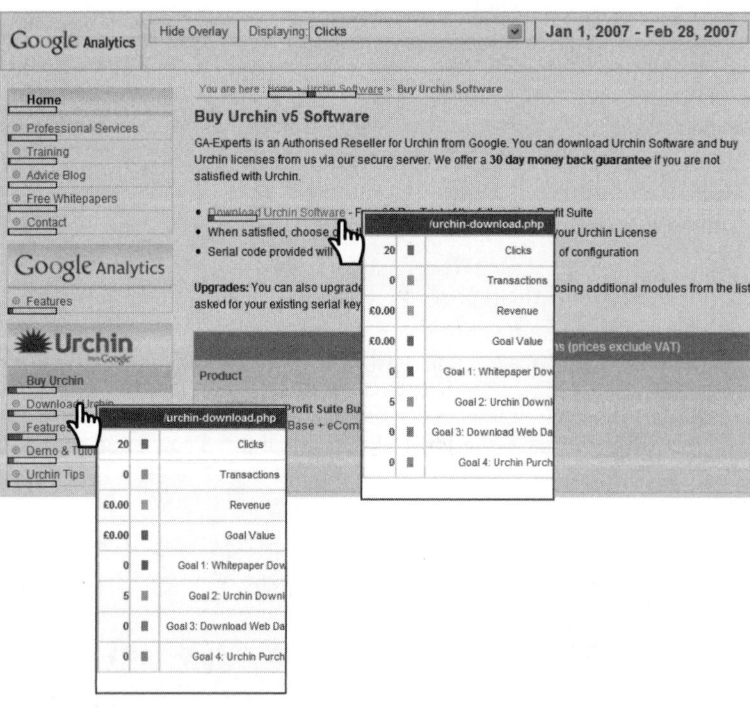

Figure 9.13 Default site overlay report for two links that point to the same page

With this method, your Site Overlay report will be able to clarify whether a text link has more of an impact than an image link or menu link to the same page.

Bear in mind that when applying this method and viewing other reports, such as the Top Content report, you will need to sum the pageview data for these links to determine the page total—that is, aggregate the index.htm pageviews as shown in Figure 9.14. The pages listed in this report are all, in fact, the same page.

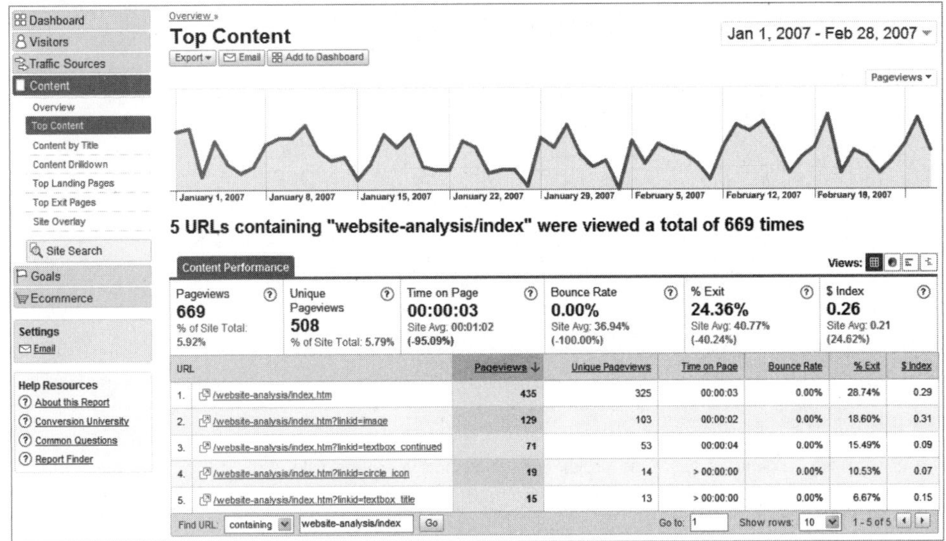

Figure 9.14 Result of adding query parameters to differentiate links to the same page

Matching Specific Transactions to Specific Keywords

As discussed in Chapter 1, web analytics is about identifying trends, so you shouldn't get hung up on precise numbers. Understand the strength and accuracy limitations of your data and get comfortable with it. For Google Analytics, Google's strong stance on privacy means that individuals are not tracked and all data is reported at the aggregate level.

However, for e-commerce transactions a little more detail is usually desired by e-commerce and marketing managers. Without identifying individuals, the following hack enables you to view your transaction list and identify which referrer source, medium, and keywords were used by the purchaser to find your website in the first place.

Note: This technique was originally discussed in an article by Shawn Purtell from ROI Revolution (www.roirevolution.com/blog/2007/05/matching_specific_transactions_to_specific_keyword.html) and is reproduced here with permission.

The hack works by cascading three advanced filters as follows:

Filter 1 Figure 9.15 shows the first filter, which grabs the campaign source and medium of a visit and places this in a custom field.

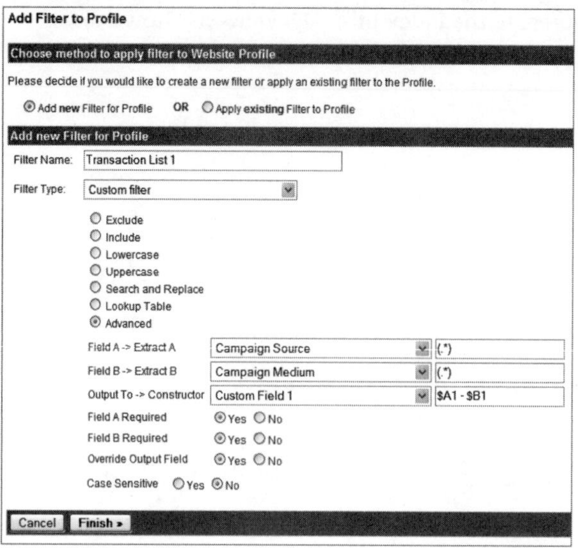

Figure 9.15 Capturing the campaign source and medium and storing these in a custom field

Filter 2 Figure 9.16 shows the second filter, which adds the keyword to the custom field. The custom field then contains the referrer source, medium, and keyword.

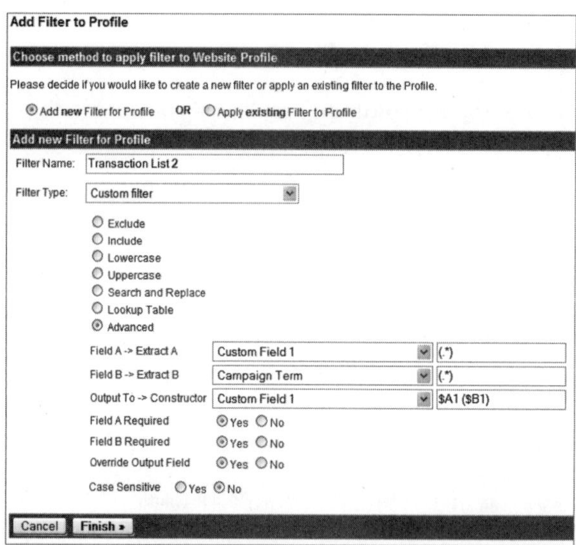

Figure 9.16 Appending the referral keyword to the custom field

Filter 3 Figure 9.17 shows the third and final filter, which takes the custom field created and appends it to the transaction order ID. This matches up sources with specific transactions.

Add Filter to Profile

Choose method to apply filter to Website Profile

Please decide if you would like to create a new filter or apply an existing filter to the Profile.

(◉) Add **new** Filter for Profile **OR** ( ) Apply **existing** Filter to Profile

Add new Filter for Profile

Filter Name: Transaction List 3

Filter Type: Custom filter

( ) Exclude
( ) Include
( ) Lowercase
( ) Uppercase
( ) Search and Replace
( ) Lookup Table
(◉) Advanced

Field A -> Extract A: E-Commerce Transaction Id (.*)
Field B -> Extract B: Custom Field 1 (.*)
Output To -> Constructor: E-Commerce Transaction Id $A1 $B1
Field A Required: (◉) Yes ( ) No
Field B Required: (◉) Yes ( ) No
Override Output Field: (◉) Yes ( ) No
Case Sensitive: ( ) Yes (◉) No

Cancel | Finish »

Figure 9.17 Appending the custom field information to the transaction ID

Of course, the order of the filters is important and these should be maintained as described. When done correctly, the cumulative result is an Ecommerce > Transaction report that is transformed from just showing the list of transaction IDs to one that includes details of these alongside the referring source, medium, and keyword, as shown in Figure 9.18. The format shown is as follows:

```
Transaction-ID referral source - medium (keywords)
```

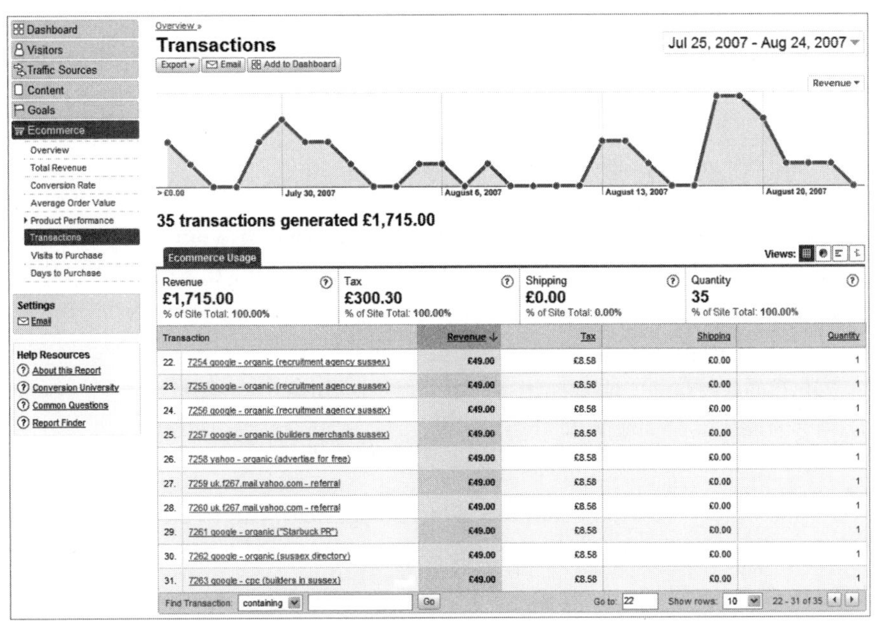

Figure 9.18 Matching specific transactions to specific keywords

Tracking Links to Direct Downloads

What if your campaign sends visitors directly to a file that does not accept the GATC JavaScript page tags? This can be the case with e-mail marketing or other specialized types of campaigns whereby visitors are referred directly to a PDF, EXE, ZIP, DOC, XLS, or PPT download—or any other file type that is not a website landing page. Without the GATC in place, Google Analytics will not detect a visitor from such a campaign. However, you can address this challenge by creating an intermediate landing page to capture the campaign variables *before* forwarding the visitor on to the actual file download.

Figure 9.19 shows an example intermediate landing page generated by a link from an e-mail message, such as a marketing campaign or even a regular e-mail signature that points to the following URL:

```
www.mysite.com/forwarder.php?file=download1.pdf&utm_source=sales&utm_campaign=
first-followup&utm_medium=email
```

The following table describes the elements of the preceding URL:

forwarder.php	Name of the page that will forward the visitor to the correct file
download1.pdf	Name of the file requested by the visitor
&utm_source=	Campaign identifier
&utm_medium=	Campaign identifier
&utm_campaign=	Campaign identifier

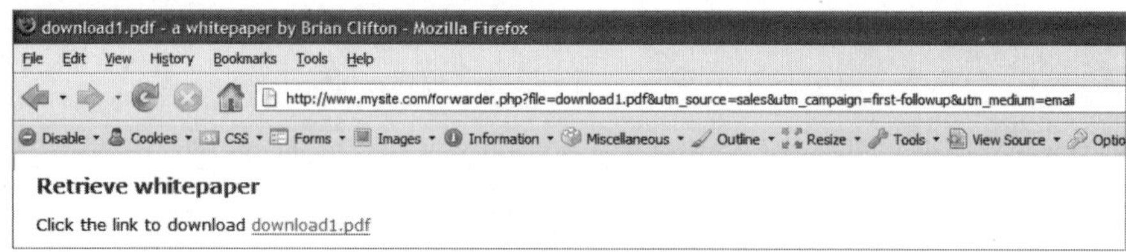

Figure 9.19 Example use of an intermediate landing page for file downloads

In this example, the forwarding page, forwarder.php, contains your GATC and the following code in the body tag—no other content is required for this page:

```
<body onLoad="var tmp='<? echo $file; ?>';if(tmp){
pageTracker._trackPageview('/downloads/<? echo
"$file?utm_source=$utm_source&utm_medium=$utm_medium&utm_campaign=$utm_campa
ign');window.location.href='$file'" ?>
}else{alert('No download file specified')}">
```

As you can see, forwarder.php contains a *_trackPageview()* call to create a virtual pageview for the requested file download, along with its associated campaign variables.

The final part is a redirect to the actual file itself: As soon as the forwarder.php page is loaded, the visitor is forwarded directly to the file in question.

The beauty of this method is that you can view each download file as a pageview in your Google Analytics reports with the referral campaign, medium, and source correctly attributed to the referring campaign. In addition, forwarder.php will be listed will all the aggregate referral information; however, you might want to remove this page from your reports with an exclude filter to prevent double counting.

Note: Although PHP is used in the example, the technique is equally applicable for any server-side web scripting language you might use, such as ASP, .NET, CGI-Perl, Python, and so on.

Changing the Referrer Credited for a Conversion

By default, Google Analytics gives credit for a conversion to the last referrer a visitor used. For example, consider the following search scenario for a user who visits your website by way of a different referrer each time:

- Google organic search, visitor leaves your website (referrer 1)
- Yahoo! paid search, visitor leaves your website (referrer 2)
- Google paid search, visitor converts (referrer 3)

All visit referrals are tracked with credit for the conversion given to referrer 3. This is the case except when the last referrer is direct—that is, the visitor uses their bookmarks or types your URL directly into their browser address bar. For example:

- Google organic search, visitor leaves your website (referrer 1)
- Yahoo! paid search, visitor leaves your website (referrer 2)
- Google paid search, visitor leaves your website (referrer 3)
- Direct (bookmark), visitor converts (referrer 4)

Credit for the conversion is still given to referrer 3. That makes sense, as it is most likely referrer 3 that led to the bookmarking (or remembering) of your website address. In the next section, you'll see what you can do if viewing the first referrer is more important to your conversions and you want to see this in your reports instead of the last referrer.

Capturing the Previous Referrer for a Conversion

For tagged landing page URLs only (that is, not organic landing pages), you can change the referrer given credit for a conversion to the previous referrer by appending your landing page URLs with the utm_nooveride=1 parameter, as shown in the following example:

```
http://www.mysite.com/product1.php?utm_source=sales&utm_campaign=first-
followup&utm_medium=email&utm_nooverride=1
```

When Google Analytics detects the utm_nooverride=1 parameter, it retains the previous referrer campaign information. Therefore only if there are no existing campaign variables will new ones be written. You can also use this parameter in a mixed environment—that is, with some landing page URLs having utm_nooveride=1 while others are not set.

Consider, for example, an online marketing campaign using AdWords to drive visitors to your site where the call to action is an e-mail subscription. You then follow up by e-mailing your newsletter to new subscribers. In this scenario, you will probably want to maintain the AdWords campaign details about how visitors came to subscribe and have these associated with any future conversions. To achieve this, prevent your e-mail marketing from overwriting the campaign details by appending your URLs within the e-mail with the utm_nooveride=1 parameter. The manual tagging of landing page URLs for e-mail is discussed in the section "Online Campaign Tracking," in Chapter 7.

For your AdWords landing pages (for which auto-tagging is enabled), you only need to append the utm_nooveride=1 parameter to your landing pages as follows:

- Example AdWords landing page URL for a static web page with auto-tagging on:

  ```
  http://www.mysite.com/product1.php?utm_nooverride=1
  ```

- Example AdWords landing page URL for a dynamic web page with auto-tagging on:

  ```
  http://www.mysite.com/product.php?id=101&lang=en&utm_nooverride=1
  ```

Note: If you are using a third-party ad tracking system with your AdWords campaigns, read "Testing After Enabling Auto-tagging," in Chapter 6.

By this method, if your e-mail recipient has not previously been associated with any other online marketing activity, then they will be reported as coming from your e-mail marketing should they click through on a link to your website. Otherwise the original referral details (AdWords in this example) will be maintained. For consistency, this is the same should a conversion take place.

Capturing the First and Last Referrer of a Visitor

The previous section describes overriding which tagged referrer is given credit for a conversion—from the last referrer (default) to the p revious referrer. The hack in this section is an extension of that; it captures both the first and last referrer together. This works whether a conversion takes place or not and will work for all referrers, including organic visitors—not just those that result from tagged landing pages.

The caveat is that it requires a little bit more work. Firstly, you have to modify your GATC, and then apply an advanced filter:

1. Capture and store the first referrer.

Modify your GATC on all pages as follows:

```
<script type="text/javascript">
   var gaJsHost = (("https:" == document.location.protocol) ?
"https://ssl." : "http://www.");
   document.write(unescape("%3Cscript src='" + gaJsHost + "google-
analytics.com/ga.js' type='text/javascript'%3E%3C/script%3E"));
</script>

function _uGC(l,n,s) {
   // used to obtain a value form a string of key=value pairs
   if (!l || l=="" || !n || n=="" || !s || s=="") return "-";
   var i,i2,i3,c="-";
   i=l.indexOf(n);
   i3=n.indexOf("=")+1;
   if (i > -1) {
      i2=l.indexOf(s,i); if (i2 < 0) { i2=l.length; }
      c=l.substring((i+i3),i2);
   }
   return c;
}

function checkFirst(){
   // check if this is a first time visitor
   newVisitor = 0;
   var myCookie = " " + document.cookie + ";";
   var searchName = "__utma=";
```

```
        var startOfCookie = myCookie.indexOf(searchName)
        if (startOfCookie == -1) {   // i.e. first time visitor
            newVisitor = 1;
        }
    }

    function grabReferrer(){
        // grab campaign and referrer info from the _utmz cookie
        if (newVisitor) {
            var z = _uGC(document.cookie, "__utmz=", ";");
            urchin_source = _uGC(z,"utmcsr=", "|");
            urchin_medium = _uGC(z,"utmcmd=", "|");
            urchin_term = _uGC(z,"utmctr=", "|");
            urchin_content = _uGC(z,"utmcct=", "|");
            urchin_campaign = _uGC(z,"utmccn=", "|");
            var gclid = _uGC(z,"utmgclid=","|");
            if (gclid) {
                urchin_source = "google";
                urchin_medium = "cpc";
            }
            pageTracker._setVar(urchin_term);
        }
    }
        var pageTracker = _gat._getTracker("UA-12345-1");
        pageTracker._initData();
        checkFirst();          // checks if this is a new visitor
        pageTracker._trackPageview();
        grabReferrer();        // Grab referrer info
</script>
```

The function checkFirst() simply checks whether this is a first-time visitor by looking for the presence of the _utma cookie. This is always set for a visitor, so its presence indicates a returning visitor. The function grabReferrer() grabs all the current first-time visitor referral information—source, medium, keyword term, campaign content, and campaign name—and stores these as local variables. The last line of this function stores the keyword term as a visitor label by calling __setVar().

Notice that in the function grabReferrer(), only the campaign term (the keywords) is stored as a visitor label. However, if you want you can store any of the campaign variables listed by modifying the __setVar() line accordingly, or use combinations of the campaign variables.

2. Use an advanced filter (see Figure 9.20).

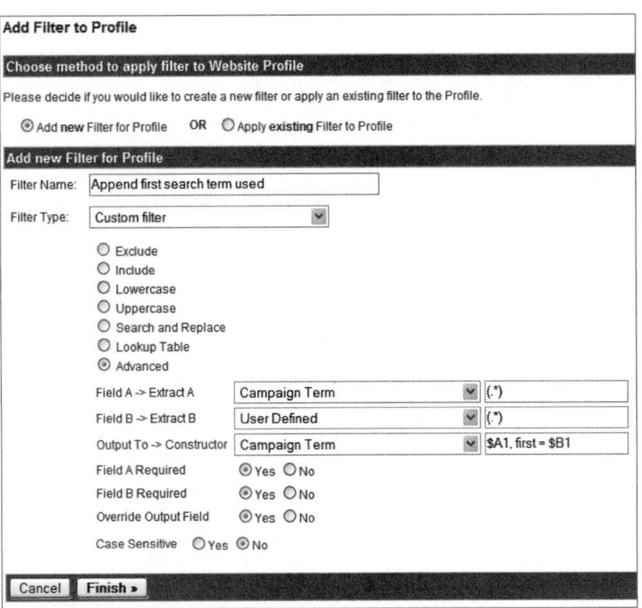

Figure 9.20 Advanced filter to combine the first and last referrer

When you implement this filter, you will see the first and last referral keywords displayed in your keywords reports, as shown in Figure 9.21. Highlighted is an interesting combination: The original referral search term was "google analytics accreditation" and the last one was "google analytics training." Perhaps there is potential in targeting those people looking for accreditation information with training courses?

Note that with this method you will always see the keyword reports in the format of *last_keyword, first = first_keyword*—even when no keyword is present for one of the visits, such as for direct visitors or for single visits. In these cases, a dash "-" is shown as the keyword. To view a list of only the first referrer keywords, view the Visitors > User Defined report.

If you want to maintain your keyword reports in their original state, you could place the *last_keyword, first = first_keyword* information in the User Defined report instead. Simply change the Output To constructor of Figure 9.20 to User Defined so that it overwrites the User Defined field rather than the Campaign Term.

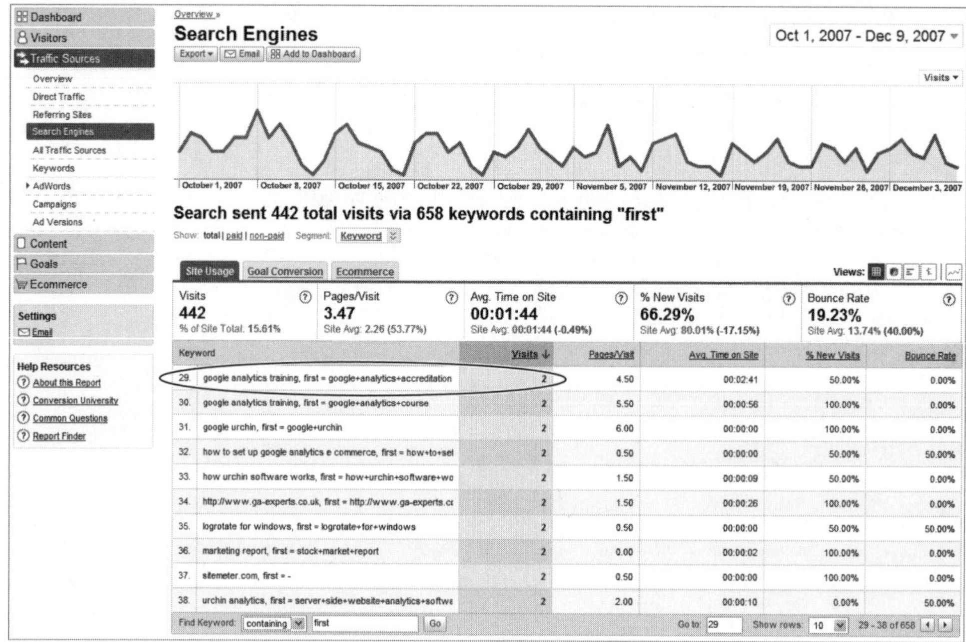

Figure 9.21 A modified Traffic Sources > Keyword report

Importing Campaign Variables into your CRM System

Campaign variables (medium, referral source, keywords, etc.) captured by Google Analytics are shown throughout your reports. For example, for any pageview, conversion, or transaction, you can cross-segment the data to view visitors' referral details. It might also be useful for you to have this information imported into your customer relationship management (CRM) system, external to Google Analytics. That way, when a visitor submits a brochure request form or makes a purchase—data that is transmitted to your CRM system—you can also transmit the campaign details along with it.

The method is demonstrated using a submit form. First, copy the following two JavaScript functions into the <head> section of the HTML page containing your form:

```
<script type="text/javascript">
 function _uGC(l,n,s) {
   // used to obtain a value form a string of key=value pairs
   if (!l || l=="" || !n || n=="" || !s || s=="") return "-";
   var i,i2,i3,c="-";
   i=l.indexOf(n);
   i3=n.indexOf("=")+1;
```

```
    if (i > -1) {
        i2=l.indexOf(s,i); if (i2 < 0) { i2=l.length; }
        c=l.substring((i+i3),i2);
    }
    return c;
}

function setHidden(f) {
    var z = _uGC(document.cookie, "utmz=",";");
    f.web_source.value = _uGC(z,"utmcsr=","|");
    f.web_medium.value = _uGC(z,"utmcmd=","|");
    f.web_term.value = _uGC(z,"utmctr=","|");
    f.web_content.value = _uGC(z,"utmcct=","|");
    f.web_campaign.value = _uGC(z,"utmccn=","|");

    var gclid = _uGC(z,"utmgclid=","|");
    if (gclid) {
        f.web_source.value = "google";
        f.web_medium.value = "cpc";
        //It is not possible to capture AdWords campaign details by this
        //method as GA processing is required for this. Therefore the
        //following lines are set to remove confusion should a visitor
        //use multiple referrals with the last one being AdWords.
    f.web_term.value = "";                  // remove previous info if any
    f.web_content.value = "";             // remove previous info if any
    f.web_campaign.value = "";          // remove previous info if any
    }
}
</script>
```

Then, within your HTML <form> tag of the same page, add the onSubmit event handler and hidden form fields as follows:

```
<form method="post" action="formhandler.cgi" onSubmit="setHidden(this);">
    <input type=hidden name=web_source value="">
    <input type=hidden name=web_medium value="">
    <input type=hidden name=web_term value="">
    <input type=hidden name=web_content value="">
    <input type=hidden name=web_campaign value="">
…etc.
</form>
```

If you already have an onSubmit event handler, simply append the setHidden(this) call, for example, as follows:

```
<form method="post" action="formhandler.cgi"➥
onSubmit="validate();setHidden(this);">
```

By this method, when a visitor submits the form to your CRM system, a call is made to the JavaScript function setHidden(this). This routine extracts the campaign variables from the Google Analytics __utmz cookie using the function _uGC. These are stored in your hidden form fields, which can then be transmitted to your CRM system.

Note: Even without a CRM system you may want to use this method. For example, most formhandler scripts allow you to log the details of a form submission. Simply append the hidden form fields to your logfile.

Summary

Google Analytics hacks help you delve deep into analysis. To do that, you need to think laterally and be creative with applying filters. Because the GATC is written in JavaScript, Google Analytics is extremely flexible in this regard. There are numerous ways it can be altered or customized, and a good webmaster should be able to do this for you without too much trouble. Custom labeling of visitors on a per-visitor or per-session basis is very powerful, as is the ability to use advanced filters to manipulate reported data, such as combining bid and search terms into one phrase.

The examples provided in this chapter are only a sample of what you can achieve. Feel free to experiment and share your own experiences on the book blog site: www.advanced-web-metrics.com/blog.

In Chapter 9, you have learned about the following:

- Customizing the list of recognized search engines
- Labeling visitors
- Changes to which referrer is given credit for a conversion
- Tracking error pages and broken links
- Tracking pay-per-click search terms as well as bid terms
- Tracking referral URLs from pay-per-click networks
- Site overlay: differentiating links to the same page
- Matching transactions to specific keywords
- Tracking links to direct downloads

Now that you have hacked your data, the next chapter describes how to take Google Analytics reports with you into your core business.

IV

Using Visitor Data to Drive Website Improvement

Reporting, although important, is only half the story. The real power of web analytics tools lies in what you do with the data. Having a clear understanding of visitor behavior enables you to identify bottlenecks in conversion processes and marketing campaigns so you can improve them. That is, turning inert data into actionable information.

Part IV is about using data, from determining the most important measures of performance, to optimizing pages, processes, and your search engine marketing campaigns. Part IV also considers how the online world ties in with the offline channel, so this part is aimed at the marketer and analyst in you.

In Part IV, you will learn about the following:

10 Focus on Key Performance Indicators

By now you understand what web analytics tools can do, how to set up Google Analytics using best practices, and how to navigate around its interface so that you feel comfortable with the data.

What has been discussed so far has been fairly straightforward—dare I say easy? The next step is the difficult part—not from a technical perspective, but purely in terms of communication.

In Chapter 10, you will learn about the following:

Setting objectives and key results
Selecting and preparing KPIs
Presenting hierarchical KPIs
Example KPIs segmented by stakeholder job roles
KPIs for a web 2.0 environment

Setting Objectives and Key Results (OKRs)

To summarize the story so far, the first best practice implementation principals are as follows:

- Tag everything—get the most complete picture of your website visitors possible.
- Clean and segment your data—apply filters.
- Define goals—distill the 80-plus reports of Google Analytics into performance benchmarks.

If you have followed these steps, that's excellent. However, the usual problem is that few other people in your organization know what you've done or appreciate your work. To many people, you have created a set of nice charts and reports. Even if they don't say it aloud, they may be thinking, "So what?"

The unfortunate truth is that you will have wasted your time unless you can get the buy-in to use the visitor data in driving business decisions and be the focal point for instigating change on your website. With your initial understanding of your visitor data, this is your next step—that is, to set *key performance indicators* for your website and align these with the objectives and key results of your organization. For this you need to bring in your key stakeholders from the other parts of the business.

> *Visitor analysis is very important, but it seems like few people are using it in an actionable way. People are beginning to discover you can dramatically improve profitability, double and triple it, just by understanding which traffic is most likely to convert, what it is people do (and don't do) on your website, and how to measure the effectiveness of changes you make on the site to improve visitor conversion.*
>
> —Jim Novo
> Co-Chair, Web Analytics Association Education Committee

Objectives and key results (OKRs) are about understanding your business goals. This is an important prerequisite before you delve into the specific key performance indicators for your website. Essentially, you need to ensure that the two are in alignment, and the setting of OKRs prepares the way. The process consists of four steps:

1. Map your stakeholders.
2. Brainstorm with your stakeholders.
3. Define your OKRs.
4. Distill and refine OKRs.

1. Map your stakeholders. Who are your stakeholder departments? These may be marketing, sales, PR, operations, web development and design agencies, e-commerce managers, content creators—even the CEO. Of course it may only be the CEO, but if not, select one person from each department as the key contact for initial discussions. Your first choice may not end up being the right person but that can be changed later. The important thing is to get people on board from those departments.

Your key contacts are the individuals who represent the interests of that department within your organization. They can canvas opinion from the rest of the organization on your behalf; in other words, they do not have to be the most senior people in their departments. Try to make this a two-way street—with you setting the scene with your initial data and thoughts on the current situation, and stakeholders providing their perspective on how it fits with their department. For example, they may provide information from CRM systems, call center figures, web server performance, and so on.

2. Brainstorm with your stakeholders. Do this by arranging regular meetings with your stakeholders. The frequency will vary depending on how significant your website is to the overall business model of your organization. However, try to meet weekly in the early days and adjust according to feedback.

Initial meetings should focus on what is currently happening—not whether it is good or bad, but rather what information is available. Often your stakeholders will ask for more information, possibly less; but usually they want to see it cross-referenced against other metrics—something to prepare for the next meeting.

As meetings progress, you and your stakeholders should start to understand each other with respect to terminology, what data can be collected, what information can be gleaned from it, and how it can be useful to the business (this usually takes from one to three meetings).

3. Define your OKRs. By week four, it can be a good idea to meet separately with each stakeholder. You should be ready to ask the question, "What is the objective of our website from your point of view?" Don't worry if you need a few more weeks to achieve this. Every organization is different. But, try not to let this drag on beyond the six-week mark or you risk losing momentum. The process taking place is not set in stone and can be reviewed and modified in six months or whenever necessary.

Encourage your stakeholders to give measurable answers to your objectives question; these form the *results* part of your OKRs. The following are measurable examples:

- Generate more sales leads.
- Download more catalog PDFs.
- Encourage more cross-selling (increase average order value).
- Create greater brand or product awareness.
- Acquire more traffic.
- Provide customer service (reduce call center volume).
- Build relationships with visitors (e.g., blog comments, forum posts).

Include anything that can be judged as a success for your website.

4. Distill and refine OKRs. With a long list of objectives and key results from your stakeholders (such lists are always long to start with), distill it down to the 10 most important for each. This should be your maximum—fewer than 10 is better because during this first phase of building your web analytics framework, managers generally cannot cope with a long list of levers to act on. Objectives can always change later, so focusing on the most important 5–10 OKRs will stand you in good stead.

Once you have your list of OKRs, the business language of your organization, you can use these to build your KPIs, the analyst language with respect to the website.

Selecting and Preparing KPIs

Google Analytics is your free data gathering and reporting tool, but it will not optimize your website for you. That requires smart people (you!) to analyze, interpret, and act on the reported findings. To act on your Google Analytics information—that is, instigate required changes—you need to present your findings in a clear, understandable format to your stakeholders. These are a diverse group of people who sit at different levels in your organization—all the way up to the board, one hopes. That's the caveat: Presenting web analytics data outside of your immediate team is a challenge because most business people simply do not have the time to understand the details that such reports offer by default.

To communicate your story effectively to your stakeholders, create reports in a format and language that business managers understand—that is, KPI reports. These are abridged versions of your web analytics reports, usually summarized in Microsoft Excel.

What Is a KPI?

Web analytics aside, organizations around the world use key performance indicators (KPI) to assess their performance. Also sometimes referred to as Key Success Indicators (KSI) or Balanced Score Cards (BSC), KPIs are used in business intelligence to appraise the state of a business. Once an organization has set its OKRs, it needs a way to measure progress. Key performance indicators are those measurements.

Similarly, in web analytics, a key performance indicator is a web metric that is essential for your organization's online success. The emphasis here is on the word *essential*. If a 10 percent change—positive or negative—in a KPI doesn't make you sit up and call someone to find out what happened, then it is not well defined. Good KPIs create expectations and drive action; and because of this, they are a small subset of information from your reports.

When considering your KPIs, bear in mind the following:

- In most cases a KPI is a ratio, percentage, or average, rather than a raw number. This allows data to be presented in context.

- A KPI needs to be temporal–that is, time-bound. This highlights change and its speed.
- A KPI drives business-critical actions. Many things are measurable, but that does not make them key to your organization's success.

Use KPIs to put your data into context. For example, saying "we had 10,000 visitors this week" provides a piece of data, but it is not a KPI because it has no context. How do you know whether this number is good or bad? A KPI based on this data could be "our visitor numbers are up 10 percent month on month." This is a temporal indication that things are looking good over the time span of one month. In this example, the raw number should still be part of the KPI report, but it is not the KPI itself.

For the reasons just given, the vast majority of KPIs are ratios, percentages, or averages. However, sometimes a raw number can have a much greater impact. Consider the following examples:

- Our website lost 15 orders yesterday because our e-commerce server was down for 34 minutes.
- We lost $10,000 in potential revenue last week because our booking system does not work for visitors who use Firefox.
- We spent $36,000 last month on PPC keywords that did not convert.

Clearly, knowing whether any of these numbers are increasing or decreasing and what fraction of the total they represent is important, but the impact of these raw numbers is far greater at obtaining action and therefore should be the KPI in these examples.

The key point is that you should develop KPIs relevant to *your* particular business and *your* stakeholders. Any metric, percentage, ratio, or average that can help your organization quickly understand visitor data and is in context and temporal should be considered as a KPI. Try to use monetary values where possible; everybody understands $$$.

Preparing KPIs

Most of the hard work of preparing KPIs consists of defining OKRs—the dialogue you had with your stakeholders in obtaining the business objectives of your company. The key results used to establish OKR success are in fact your KPIs; you just need to turn these into actual web metrics that are available to you.

Sometimes (actually, quite often) discussing KPIs with stakeholders instills fear in your colleagues. They think you are performing the web equivalent of a time and motion study that is going to spotlight their deficiencies and single them out as not doing a good job. That fear is understandable: Being measured is not a comfortable feeling. However, my approach has always been to dispel that image. Evangelize web analytics KPIs as the tools to help your stakeholders shine and be rewarded for their efforts. Wield a carrot, not a stick.

The art of building and presenting a KPI report lies in being able to distill the plethora of website visitor data into metrics that align with your OKRs. For small organizations, having a report of 10 KPIs aligning with 10 OKRs is usually sufficient. For organizations with many different stakeholders, having only one KPI report will not cover the requirements of your entire business. There are simply too many stakeholders to reach a consensus about what the KPI short list should contain. Therefore, ensure that you tailor your KPI reports to specific needs by having individual stakeholder and hierarchical KPI reports.

Here is a six-point KPI preparation checklist:

1. Set your OKRs.

I repeat this here because of its importance. Identifying your stakeholders, discussing their needs, and being aware of the overall business plan for your organization enables you to put in place relevant metrics. This is an essential first step to ensure that your KPIs align with the business objectives of your organization.

2. Translate OKRs into KPIs.

This means setting specific web metrics against the business OKRs. Some metrics will be directly accessible from your Google Analytics reports; for example, if your e-commerce department says they want to "increase the amount of money each customer spends," then you will look for the average order value (AOV) from within the e-commerce section and monitor this over time. However, not all KPI metrics can be obtained in this way; sometimes segmentation is required or the multiplication or division of one number by another. Table 10.1 is a useful translation tool.

3. Ensure KPIs are actionable and accountable.

For each translated KPI, always go back and ask the stakeholder, "Who would you contact if this metric fell by 10 percent?" and "Who would you *formally* congratulate if it rose by 10 percent?" If a good answer for both is not forthcoming, then the suggested KPI is probably not a good one to include in your short list. I emphasize the word *formally* as this is a good way to focus the minds of your stakeholders on KPIs that lead to actions. A formal recognition could be a department-wide e-mail bulletin or a performance bonus—that usually does the trick.

4. Create hierarchical KPI reports.

Ensure that each recipient of your KPI report receives only the data he or she needs; the more relevant the information presented, the more attention and buy-in you will gain. It follows that a chief marketing officer will need a different, though similar, KPI report than a marketing strategist.

5. Define partial KPIs.

A frequently requested OKR is to increase the website conversion rate. This is usually straightforward to measure, but it is also black and white—the visitor either converts or doesn't. By providing partial KPIs, you can preempt your stakeholder's next question: "Why is the conversion rate so low?" I refer to these as partial KPIs because they relate to the partial completion of a full KPI. For example, if the conversion is to download a file, then navigating to the download page could be the partial KPI. Similar partial KPIs include the following:

- Navigating to the Contact Us page
- For a multi-page request form, the completion of the first page
- Reaching a certain point in a form-completion process
- Adding items to the shopping cart
- Navigating to the Special Offers page
- Completing an onsite search query

Tip: Tracking partially completed forms is discussed in the section "Virtual Pageviews for Tracking Partially Completed Forms," in Chapter 7.

► **Table 10.1** Sample OKR-to-KPI Translation Table

Stakeholder OKR	Suggested KPIs
To see more visitors access our site from search engines	Percentage of visits from search engines Percentage of conversions from search engine visitors
To sell more products	Percentage of visits that add to shopping cart Percentage of visits that complete the shopping cart Percentage of visits in which shopping cart is abandoned
To see visitors engaging with our website more	Percentage of visits that leave a blog comment or download a document Percentage of visits that complete a Contact Us form or click on a mailto: link Average time on site per visit Average page depth per visit
To cross-sell more products to our customers	Average order value Average number of items per transaction
Improve the customer experience	Percentage of visits who bounce (single-page visits) Percentage of internal site searches that produce zero results Percentage of visits that result in a support ticket being submitted

6. Consolidate.

   After forming a list of required KPIs for each stakeholder, consolidate them by looking for overlaps. The point of KPIs is to focus on the important metrics to your business. If your KPI report represents all the key factors that you need to measure success, each KPI should represent at least 10 percent of the whole (so no more than 10 KPIs are allowed). If a single KPI is much less than 10 percent in importance, then drop it or consolidate it into a more important KPI.

Remember that KPI reports are not set in stone—they can and should evolve as your audience learns to understand the metrics of their website and develop their actions to effect change. Review your KPI short list quarterly, at the very least.

Tip: As part of your role as a web analyst, you might also want to include KPIs that are not part of your Google Analytics reports—for example, server uptime, server response speed, notes of any offline campaigns or PR that could influence numbers, changes made to the website, new product launches, or user feedback. All of these can help explain what you see and therefore add value to your data.

Presenting Your KPIs

The best way to present KPI reports is by using Microsoft Excel. Every strategist, manager, and executive is familiar with the Excel format and recognizes its layout immediately. It is far better to present your KPI reports using Excel than to try to teach a new interface (Google Analytics) to old hands. In addition to Google Analytics, you may be collecting data from different sources, such as your own web server log files or offsite metrics such as search engine ranking reports. Combining all of them into one familiar interface will make it easy for everyone to understand the material you are presenting.

Figure 10.1 is an example KPI report for an online marketing executive containing 10 key metrics. Color coding (using Excel's conditional formatting) and arrows have been used to highlight positive and negative changes, with a threshold of 5 percent used to "double highlight" values.

All the data shown in Figure 10.1 is readily available from within Google Analytics, but using Excel to combine exactly what data elements your stakeholder wants to see enables you to deliver a concise report within a familiar interface.

Tip: You can download the example spreadsheet used in Figure 10.1 from the book blog site at www.advanced-web-metrics.com/scripts.

	A	B	C	D	E	F	G	H	
1	**KPI Report**								
2		May	June	% Change		July	% Change		
3	Conversion rate	n/a	3.7%			4.1%	10.8%	▲	1
4	Booking income	n/a	£464,823			£377,995	-18.7%	▼▼	
5									
6									
7	SE Visitors as a % of total	90.4%	90.1%	-0.3%	▼	87.6%	-2.8%	▼	2
8	Non SE Visitors as % of total	9.6%	9.9%	2.9%	▲	12.4%	25.6%	▲	
9									
10									
11	Quality of SE Visitors (% entering booking system)	n/a	26.7%			26.5%	-0.5%	▼	3
12	Quality of Non-SE Visitors (% entering booking system)	n/a	5.1%			6.8%	34.3%	▲	
13									
14									
15	Quality of PPC Visitors (% entering booking system)	n/a	34.8%			33.8%	-3.0%	▼	4
16	Quality of Organic Visitors (% entering booking system)	n/a	23.3%			27.2%	16.7%	▲	
17									
18									
19	% unable to book (non-IE browser)	n/a	5.0%			4.7%	-6.0%	▼▼	5
20	Money lost from non-IE visitors entering booking system	n/a	£23,078			£17,638	-23.6%	▼▼	
21									
22	**Key & Defintions:**							*time* →	
23	▲ = An increase								
24	▲▲ = An increase of 5%+								
25	▼ = A fall of 0% to 5%								
26	▼▼ = A fall of greater than 5%								

Figure 10.1 Example KPI report using Excel

The stakeholder (online marketer) who receives the KPI report shown in Figure 10.1 is clearly interested in the difference between visitors from search engines (SE) and non-search traffic and how likely they are to convert—in this case, to book a holiday.

Interpreting the KPI report from Figure 10.1:

1. Online revenue is down 18.7 percent for July compared with June.
2. Approximately 90 percent of all visitors who arrive at the website do so from search engines.
3. Visitors from search engines are almost five times more likely to enter the booking system than non-search-engine visitors.
4. Visitors from pay-per-click sources are 24–49 percent more likely to enter the booking system than organic search engine visitors.
5. Because the website booking engine does not work with non-Internet Explorer web browsers, the website is losing £17,000–23,000 per month.

Action points for stakeholders of this KPI report:

- Check whether the drop in online revenue is a seasonal fluctuation experienced across the whole business or unique to the online channel.
- Ninety percent of visitors arriving via a search engine appears at first glance to be high; share with the rest of the marketing department for discussion. Is this the result of a great search engine marketing strategy or are other channels not working very well?

- Increase the budget for pay-per-click campaigns—they work! However, pay-per-click may be working better here because of failings with organic search optimization, so this should be investigated further. Regardless, in the short term, raising the pay-per-click budget makes sense.
- Set up a meeting with the web development department to investigate an improved booking engine that will work for Firefox users.

As you can see, significant action points are required as a result of the KPI report presented. Without this data being shown in such a clear and concise way, discovering the action points from the wealth of Google Analytics reports available would be like finding a needle in a haystack and could even be missed.

Tip: Consider delivering your KPI reports at least on a monthly basis. If you are a transactional e-commerce site, certain stakeholders will want to receive reports weekly, even daily for very high volume websites. Consider which report frequency is realistic for you. If your organization cannot take action on a daily basis, particularly your web development and design team, then daily KPI reports do not make sense. Bear in mind the issues discussed under "Getting Comfortable with Your Data and Its Accuracy," in Chapter 2.

Presenting Hierarchical KPIs via Segmentation

There are hundreds of potential KPIs for your business. Which ones are relevant to your organization is an important discussion you will need to have with your company stakeholders. A key point stressed earlier is that you must deliver hierarchical KPI reports. That is, KPI reports for the chief marketing officer will differ from those for departmental managers, and those will differ from those for the account managers and strategists within each department.

For example, the CMO of a retail site would want to see the average conversion rate, average order value, and cost per acquisition. A marketing strategist would also like to see this same information segmented by referral medium type (paid search versus organic search versus e-mail marketing versus display banners, etc.). Without wishing to insult any chief marketing officer's intelligence, segmentation detail is generally too much information and is not required in order to give direction to the team. However, it is required for the strategists to be effective in their role. Detailed KPIs are usually obtained by segmentation.

A great deal of segmentation is available within the Google Analytics interface. As described in Chapter 4, rather than use a menu-style navigation system, Google Analytics encourages you to drill down through the data itself, automatically cross-segmenting by each click-through of the reports. In addition, where applicable, you will often see a drop-down menu for further analysis. For example, Figure 10.2 highlights the 24 ways to cross-segment visitors for medium = organic.

Segmenting on-the-fly with this drill-down method is a great tactic for quickly understanding the behavior of different visitor segments. Once you have identified the key segments that affect your website, you may wish to create specific profiles that report on only these. Having dedicated segmented reports enables you to investigate visitor behavior in greater detail, more efficiently, and more quickly.

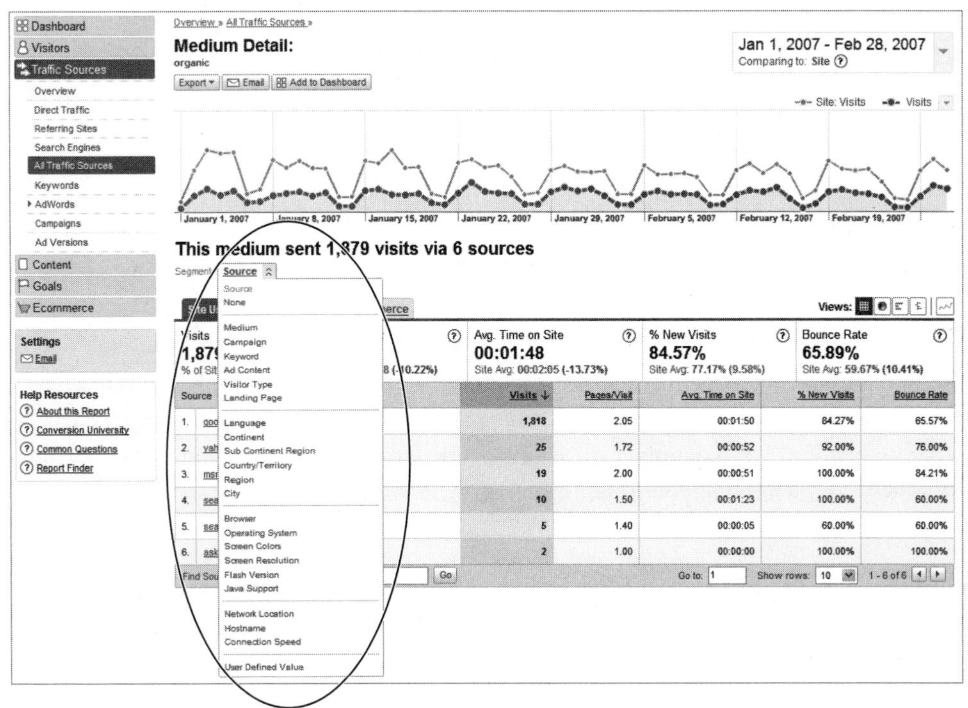

Figure 10.2 The 24 cross-segmentation options for organic search visitors

To create separate reports of your visitors, apply the segmenting filters as described in Chapter 8. Most segmentation involves the visitor type, referring source, or visitor geography.

Example visitor type segments:

- New visitors (or returning visitors)
- Customers (or non-customers)
- New visitors who are customers (or returning visitors who are customers)
- New visitors who are non-customers (or returning visitors who are non-customers)

Example visitor source segments:

- Search visitors (or non-search visitors)
- Affiliate visitors (or non-affiliate visitors)
- Paid search visitors only
- Organic search visitors only
- E-mail visitors only

Example visitor geographic segments:

- California visitors only, U.S.-only visitors, etc.
- Regional visitors only (Europe, North America, Asia, Latin America, Africa, Far East, Oceania, etc.)
- English language only (or rest of world language visitors)
- Language type XXX (or rest of world language visitors)

Performing segmentation is a fine balance between obtaining clarity about visitor behavior and obtaining information overload. Clearly, Google Analytics offers a great number of segmentation options. However, whenever you cross-segment data, you double the information reported. This is clearly contrary to the purpose of KPI reporting. Therefore, a good deal of thought and investigation should be applied prior to creating separate profiles. For example, how is this profile going to enhance your understanding of visitors and what will you do with such information?

Tip: When filtering visitors to create segmented profiles, always apply your filters to an additional profile. That is, keep a profile with no filters applied and work with a copy. This enables you to refer back to your default profile should there be any discrepancies or errors in your filtering.

Benchmark Considerations

Keep in mind that KPIs are important to drive improvement for your *own* website. Although it is obviously interesting and insightful to compare how your website is performing against your peers and competitors, in my opinion it is a mistake to place too much emphasis on external industry benchmarks. These can be misleading and often end up with you finding the benchmark that fits your story—giving a false impression of success.

KPIs vary greatly by business sector—for example, retail, travel, technology, B2B, finance, and so on. Even within subsectors there is wide variance: think flights versus holidays or food retail versus clothing retail. Even comparing against your competitors with *identically defined goals* is fraught with gross approximations. The exact path that visitors will take to complete a goal and the quality of their user experience along the way will vary for every website. Slight changes in these can have a major impact on conversion rates. I deliberately emphasize the phrase *identically defined goals* here, as definitions from different organizations can become blurred. For example, retail managers will often wish to differentiate existing customer visits from non-customer visits. Quoting a standard conversion rate across an industry can therefore be misleading.

Also, consider that e-commerce conversion rates can be measured in a variety of ways:

- The number of conversions / total number of visits to the website
- The number of conversions / total number of visitors to the website
- The number of conversions / total number of visits that add to cart
- The number of conversions / total number of visitors that add to cart

In the preceding list you can also substitute the word "transactions" for "conversions." That is, a visitor may complete a purchase and enjoy the experience so much that they return to make an additional purchase within the same visit session. Depending on the web analytics tool used and the preference of the organization, that can be defined as one conversion with two transactions, or two conversions with two transactions.

Note: For the preceding scenario, Google Analytics would show one conversion and two transactions, as the visitor has converted to a customer and this can happen only once during their session.

Other onsite factors that can greatly affect conversion rates, and therefore muddy the waters for benchmarking, include the following:

- Your website's search engine visibility (organic and paid search listings)
- You website's usability and accessibility (is your site easy to navigate?)
- Whether your website works in all major browsers
- Whether a purchase requires registration up front
- Your page response and download times
- Page content quality and imagery
- The use of trust factors such as safe shopping logos, a privacy policy, a warranty, use of encryption for payment pages, client testimonials, etc.
- The existence of broken links or broken images
- Quick and accurate onsite product searching

As you can see, comparing apples with apples is complicated. By all means benchmark yourself against your peers. It can be an interesting and energizing comparison. However, I emphasize the need for internal benchmarking as the main drivers for your website's success.

KPI Examples by Job Role

The following are KPIs I have employed when using Google Analytics. This is not intended to be an exhaustive list; rather, it is a sample to demonstrate how KPIs are defined and used. KPIs tell a story; therefore, to maintain continuity, I have chosen a single website to illustrate the value of the KPI metrics throughout. The website chosen for the examples is a partner of Google, based in the U.K. The business objectives of this partner are twofold: to sell software and to solicit an inquiry for professional services. For this, the website incorporates a number of key areas:

E-commerce section To sell product software, the value of which is relatively high compared to most e-commerce websites (from $695)

Lead generation section For visitors to inquire about professional services (training, implementation advice, strategic consultation). This is also of a high value.

Brand promotion area Writing of blog articles providing best-practice implementation advice

For job roles, I have grouped and differentiated the KPIs into four stakeholders: e-commerce manager, marketing manager, content creator, and webmaster. These should not be considered mutually exclusive, though. As discussed, the level of segmentation applied will determine the hierarchy.

Lastly, there is almost always more than one way to discover the KPI information within Google Analytics, and quite often the data points lie within several overlapping reports. I list the most obvious or most likely way to access the data.

Note: In Google Analytics, goal conversions and revenue (if you have monetized your goals) are reported separately from purchaser (e-commerce) conversions and revenue. Metrics that require the total revenue use the e-commerce plus goal revenue amounts.

E-Commerce Manager KPI Examples

An e-commerce site probably has the most potential KPIs to choose from, as the main goal (purchase) is relatively easy to measure and the site objective (driving visitors into the shopping cart system) is so clearly defined. Google Analytics has an entire section dedicated to the reporting of e-commerce activity. However, most of my KPIs come from other reporting areas.

Looking beyond visitor volume, some suggested KPIs for an e-commerce manager include the following:

- Average conversion rate
- Average order value
- Average per visit value
- Average ROI
- Percentage revenue from new visitors
- New customer on first visit index—new KPI defined

Average Conversion Rate

This is a high-level metric that every retailer watches with a keen eye in the offline world and is very easy to identify for online transactions. View the Ecommerce > Conversion Rate report or the Ecommerce > Overview report, as shown in Figure 10.3.

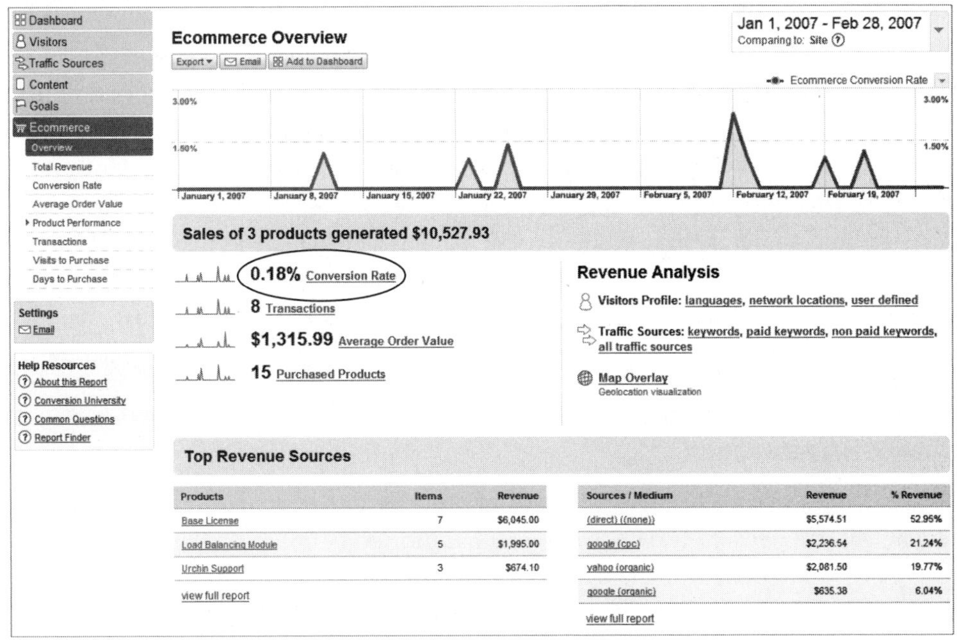

Figure 10.3 E-commerce Overview report graphing the conversion rate KPI over time

At first glance the average conversion rate is quite low at 0.18 percent. However, each purchase item has a high value, so this could be expected. In addition, without segmentation, this catch-all report includes blog visitors. Because of the nature of this website's blog (product advice), it is unlikely blog visitors would purchase. In fact, they are more likely to be existing customers. Therefore, before reading too much into this metric, blog visitors should be removed (filtered out) from this report. See Chapter 8 for more information about applying filters.

Average Order Value

Like the average conversion rate, the average order value is an important high-level KPI that retailers watch closely. The value (£1,315.99) is shown in Figure 10.3 and can be plotted against time by changing the chart options.

Average Per Visit Value

Some visitors become purchasers and some do not. What is the average value per visit on your website? By default, Google Analytics measures two types of per visit value: per visit goal value (based on the value of your goals) and per visit value (based on e-commerce transaction data). See Figure 10.4a and b. Add the two together for the overall average per visit value KPI.

Figure 10.4a and b shows the per visit goal value and per visit value, respectively. These are segmented by medium in the tables and compared with the overall visit volume over time in the graphs. From Figure 10.4a, it is Forum visitors who have the highest per visit goal value—that is, 10 times higher than the average for all media. However, comparing this with Figure 10.4b, we can see that Forum visitors do not purchase in this time frame—that is, the Forum per visit value is zero. The report shows that it is Google AdWords (medium = cpc), direct access (medium = none), and, to a lesser extent, organic visitors that are driving sales. Because Forum visitors are driving goal conversions but not transactions, it would be best to segment these visitors into a separate profile.

Average Return on Investment

Return on investment (ROI) is a KPI that all business managers understand. It tells you how much, as a percentage, you are getting back for every dollar you spend acquiring visitors. For clarity, the formula used for calculating return on investment in Google Analytics, expressed as a percentage, is as follows:

ROI = (Revenue – Cost)/Cost

For example, if for every $1 you spend on AdWords, you get $2 back in sales from your website, your ROI would be 100 percent. If you received $3 back for the same outlay, your ROI = 200 percent, and so forth. Obviously, you want to maximize your ROI—the greater this number, the better.

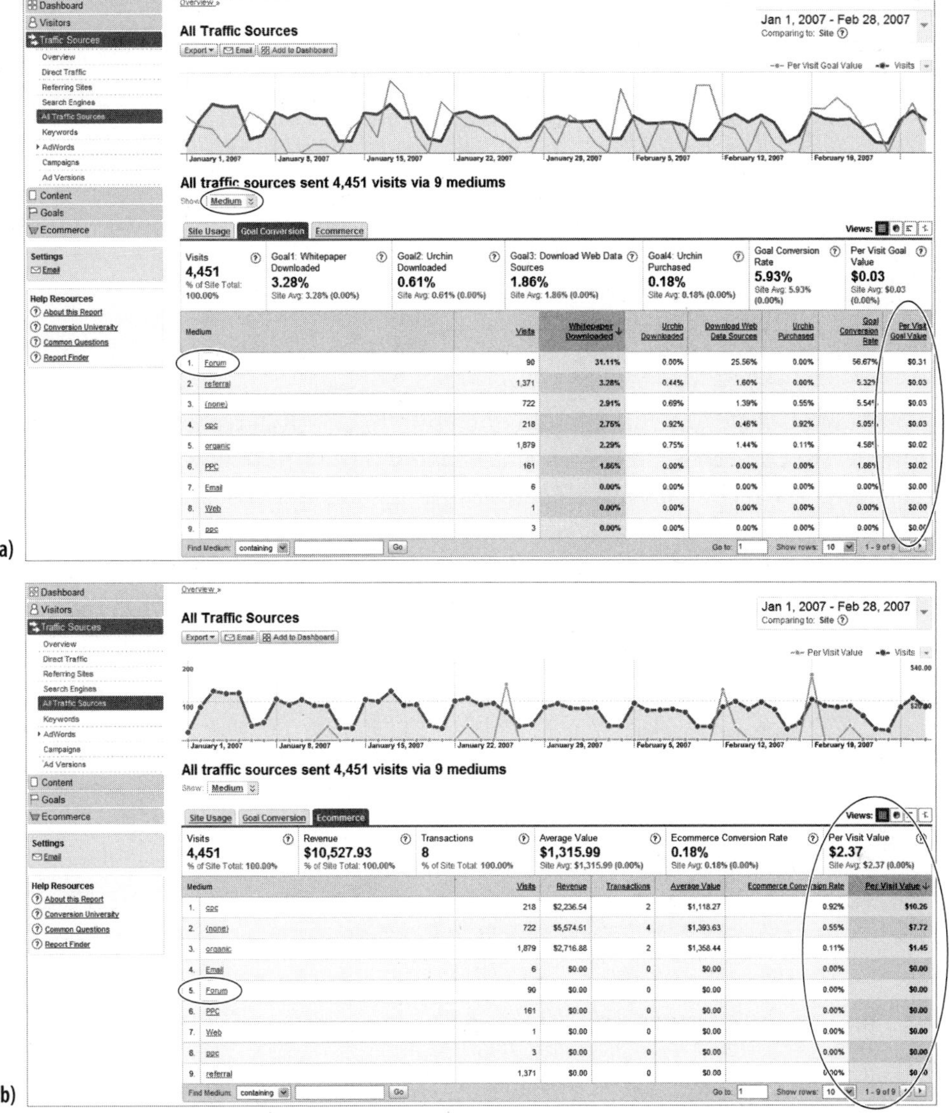

Figure 10.4 Obtaining (a) the per-visit goal value; (b) the per-visit value

A negative ROI means you are losing money: Your costs of acquisition are greater than your returns. However, bear in mind that when launching a new campaign, ROI is likely to be negative until repeat visitors or brand awareness starts to grow and lead to more conversions (see Figure 10.5). The breakeven point (zero percent ROI) could be hours, days, weeks, or even months, depending on many (visitor-centric, online, off-line) factors. For mature campaigns, keep your ROI above zero percent unless there is a clear reason not to do so. For example, you may be a new entry in the market and want to buy market share to gain customers at a later date.

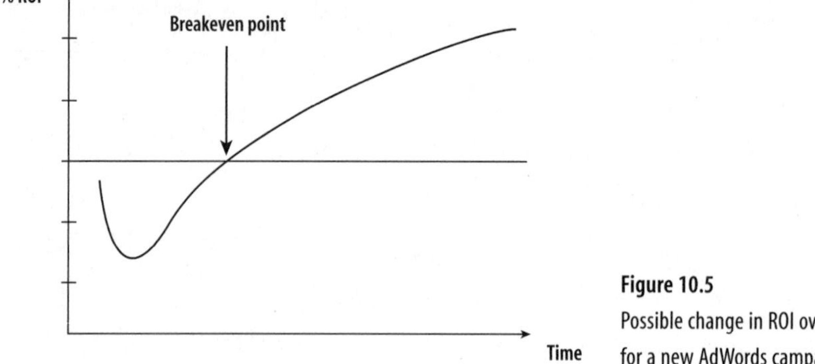

Figure 10.5 Possible change in ROI over time for a new AdWords campaign

Within Google Analytics you can drill down to view ROI reports for AdWords at three levels: Campaign, Ad Group, and Keyword. Figure 10.6 shows data at the Ad Group level. The report table clearly shows a large amount of variance reported for ROI. This is due to the use of generic keywords in these ad groups. For example, there will be many people looking for all sorts of information around the keyword "Urchin software," some of whom have no intention of purchasing the product when they see that its sibling product (Google Analytics) is free.

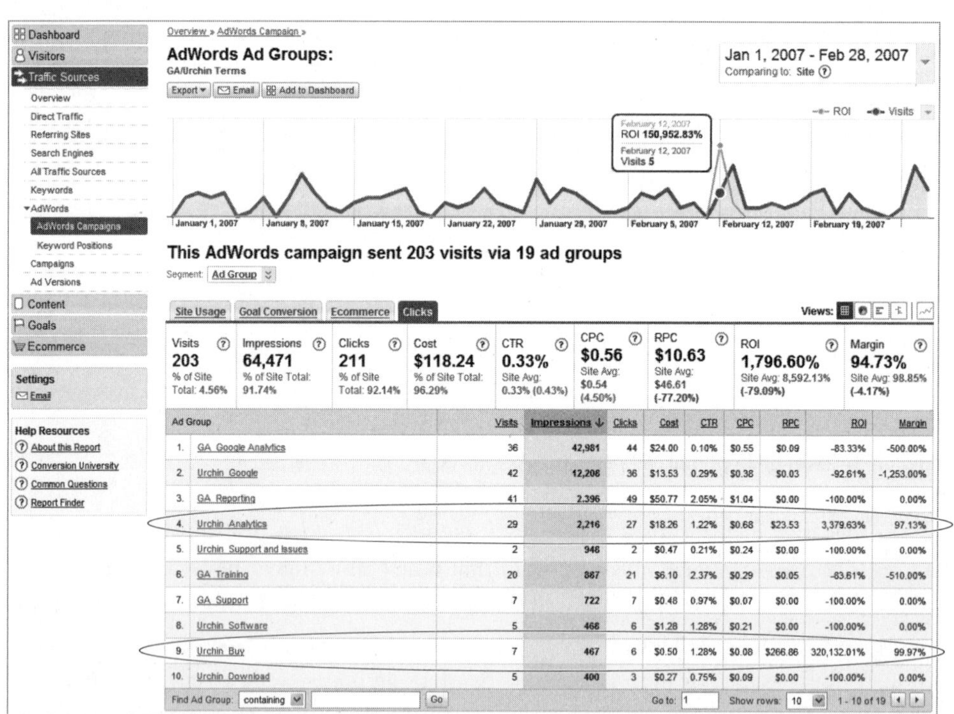

	Ad Group	Visits	Impressions ↓	Clicks	Cost	CTR	CPC	RPC	ROI	Margin
1.	GA Google Analytics	36	42,981	44	$24.00	0.10%	$0.55	$0.09	-83.33%	-500.00%
2.	Urchin Google	42	12,206	36	$13.53	0.29%	$0.38	$0.03	-92.61%	-1,253.00%
3.	GA Reporting	41	2,396	49	$50.77	2.05%	$1.04	$0.00	-100.00%	0.00%
4.	Urchin Analytics	29	2,216	27	$18.26	1.22%	$0.68	$23.53	3,379.63%	97.13%
5.	Urchin Support and Issues	2	948	2	$0.47	0.21%	$0.24	$0.00	-100.00%	0.00%
6.	GA Training	20	887	21	$6.10	2.37%	$0.29	$0.05	-83.61%	-510.00%
7.	GA Support	7	722	7	$0.48	0.97%	$0.07	$0.00	-100.00%	0.00%
8.	Urchin Software	5	468	6	$1.28	1.28%	$0.21	$0.00	-100.00%	0.00%
9.	Urchin Buy	7	467	6	$0.50	1.28%	$0.08	$266.86	320,132.01%	99.97%
10.	Urchin Download	5	400	3	$0.27	0.75%	$0.09	$0.00	-100.00%	0.00%

Figure 10.6 AdWords ROI report

As shown in Figure 10.6, only two ad groups are driving a positive ROI, but the ROI for these is so high that it drives up the average AdWords ROI for the site to a massive 1,796.60 percent. That is to say, for every $1 invested in AdWords, an average of nearly $18 is returned—a pretty good investment! By graphing two metrics (AdWords visitors and ROI), Figure 10.6 also shows when the vast majority of ROI was earned—February 12th and 13th.

Of course, ROI is a top-level indication of performance from your gross revenue. It does not take into account what profit you make on your sales. Nor does it take into account the volume of transactions or visitors received. For example, a high ROI campaign may be so specific that it generates only a small revenue. A lower ROI (less specific) campaign may in fact produce greater revenue due to the higher visitor volume it generates.

Average Margin

The last column in Figure 10.6 shows the margin KPI. Expressed as a percentage,

Margin = (Revenue – Cost) / Revenue

This is similar to the ROI calculation described previously, and the two metrics are closely related. For the margin, your aim is to keep the costs of acquiring new visitors to a minimum. Hence, trying to keep the margin as close to 100 percent as possible is your objective. Unlike ROI, margin can never be greater than 100 percent.

The smaller your margin percentage, the less profit you make. At zero percent, you break even. At a negative percentage you are losing money: Your costs of acquisition are higher than your returns. As with ROI, bear in mind when launching a new campaign that your margin may well be negative until repeat visitors or brand awareness starts to grow and leads to more conversions (refer to Figure 10.5).

Note: Margin and ROI reports are currently available for AdWords campaigns only.

Percentage Revenue from New Visitors

Most e-commerce managers would like their visitors to become purchasers as soon as possible. That is, when new visitors arrive, they are so convinced by the value proposition they make their purchase on that first visit. Whether visitors convert on their first visit or not depends on many factors, as discussed earlier in the section "Benchmark Considerations." One major influence is pricing. High-value products usually require a longer consideration period, which often equates to more visits in order to convince the visitor to buy than a lower value item would require. The percentage revenue from new visitors KPI enables you to ascertain whether this is the case for your website.

Consider Figure 10.7. Even though the average order value is high, at $1,315.99, most of the revenue generated in this period is from first-time visitors (68.16 percent). This indicates that the value proposition and other factors such as trust, page quality, and so on are high for this website.

Note: A caveat here when interpreting these particular example metrics: the number of transactions is low (8). As the average order value KPI is high, it is entirely possible that one single transaction from one new visitor could be skewing the data. Before taking action on this KPI, collecting a larger sample of data is recommended—at least hundreds of transactions.

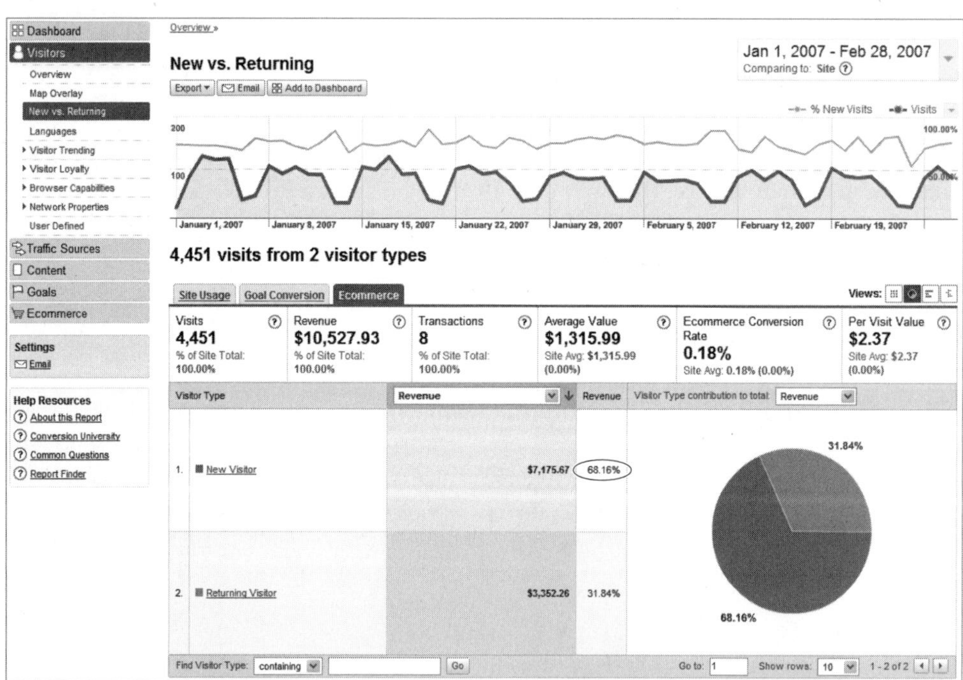

Figure 10.7 Percentage revenue from new visitors

New Customer on First Visit Index—A New KPI Defined

What is the likelihood of new visitors becoming new customers on their first visit? You saw in Figure 10.7 that a high proportion of revenue is generated by first-time visitors, but how does that relate to the number of first-time visitors to the website?

The new customer on first visit index KPI can tell you that. It is defined as follows:

$$\text{New Customer on First Visit Index} = \frac{\text{percent transactions from new visitors}}{\text{site percentage of new visitors}}$$

From the data in Figure 10.8 and Figure 10.13 (marketing KPIs), the value is calculated as follows:

New Customer Index = 62.50 / 77.20
New Customer Index = 0.81

Interpretation:

A value of 1.0 indicates that a new visitor is equally likely to become a customer as a returning visitor. A value less than 1.0 indicates that a new visitor is less likely to become a customer than a returning visitor, and a value greater than 1.0 indicates that a new visitor is more likely to become a customer than a returning visitor.

Figure 10.8 Percent transactions from new visitors

For the example website, this KPI shows that a new visitor is less likely to purchase than a returning visitor. This is not surprising, as the average order value KPI is high. However, considering this, what is surprising is that the new customer on first visit index is as high as 0.81. As for the percentage revenue from new visitors KPI, this indicates that the value proposition and other onsite factors such as trust and page content quality are very high for this website.

Marketer KPI Examples

Bringing good quality visitors—that is, qualified leads—to your website is the bread and butter of your marketing department. Putting offline marketing to one side, the "bringing" part is achieved with online marketing and may include any or all of the following sources: search engine optimization (free search rankings), pay-per-click advertising (paid search), banner advertising, affiliate networks, blog marketing, links from site referrals, and e-mail marketing.

Determining which traffic is qualified means looking at the conversion rates, campaign costs, revenue generated, and ROI. KPIs for the marketer therefore overlap strongly with KPIs for the e-commerce manager. One important difference is that marketers not only look for purchaser conversion rates, but also goal conversions, as these build visitor relationships that, it is hoped, will later lead to purchases. As e-commerce conversions have been discussed in the previous section, only KPIs related to goal conversions are considered here.

In most cases, online marketing is grouped under the general marketing department. It is therefore critical here to use hierarchical KPIs to differentiate those members of your audience familiar with the online channel from those who are not.

Looking beyond the overall visitor volume to a site, some suggested KPIs for e-commerce managers include the following:

- Percentage visits by medium type
- Percentage goal conversion rate by medium type
- Percentage visits by campaign type
- Percentage goal conversion by campaign type
- Goal conversion index by campaign
- Average ROI by campaign type
- Percentage of new versus returning visitors
- Percentage of new versus returning customers
- Percentage brand engagement

Percentage Visits by Medium Type

Viewing a breakdown of visitor source by medium is an extremely effective KPI for the marketer. It is shown in Figure 10.9. For example, what's driving your visitor acquisition—e-mail marketing, organic search, paid advertising, affiliates, or your offline marketing?

In the example of Figure 10.9, organic search engine traffic is clearly driving the majority of visitors to the website in question (42 percent), though referral links are a close second. Given this at-a-glance KPI report, managers can immediately start to ask themselves, "Does the distribution of our marketing budget match the received visitors?"

If, for example, little of your budget is being spent on acquiring organic search visitors, then you know from Figure 10.9 that this source provides a great ROI for you and perhaps should be exploited further.

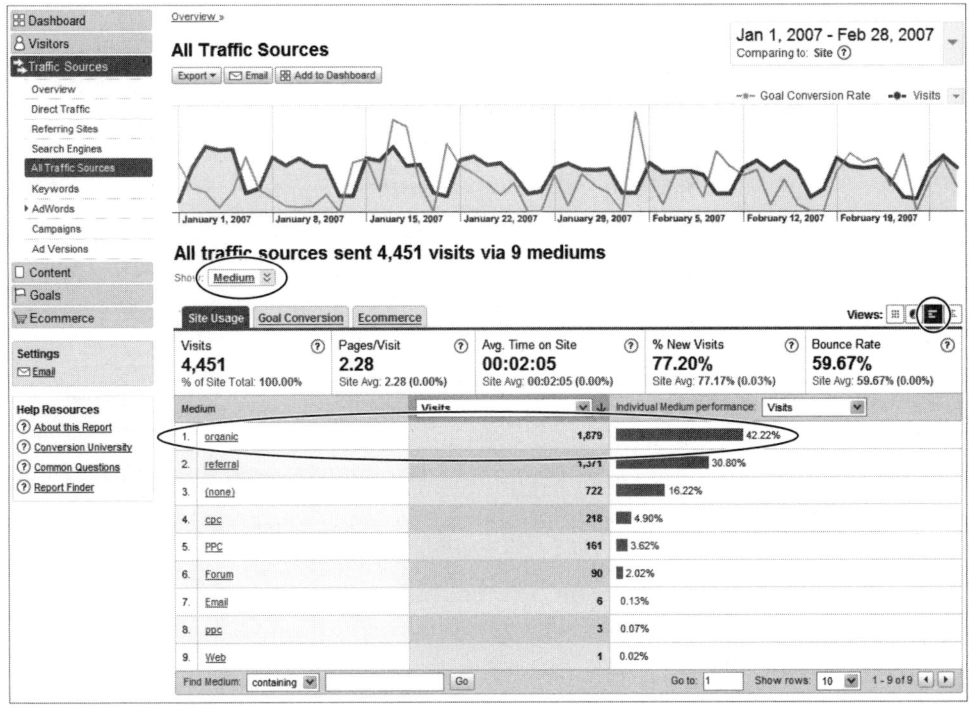

Figure 10.9 Visitors by referral medium

Percentage Goal Conversion Rate by Medium Type

Once you understand which media and which campaigns are driving traffic to your website, the next logical question is to consider how well such visitors convert; from Figure 10.9, click the Goal Conversion tab to get to Figure 10.10.

As shown in Figure 10.9, organic sources provide the highest volume of visitors. However, Figure 10.10 indicates that goal conversions from organic sources are comparatively low (4.6 percent of all organic visits convert, equating to 86 conversions). The most qualified traffic—that is, those who are most likely to convert—come from "Forum" (57% of all forum visits convert, equating to 51 conversions).

"Forum" reflects employees of the example website who participate in online forums and blogs, leaving backlinks to their company website where appropriate. It makes sense that people following such links would be highly qualified and more likely to convert. However, visitor volume from this medium is low compared to others.

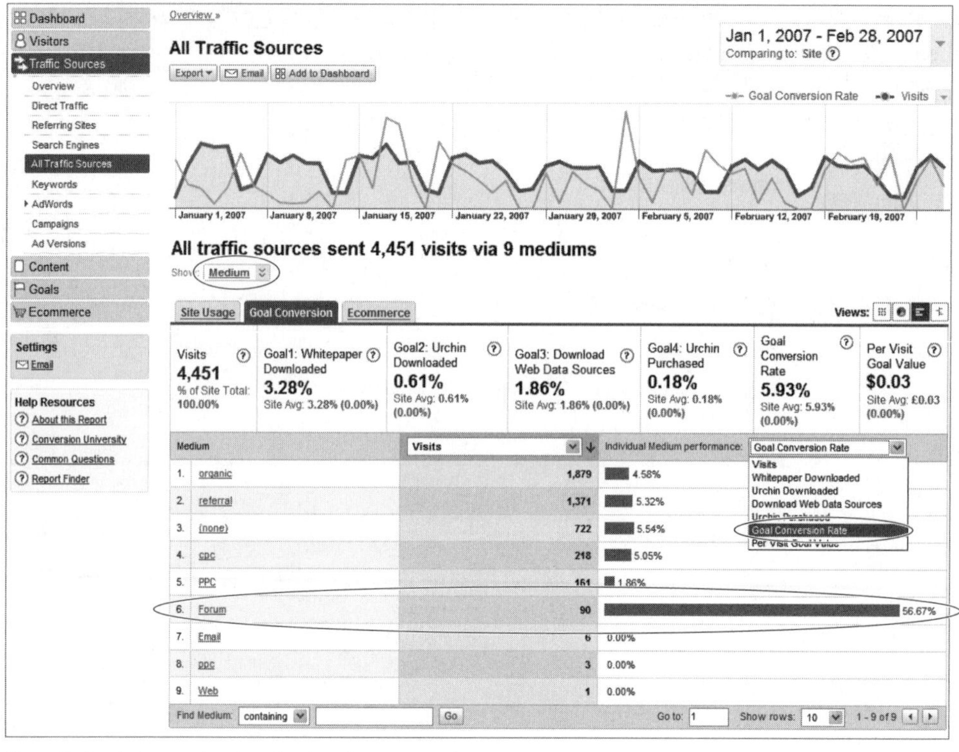

Figure 10.10 Goal conversions by referral medium

With such information, the marketing manager and strategist can then determine (or even better, test) whether it is worth increasing their budget to try to acquire more forum visitors.

Tip: To determine which individual goal is providing the high conversions, select the relevant Individual Medium performance report from the drop-down menu, as highlighted in Figure 10.10. You can also plot individual goal conversions over time by selecting the appropriate goal in the graph mode (the drop down option at the top right of the chart).

Percentage Visits by Campaign Type

Once you know which media are driving traffic and goal conversions to your website, your strategist will also want to drill down into specific campaigns. I use the term *campaign* here loosely to mean any online channel from which visitors arrive at your website, rather than specific campaign-labeled visitors (see corresponding note). For example, which particular e-mail marketing campaign is driving visitors to your website;

which search engine; and which pay-per-click sources? These can be quickly obtained from the Traffic Sources report section.

As shown in Figure 10.11, from a search engine perspective, Google is driving the most traffic to this website. You can further segment this by clicking the non-paid or paid text menu items. Similar reports are viewed by selecting Traffic Sources > Direct Traffic or Traffic Sources > Campaigns.

Figure 10.11 Percentage visitors by campaign type

Note: Campaign-labeled visitors are those who have arrived either from AdWords or via a landing page URL tagged with the utm_campaign tracking variable, as described in "Online Campaign Tracking," in Chapter 7. Landing page URLs are used in pay-per-click marketing, affiliate, e-mail, digital collateral and banner advertisements.

Percentage Goal Conversion by Campaign Type

As with percentage visitors by medium, once you drill down into percentage visitors by campaign, you will also wish to see this KPI by goal conversions. For each campaign type within the Traffic Sources section, you can view the corresponding conversions by clicking the Goal Conversion tab, as shown in Figure 10.12. In this case, for goal conversions, Yahoo! visitors are more qualified than Google visitors.

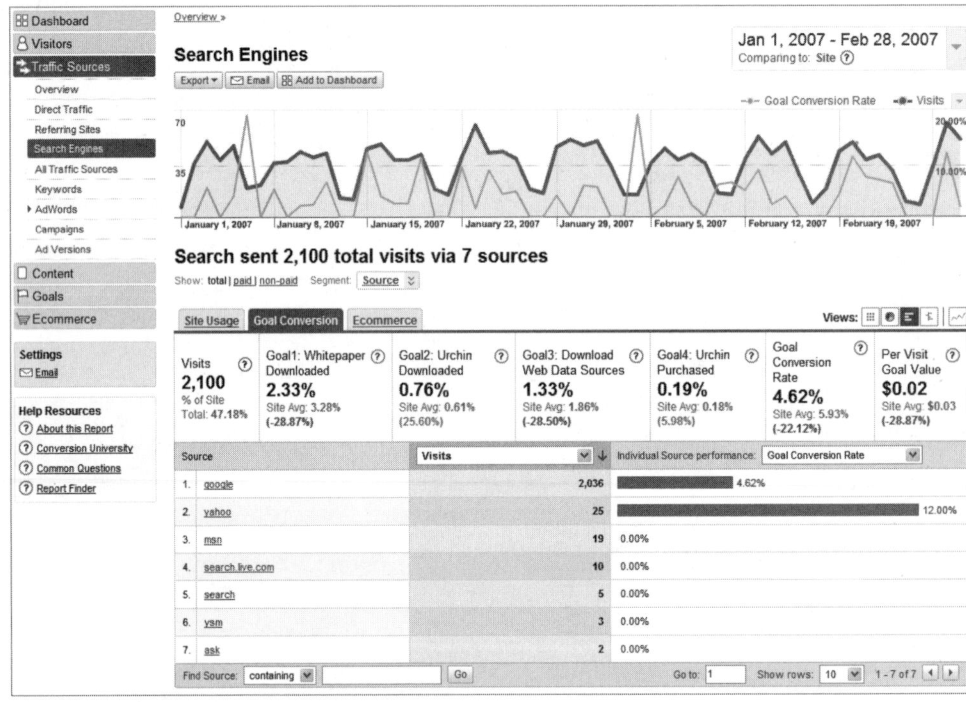

Figure 10.12 Percentage goal conversion by campaign type

Campaign Quality Index—A New KPI Defined

The campaign quality index is all about measuring how well targeted your campaigns are at driving qualified leads to your website. For example, suppose 50 percent of your visitors are from AdWords (labeled in your reports as "google / cpc") but only 20 percent of conversions are from this campaign source. That's an underperforming campaign because given two equally targeted campaigns, each producing 50 percent of your visitor traffic, both should produce 50 percent of your conversions. If one outperforms the other by generating more than its share of conversions, then by definition that campaign must be better targeted.

The campaign quality index, shown here, enables you to view these differences so you can better understand the effectiveness of your visitor acquisition targeting.

$$\text{Campaign Quality Index (for campaign } X) = \frac{\text{percent goal conversions from campaign } X \text{ visits}}{\text{percentage visits from campaign } X}$$

This report does not yet exist in Google Analytics. However, it is easy to calculate from the available reports using the data in Figure 10.13a and b, plus knowing the total number of conversions from the Goals > Overview report (264—not shown here). The only difference between Figure 10.13a and b is the data displayed in the bar chart, as determined in the drop-down menu (highlighted).

The values from these reports can then be used to populate the rows of Table 10.2. If individual referral site detail is unimportant to you, obtain the total for these by changing the Show value (cross-segmentation drop-down menu under each chart) from Source Medium to Medium in each of the reports. This will quickly provide the total number of referral site visits.

Interpretation for the Campaign Quality Index KPI:

A value of 1.0 tells us that a visitor from said campaign is as likely to convert as a visitor from any other campaign. A value less than 1.0 indicates that a visitor is less likely to convert than a visitor from any other campaign, and a value greater than 1.0 indicates that a visitor is more likely to convert than a visitor from any other campaign. As a marketer, you should be aiming for a value of 1.0 for each campaign you set up, with a margin of ±0.1.

Table 10.2 indicates three distinct categories of visit sources for the campaign quality index:

- **High:** Forum, Yahoo! organic
- **Expected:** Referrals, Direct, Google cpc, Google organic
- **Low:** YSM ppc, other

▸ **Table 10.2** Campaign Quality Index (CQI)

	A	B	C	D
Campaign	% Visits (Figure 10.13a)	# Conversions (Figure 10.13b)	% of Total Conversions (B/264)	Campaign Quality Index (C/B)
Forum	2.02	51	19.32	9.56
Google cpc	4.90	11	4.17	0.85
Google organic	40.84	83	31.44	0.77
YSM ppc	3.62	3	1.14	0.31
Yahoo! organic	0.56	3	1.14	2.04
Referral	29.59	73	27.65	0.93
Direct	16.22	39	14.77	0.91
Other	2.25	1	0.38	0.17
TOTAL	100.00	264.00	100.01	

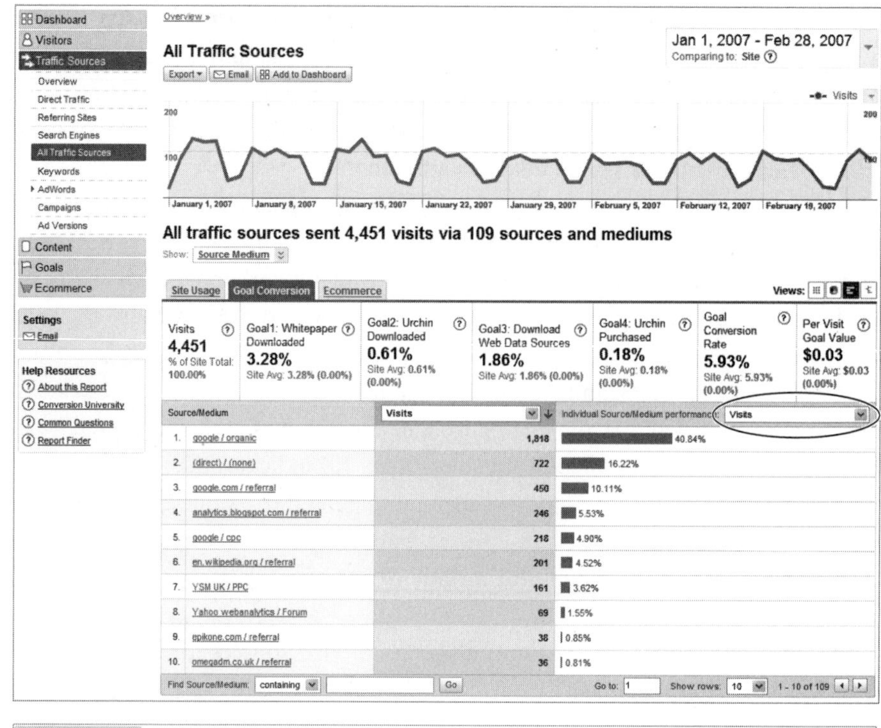

a)

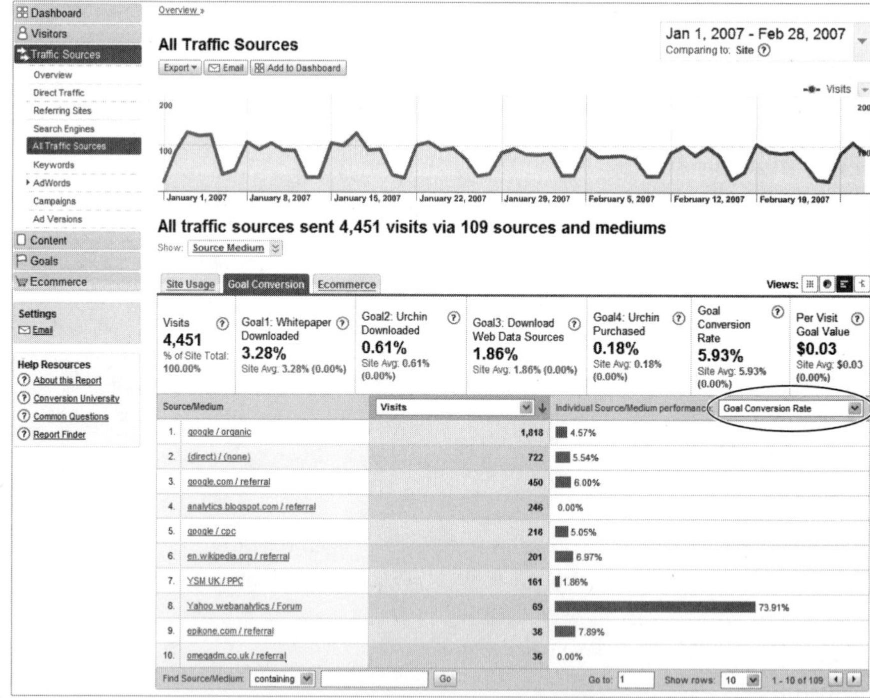

b)

Figure 10.13 (a) Number of visits and percentage by referral source; (b) Conversion rates as a percentage. Multiply each conversion rate (as a decimal) by the number of visits to obtain the number of conversions.

I single out forum visits because these are nearly 10 times more likely to convert than a visit from any other campaign or source (except that a forum visitor is nearly five times more likely to convert than a Yahoo! organic visitor). This indicates that the forum campaign is an extremely well targeted campaign. As shown in the previous section, forum visitors are visitors who have followed links left by employees of this example website, so it makes sense that such visitors are much more qualified.

Yahoo! organic visitors also appear highly targeted (CQI = 2.04), but the number of conversions is very low, just three, so this should be disregarded until more data is collected. In fact, this highlights the need to keep raw numbers close at hand when calculating KPIs that are averages, ratios, or percentages. Taking the CQI KPI at face value, a great deal of time and effort could be wasted on investigating why a Yahoo! organic visit is nearly three times more qualified than a Google organic visit, when in fact such a small sample size is statistically meaningless.

Tip: For further reading on taking care with averages, see "Why Segmentation Is Important," in Chapter 8.

Why is Google organic showing as a relatively low CQI? This may be because of the ubiquitous nature of search and the current popularity of Google as a search engine. For example, people will arrive at your website through a Google search for all sorts of reasons that may not be relevant to your business, including job search, competitive research, clients searching for your contact details, spammers, misspellings, and mis-associations (Omega watches versus Omega Couriers, for example). This can lead to high volumes of organic referral traffic from Google that is not qualified. If possible, filter out such nonqualified visitors based on the Google search terms used.

Average ROI by Campaign Type

This KPI is the same as the one discussed for e-commerce managers and shown in Figure 10.6. I list it here for completeness.

Percentage of New Versus Returning Visitors

Knowing whether it is new or returning visitors who are driving your website metrics is an important top-level guide to the success of your online marketing strategy (see Figure 10.14). If your marketing focus is on acquiring new visitors, then you would expect a greater proportion of these. If you focus on visitor retention, then you would expect returning visitors to be higher.

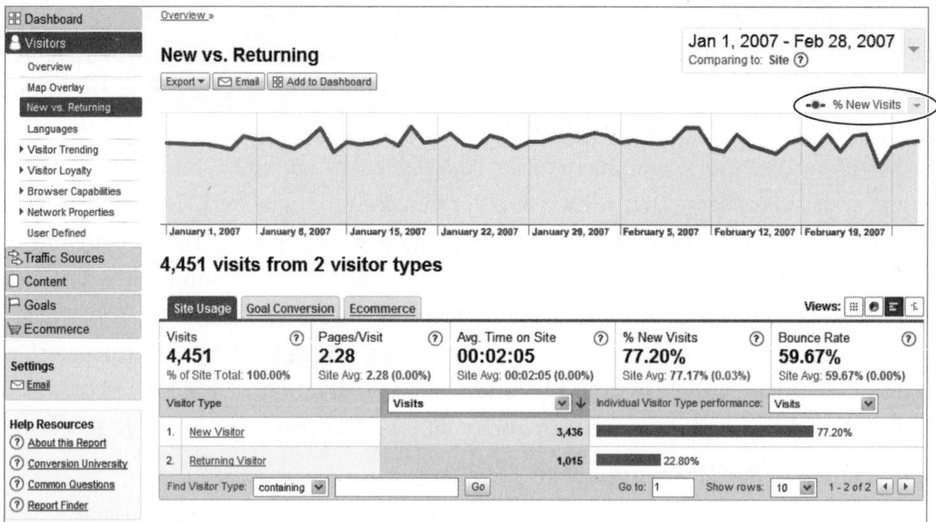

Figure 10.14 Understanding new versus returning visitors

Unless you are embarking on a new online marketing initiative, these metrics should remain fairly stable. Generally speaking, the more proactive your organization is at search engine marketing, the higher the percentage of new visitors—typically, 70 percent plus. Exceptions to this are customer support websites and content publishing websites that have a more even mix of new versus returning visitors.

Be careful when interpreting changes in percentage of visitor types. For example, a decrease in percentage of new visitors could in fact be due to an increase in percentage of returning visitors, rather than any change in your new visitor acquisition strategy. To check, compare different date ranges and examine the raw numbers (see Figure 10.15).

By viewing the raw visit numbers of Figure 10.15, you can see that both new and returning visitor numbers have increased. Therefore, the slight increase in the percentage new visitors (0.83 percent) does not appear to be due to any lack of activity from returning visitors.

Percentage of New Versus Returning Customers

Providing you are using the Google Analytics function _setVar() to label visitors who purchase as customers (as described in Chapter 9), you can view the ratio of new versus returning customers by viewing the Visitors > User Defined report. Click the Customers table entry, and then cross-segment by Visitor Type, as shown in Figure 10.16.

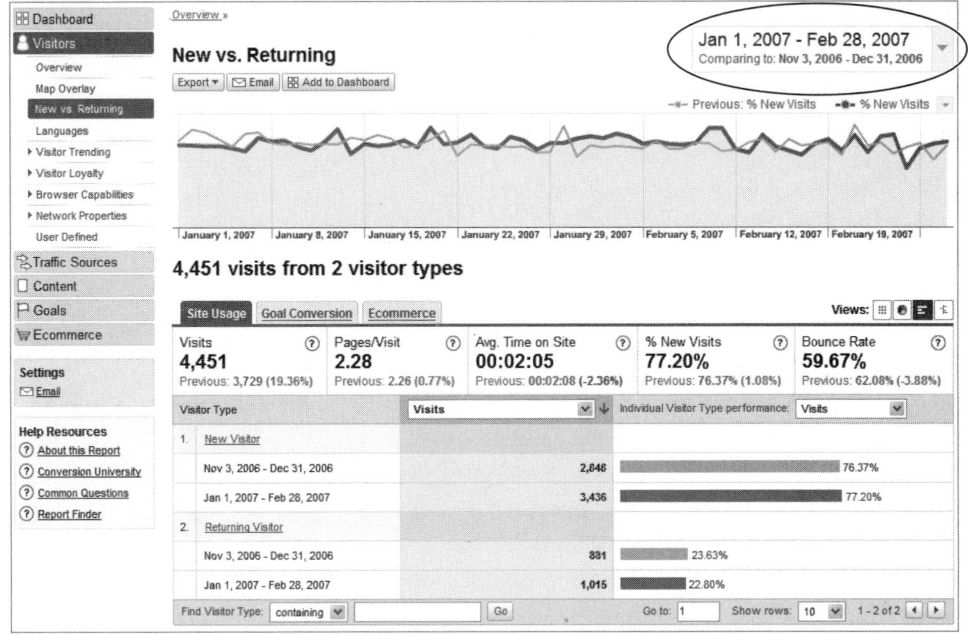

Figure 10.15 Comparing new versus returning visitors over time

Figure 10.16 Visitor types for the custom label segment customer

Percentage Brand Engagement

In his blog www.webanalyticsdemystified.com, Eric T. Peterson describes brand engagement as the brand index KPI. Visitors who know your brand and have arrived at your site because of it have, by definition, engaged with you. This KPI is defined as follows:

$$\text{Percentage brand engagement} = \frac{\text{number of search terms containing your brand names + number of direct access visits}}{\text{total number of search terms + total number of direct access visits}}$$

Note that when referring to search terms here, I am referring to search engine referral keywords. Direct access visits are also included because these are people who know your website address and therefore your brand—assuming you have excluded the access of your own company employees from your reports (see "Filtering: Segmenting Visitors Using Filters," in Chapter 8).

A percentage brand engagement report is not yet directly available within Google Analytics, but it is straightforward to calculate from two other reports. First, from the Traffic Sources > Keywords report, use the inline filter to enter your regular expression of brand keywords (see Figure 10.17a). The number of direct visits is taken from the Traffic Sources > Direct Traffic report (see Figure 10.17b).

Constructing regular expressions

Because a maximum of 256 characters is allowed within the in-line filter box, construct your regular expression with some thought. For example, in Figure 10.17a, the brand term I am actually looking for is "GA Experts" or "GA-Experts"—the brand name of the website. I only require the term "experts" in this case because this will pick up both terms (and other brand terms) and is unlikely to match non-brand terms.

The same technique is used for finding the brand terms "Google Analytics" and "Urchin Software." The terms "urchin" and "google" are all that is required to match the full terms.

Using the data from Figures 10.17a and b:

Percentage brand index = (1097 + 722) / (1511 + 722)
Percentage brand index = 81.46%

This illustrates how important branding is for the site in question.

By selecting the Goal Conversion tab within the reports of Figures 10.17a and b, the analyst or strategist can also quickly calculate the brand index KPI on a per-goal basis.

a)

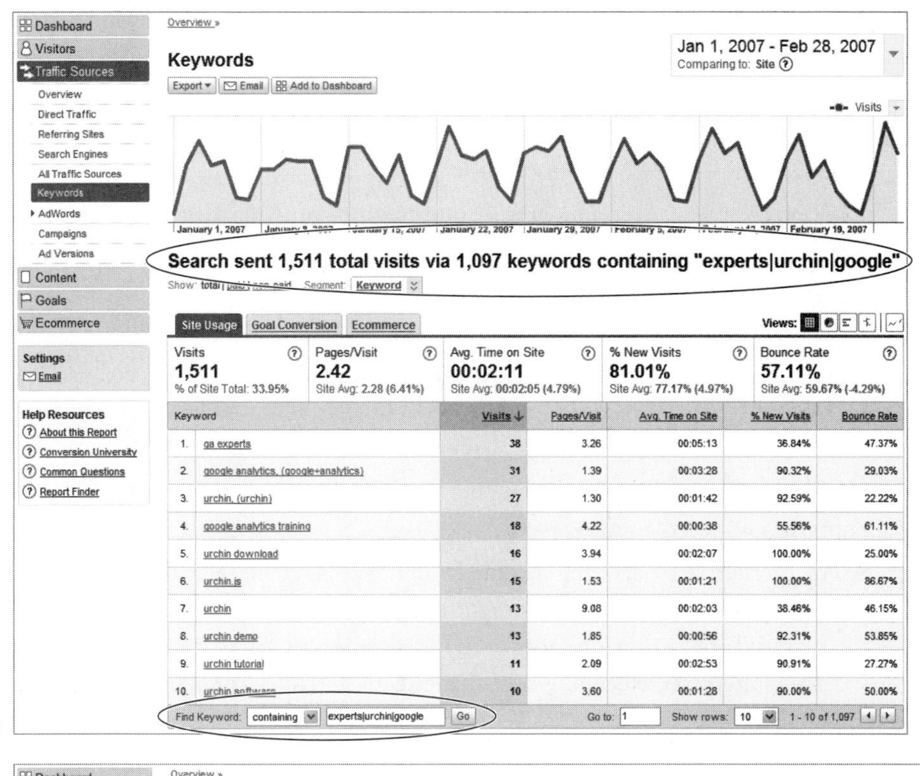

b)

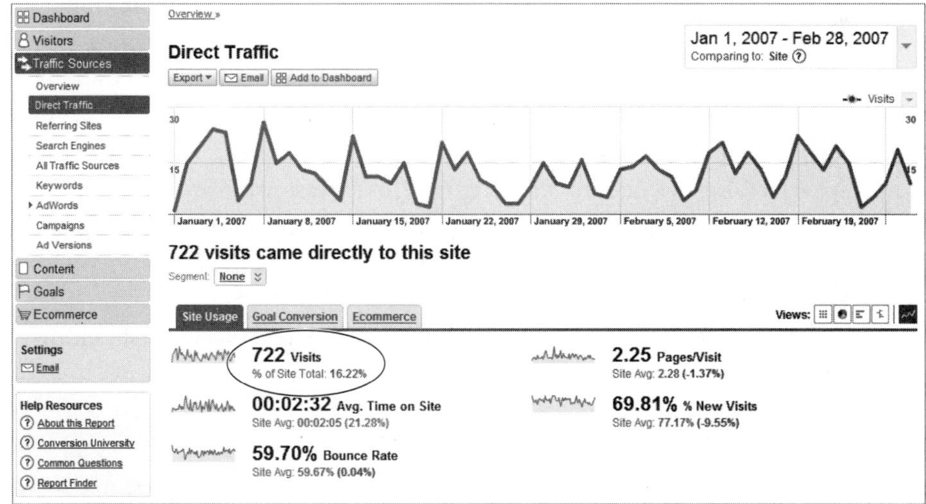

Figure 10.17 (a) Search keywords used by visitors; (b) Direct traffic metrics

Content Creator KPI Examples

If you create content—that is, you are an author, journalist, or copywriter for a content-driven website—then audience engagement is your goal. How long people spend reading your content and how much of it they consume are key indicators for measuring engagement.

Essentially, there are three categories of content-driven websites:

1. Product and organization information

   Examples include corporate website information, product review sites, blogs, help-desk support, online training sites, and so on.

2. Advertising-based content

   Free-to-read content websites that derive revenue from selling advertisements (banner or text ads) alongside content. Examples include cnet.com, myspace.com, and most TV, newspaper, and magazine websites such as nytimes.com, ft.com, cnn.com, and so on. Some blogs also embed contextual advertising within their articles—for example, using AdSense.

3. Subscription-based content

   As an alternative to deriving income from advertising, content-driven websites can offer subscription-based content; that is, you pay as a subscriber to access the material (or perhaps a more complete version of an article). Examples include jupiterresearch.com, e-consultancy.com, forrester.com, and many daily newspaper sites.

The latter two categories I classify as "publishers," and they usually employ both methods of generating income. As a publisher, if you provide advertising-based content, then you have a dilemma: If you write the perfect article to fit on one page, visitors will read that single page, be satisfied, and move on to another site or activity. They will be single-page visitors. However, single-page visits are not good for business when you derive your revenue from advertising. To increase your revenue, you want visitors to read more pages so that they are exposed to more advertisements (greater inventory), increasing the likelihood that they will click on one. That makes your website more attractive to advertisers.

Regardless of your content site's business model, greater engagement with your visitors is the key. Consequently, content managers are always looking at ways to include complementary subject matter with each article or page to encourage this. Clearly for content sites, visit volume—the number of visits per day, week, or month—is an important KPI, along with how this varies over time. However, the following sample KPIs focus on helping you measure engagement:

- Average time on site
- Average pageviews per visit

- Percentage bounce rate
- Average number of advertisements clicked
- Percentage engagement
- Percentage new versus returning visitors
- Percentage brand engagement

Average Time on Site

The average time on site is the length of time visitors spend interacting with your website, and it is a great base metric to help you understand whether your visitors are engaging with your site. The example shown in Figure 10.18 indicates that the average time on site taken from all visitors is 2 minutes and 5 seconds.

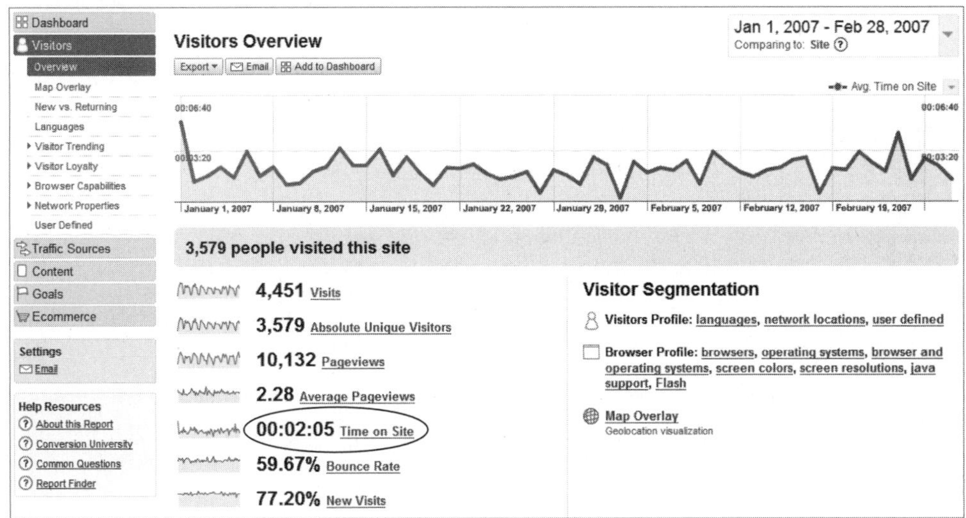

Figure 10.18 Visitor Overview is a key report for KPI metrics.

Although it's a great base metric, the overall average time on site is a blunt metric—it's an average of all visitor types. A more informative view is to compare how this varies by visitor segment. For example, compare average time on site for new versus returning visitors, or by referring traffic sources. To illustrate this, Figure 10.19 shows how the average time on site varies by referring source medium. An interesting observation is that visitors from e-mail links spend three times as long on the site as the site average. In addition, paid visitors from networks other than Google AdWords (labeled "ppc") spend 20 percent less time on the site than Google AdWords visitors (labeled "cpc").

Note: Averages and their limitations are discussed in "Why Segmentation Is Important," in Chapter 8.

Regardless of visitor segment, all content creators want to increase the average time on site KPI. By comparing segments, you can better tailor your website content, advertising, and overall usability for each visitor type. If you believe your content is already well structured in this way, yet the average time on site is relatively low, then consider how you are acquiring your visitors. Examine whether they are qualified visitors and whether the landing page they first arrive at is suitable for them.

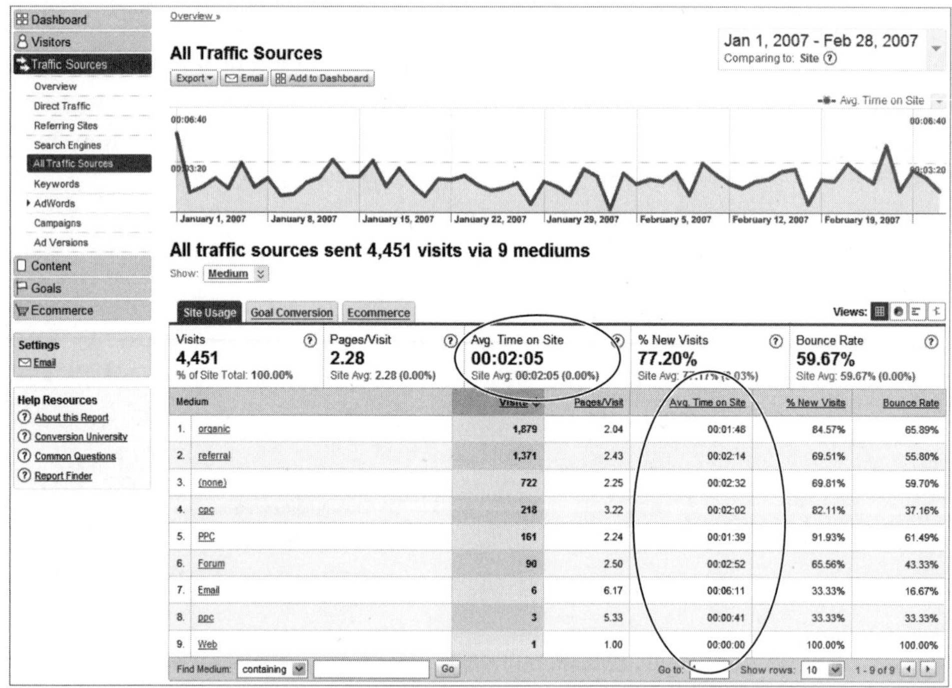

Figure 10.19 Average time on site by referring source medium

Average Pages Per Visit (Depth of Visit)

As with the time spent on site KPI, knowing the depth of visit—that is, the average pages per visit—is another excellent way to gauge how good your content is at engaging visitors. These two KPIs are closely related and are displayed together when you are viewing Google Analytics reports (refer to Figures 10.18 and 10.19). For example, if your depth of visit KPI causes you to ask further questions or instigate action, then you should also refer to the time on site. It could be that a low average pages per visit KPI is a bad thing. However, if these visitors also display a high time on site or trigger other on-page events such as watching a Flash movie clip, then it could be good thing.

As with the average time on site, the average pages per visit KPI is much more informative when you consider how this metric varies for different visitor segments (refer to Figure 10.19).

Percentage Bounce Rate

A *bounce* in Google Analytics terminology is a one-page visit—that is, a visitor arrives on your website, views one page, and then bounces off to another site or closes the browser. This calculation can vary for different web analytics vendors, so I clarify the formula here for Google Analytics:

$$\text{Percentage bounce rate for a page} = \frac{\text{number of single page visits to that page}}{\text{number of times that page was an entry page}} \times 100$$

The average website bounce rate (an average of all your page bounce rates) is quoted in numerous places throughout Google Analytics reports (e.g., Figure 10.18). To view the bounce rate for a particular page, select the page in question from the Content > Top Content report, as shown in Figure 10.20.

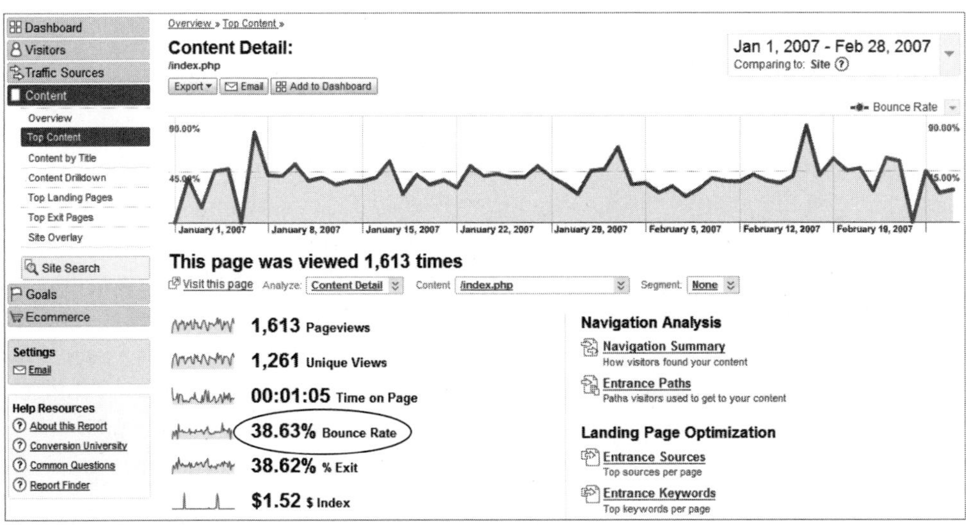

Figure 10.20 Top Content report with drill-down information on the file `index.php`

From a content creator's point of view, a high percentage of bounced visitors means poor engagement (with the caveat that content creators should not try to produce the perfect one-page article!). As with the other KPIs for this job role, segmentation is the key to making informed decisions, as per Figure 10.19.

In addition to the segmentation suggested earlier, consider using a profile that excludes visits to your home page. I have found that websites proactive with search engine optimization (SEO) can receive one-page visits from organic sources for reasons that are not relevant to their business. For example, organic visitors might click on your link in the search engine result listings because you have a high ranking—without qualifying themselves by reading your listing's description snippet. This is quite common, either because it is so easy to do, or because of brand confusion (think of the number of companies that are named Alpha).

In addition, a significant number of existing customers use their supplier's website to quickly look up an e-mail address or telephone number; often your home page is faster and easier for finding contact information than internal address books. For these reasons, experiment using this KPI with and without your home page included.

Number of Advertisements Clicked

For content-driven websites that derive their income from visitors clicking on advertisements, increasing the number of these click-throughs is an important KPI (assuming advertisements are well targeted). Advertisements generally lead a visitor to an external website, so you need to track these outbound links as discussed in "Event Tracking," in Chapter 7. With this in place, performing the calculation is straightforward from the Content > Top Content report:

$$\text{Number of advertisements clicked per 1000 visits} = \frac{\text{total number of advertisements clicked}}{\text{total number of visits}} \times 1000$$

The reason to multiply the average by 1000 is that this metric is usually very small and does not convey the information well as a KPI. In addition, advertising rate cards for content and media sites are usually priced according to a cost-per-thousand-impressions model (CPM—cost per mille; mille is Latin for thousand). Having this KPI with the same multiplier is clearly beneficial to help establish your rate card.

Taking the total number of visits from Figure 10.18 shows that in this example the number of advertisements clicked per 1000 visits is 12, very low. If this were a content media site deriving its income from advertisements, then the quality, quantity, relevance, and placement of advertisements would need to be investigated as shown by this KPI. This calculation does not take into account that a single visit could have produced all 55 advertisement click-throughs—an unlikely scenario, but possible.

Note: In Figure 10.21, tracking the 55 outbound links was performed using virtual pageviews, rather than event tracking; both methods are valid, though event tracking should be used when the occurrence of these is high (as a rule of thumb, I use greater than 10 percent of the total number of pageviews as my guide). For the virtual pageview method, the number of advertisements clicked is equal to the total number of pageviews shown in the virtual directory `external/`. Using virtual pageviews is discussed in "trackPageview(): The Google Analytics Workhorse," in Chapter 7.

This KPI has so far assumed that your advertisements are well targeted and relevant to the content your visitors are reading. To test this theory, you should calculate this metric for different visitor segments.

Figure 10.21 Number of advertisements visitors clicked

Note: At the time of writing, AdSense click-throughs cannot be tracked using Google Analytics. 

Percentage Engagement

Apart from visitors reading your content, how else could you determine engagement? Perhaps it is the number of subscriptions you receive, the number of people who proceed to read your blog, go on to contribute to your blog with a comment or rating, or provide unsolicited feedback in some other way. Whatever the method, visitors who contact you or leave a comment on your website are a valuable metric of engagement. Expressed as a percentage, the calculation is as follows:

$$\text{Percentage engaged visits} = \frac{\text{total number of engagements}}{\text{total number of visits}}$$

Google Analytics tracks all data at the aggregate level, so it is best to track this KPI on a per-visit basis, rather than a per-visitor. Hence, it is not possible to determine whether one visitor is making all the engagements (see the sidebar "Percent engaged visitors").

A simple way to obtain this KPI is to view the Goals > Overview report, shown in Figure 10.22. This assumes all of your engagements are defined as goals. If some of your engagements are not defined as goals, use the in-line filter technique (refer to

Figure 10.21) to determine the number of engagements. However do not mix both methods as the goal conversion rate is calculated on a per visit basis. That is, if a goal is defined as the download of a PDF file and a visitor downloads five of these in the same visit, Google Analytics counts this as one goal conversion. Conversely, the inline filter technique of Figure 10.21 would show five download engagements. Either calculation is valid—you just need to be aware of the difference.

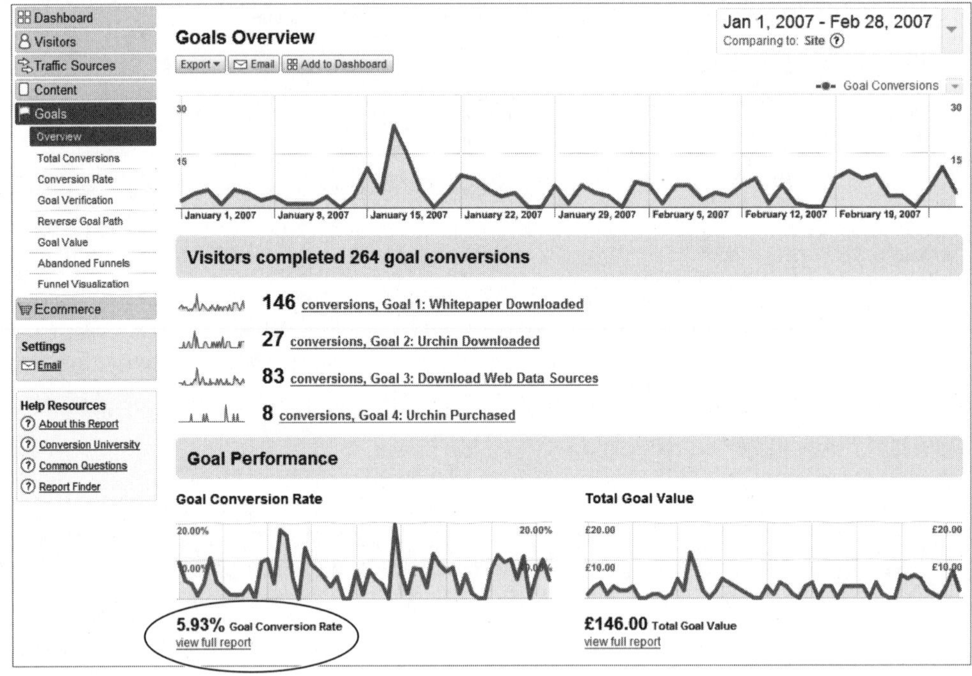

Figure 10.22 Goal conversion rates

Percent engaged visitors

It is possible to be clever here and use the _setVar() function as a label to track whether a visitor has engaged with your website (see "Labeling Visitors," in Chapter 9 for the use of _setVar()). The KPI could then be changed to percentage engaged visitors by substituting for the number of visits:

$$\text{Percentage engaged visitors} = \frac{\text{total number of engaged visitors}}{\text{total number of visitors}}$$

The total number of engaged visitors would then show in the Visitors > User Defined report.

By incrementing _setVar() for each visitor's engagement, you could also track the distribution of engagements—one of those challenges for a rainy day!

Percent New versus Returning Visitors

This is an important KPI for gauging your online business; it overlaps with marketing department KPIs (see "Example Marketer KPIs," and Figures 10.14 and 10.15). If your content is good, unique, and compelling, then you would expect a significant proportion of your visitors to be return visitors.

You need a way to separate these returning visitors from visitors attracted through your online marketing department's efforts. They will be acquiring new visitors via pay-per-click advertising, banners, and so on, as well as retaining visitors via e-mail marketing follow-ups and newsletters. It is therefore critically important for the content creator that this KPI is segmented by referral medium, as per Figure 10.23.

As shown in Figure 10.23, there are relatively few return visits for all media. However, as mentioned earlier, the example site has a mixture of e-commerce, lead generation, and blog content. The blog content has very different objectives from the other two—providing post-sales product support. To understand this KPI for the example website, it would be better to segment blog visitors into a separate profile (see "Filtering: Segmenting Visitors Using Filters," in Chapter 8).

Medium	Visits ↓	Pages/Visit	Avg. Time on Site	% New Visits	Bounce Rate
1. organic	1,879	2.04	00:01:48	84.57%	65.89%
2. referral	1,371	2.43	00:02:14	69.51%	55.80%
3. (none)	722	2.25	00:02:32	69.81%	59.70%
4. cpc	218	3.22	00:02:02	82.11%	37.16%
5. PPC	161	2.24	00:01:39	91.93%	61.49%
6. Forum	90	2.50	00:02:52	65.56%	43.33%
7. Email	6	6.17	00:06:11	33.33%	16.67%
8. ppc	3	5.33	00:00:41	33.33%	33.33%
9. Web	1	1.00	00:00:00	100.00%	100.00%

Figure 10.23 New versus returning visitors by medium

Be careful when interpreting changes in percentage of visitor types. For example, a decrease in percentage of new visitors could in fact be due to an increase in percentage of returning visitors, rather than any change in your new visitor acquisition strategy. To determine this, compare different date ranges and examine the raw visitor numbers.

Percentage Brand Engagement

See "Example Marketers KPIs," above. This is really a marketer's metric. I include this KPI here for reference. Figures 10.15a and b illustrate how to obtain this KPI.

Percent High, Medium, Low Visitor Recency

Recency is defined as the amount of time that passes between sequential visits—that is, when were the current visitors last on your site? The report in Figure 10.24 illustrates this. I have found from experience that many people struggle to understand what recency is telling them or how to interpret the chart. Maybe it is because the terminology is not widely used in business. Nonetheless, it is an essential metric for measuring engagement.

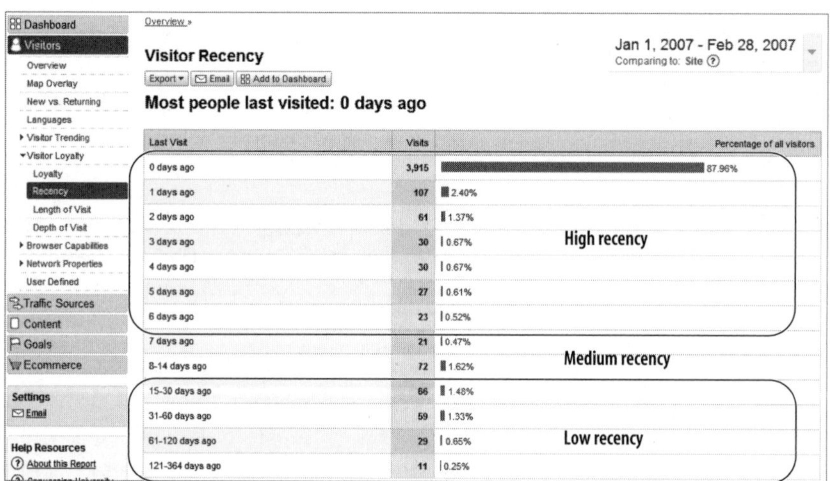

Figure 10.24 Visitor recency chart

Chart interpretation:

Of the visits made in the period shown, the vast majority (87.96% percent) of them are first-time visits (to be statistically correct, these will be a mix of first-time visits and any same-day repeat visits); 107 visitors (2.40 percent) also visited one day before, 61 visited two days ago, 72 visited 8–14 days ago, and so on.

Exercise

To check your understanding of this report, consider the following example. If an additional new visitor came to your site on 21 February 2007—that is, one week before the end of the date range shown in Figure 10.24—where will the visitor appear in the report table?

- 0 days ago
- 7 days ago
- Not shown, as they did not make a repeat visit

Answer: 0 days ago, because they were a first-time visitor who did not return.

For visitor recency KPI reports, group your metrics into high, medium, and low categories. The boundaries for each group will depend on your business model. For the example shown, it would make sense to set the following:

High = less than 7 days

Medium = between 7 and 14 days

Low = 15 or more days

In other words, collect a good sample of data (several weeks) and then define your boundaries based on the observations from this report.

In all examples, the shorter the recency value the better, excluding the first entry, "0 days ago," which represents first-time visits (and same day repeat visits). Short recency values mean fewer days between previous visits, and therefore the more engagement you have. For e-commerce websites this could be the amount of time between visit and purchase. However, not all sites exhibit this behavior; and high-value purchase items tend to have long visitor recency, as visitors take longer to consider their purchase decisions. If the example site in question were a content or media site, I would be concerned that visitors are not coming back in significant numbers. As discussed with Figure 10.23, a separate profile for just blog visitors should be used to assess this KPI in detail.

Note: According to the July 2007 ScanAlert report (www.scanalert.com/site/en/certification/moreinfo/?interest=windowshopping2007), online shoppers take an average of 34 hours and 19 minutes from their first visit to purchase.

Webmaster KPI Examples

Your webmaster department represents the people responsible for keeping your website up and running smoothly. As such, they need to know the expected visitor load on their servers. They also need to advise your design and content creation departments on visitor profiles from a technical perspective, such as what browsers are most commonly used and what language settings visitors have on their computers. This is how the industry of web analytics got started—webmasters wanting to know "how many?"

Webmaster KPIs are usually non-hierarchical because of their technical importance and intended audience—technical people for whom high-level summary indicators raise more questions. For this audience, you may also consider bringing in other non-visitor metrics to supplement the Google Analytics pageview data, such as web server uptime, server response speed, bandwidth used, and so on. These are not considered here.

Sample KPIs for webmasters include the following:

- Volume of visitors, visits, and pageviews
- Percentage of visitors with English language settings
- Percentage of visitors not using MS Internet Explorer
- Percentage of visitors with non-Windows platforms
- Percentage of visitors with high, medium, low screen resolutions
- Percentage of visitors with a broadband connection speed
- Percentage of visitors receiving an error page
- Internal search quality

Volume of Visitors, Visits, and Pageviews

This is a classic base metric that enables webmasters to quickly get a handle on the volume of traffic the website receives. Such metrics are important in determining the load on your web servers and network infrastructure, and the potential importance of your website compared to other parts of your business. The following metrics can be obtained directly from the Visitors > Overview report (refer to Figure 10.16):

- Average number of visits per time frame
- Average number of unique visitors per time frame
- Average number of pageviews per time frame
- Average pageviews per visit

For such metrics, collect data over long periods to diminish the effects of large fluctuations. If you are a B2B website, the number of visits per day averaged over a week will be skewed by the weekend. In this case, it would be better to consider the average over the working week (Monday–Friday).

Percentage of visits with English Language Settings

The more insight you have about your website visitor demographics the better, and this KPI strongly overlaps with the marketing department. The visitor language setting is an excellent way of determining your international reach and whether your content matches this. Of course, if your main website language is not English, then simply replace the KPI name "English" with the appropriate language.

You can view the distribution of visitor languages directly from the Visitors > Languages report (see Figure 10.25). You will need to do some grouping here, as all language types are reported. For example, British English (en-gb) is reported separately from American English (en-us). Similarly, Spanish, Portuguese, and French have different varieties, as do many other languages. It is therefore important to group (or not) different language versions according to your requirements.

Note: Don't infer too much from the difference between en-gb and en-us, as a great many non-U.S. users have their browser settings set as en-us by default and never bother to change this. For example, I noticed that when I access my Google Analytics reports, I do so in U.S. English. In over two years I did not bother to change this for U.K. English.

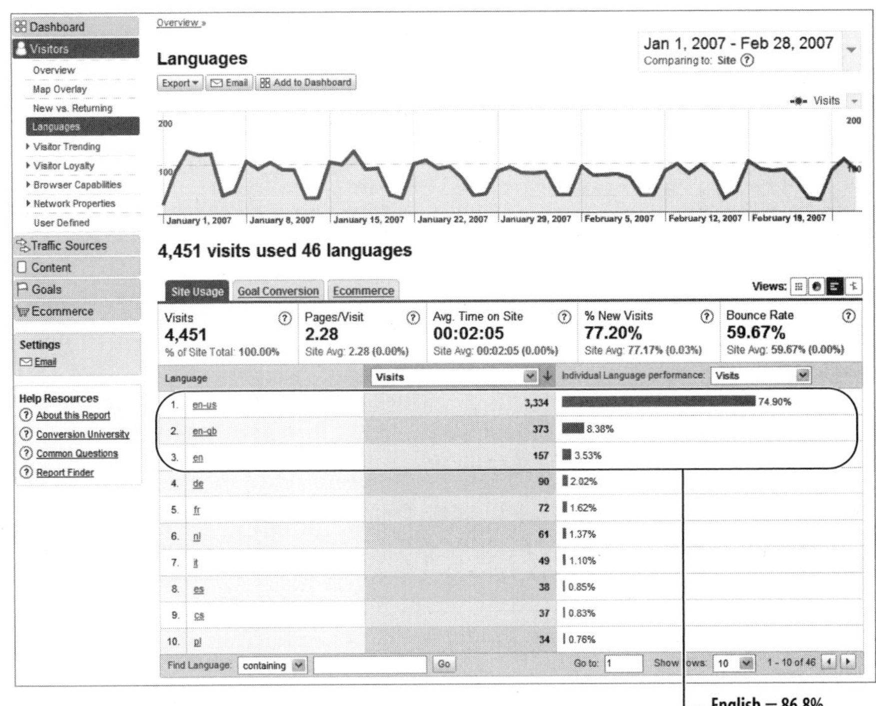

Figure 10.25 Distribution of visitor language settings

From Figure 10.25 you might assume that the vast majority of visitor language requirements (almost 87 percent) are accounted for. However, this should always be assessed further by viewing the Goal Conversions tab. You would expect that, all things being equal, the same proportion of conversions should occur for English visitors as for non-English visitors (if not higher).

Note that this is not the case for this example website, as Figure 10.26 shows. The data suggests that foreign-language visitors are more likely to convert than English-speaking visitors. Perhaps there is an opportunity for this company to market its services in other countries—in English or not.

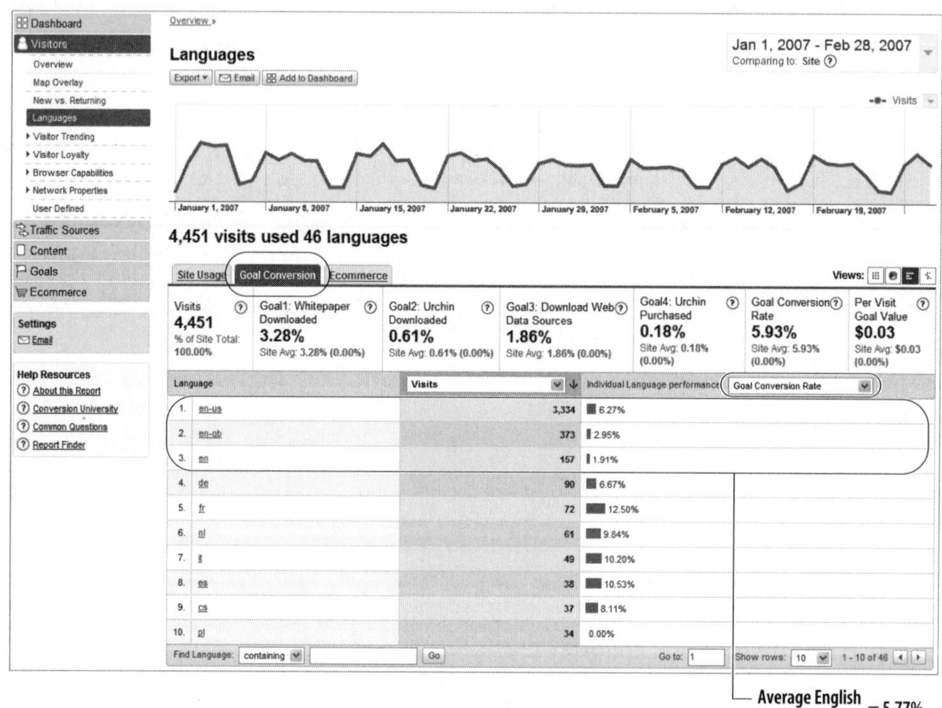

Figure 10.26 Distribution of visitor language settings by conversion rate

Note: An excellent resource for comparing Internet world statistics is: www.internetworldstats.com. See for example http://www.internetworldstats.com/stats7.htm where English accounts for 31.1 percent of world Internet usage (November 2007).

Percentage of visits Not Using MS Internet Explorer

Different web browsers (Internet Explorer, Firefox, Opera, Netscape, etc.) render web pages slightly differently. This means pages may look different from that intended or not even work at all in different browser types. Despite the vast majority of web users currently using Microsoft Internet Explorer (globally estimated at over 80 percent), if your e-commerce or booking engine can not process orders from non–Internet Explorer visitors, you could be losing out on significant revenue and damaging your brand reputation to boot.

Testing web pages in different browser windows is a laborious job for webmasters, so knowing what proportion of visitors use which browser types will enable you to prioritize resources effectively. This KPI can be accessed at a glance from the Visitors > Browser Capabilities > Browsers report, shown in Figure 10.27. As you can see, having the website working well in both MS Internet Explorer and Firefox is important (the ratio is almost 50:50). This should be assessed further by viewing the Goal Conversions tab. That is, visitors from MS Internet Explorer and visitors from Firefox should result in approximately the same proportion of conversions. If not, then it may be that your website does not work equally well for both browsers.

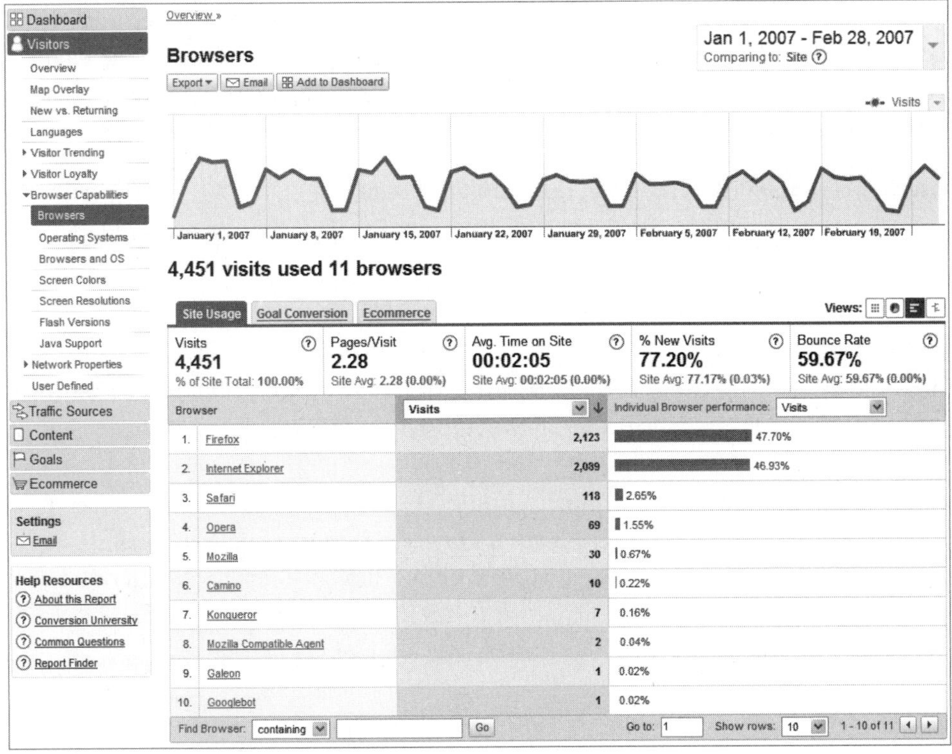

Figure 10.27 Visitor browser types

The price of incompatibility

Assuming that the estimated 20 percent of non-Internet Explorer visitors behave in the same way as Internet Explorer visitors (there is no reason to suppose otherwise for the same website), that equates to a 25 percent loss of revenue if your website cannot work in these browsers—money left on the table.

Even if your percentage of visitors not using Internet Explorer is lower, consider this value against the percentage gains your marketing department is trying to achieve from optimizing your online marketing campaigns. With browser standards now well established, there really is no excuse for not making your website work well in at least two of your visitor's most popular browsers.

Percentage of visits with Non-Windows Platforms

Similar to the issue of visitors using different web browsers, visitors with different computer operating systems (Windows, Unix, Mac, etc.) can also cause website pages to be rendered differently and may break functionality. Knowing what proportion of visitors use which operating system platform enables you to allocate resources efficiently.

In addition, operating systems provide information about the devices used to access your website, enabling you to cater to mobile phones, for example. The report is accessed from the Visitors > Browser Capabilities > Operating Systems report, shown in Figure 10.28.

Percentage of visits with High-, Medium-, Low-Resolution Screens

Are your web pages designed for 800 × 600 screen resolution? Modern LCD displays are set by default to 1024 × 768 and can go to much higher resolutions, including wide-screen formats. If you are designing web pages for a width of only 800 pixels, content that could fit on a wider resolution without the need to scroll ends up being pushed down the page. Consider also that side-menu navigation and margins can consume 150–200 pixels of your HTML page. As a result, valuable messaging and calls to action fall below the fold of the page (meaning they are not initially viewable), forcing visitors to scroll down to view them. Of course, if your visitors only have a screen resolution of 800 × 600, that cannot be helped. However, if most of your visitors are viewing wider resolutions, then you are wasting an opportunity—they may not scroll down to see your content. Rather, they may click to another part of your website, or click away, believing you do not have the content they are looking for.

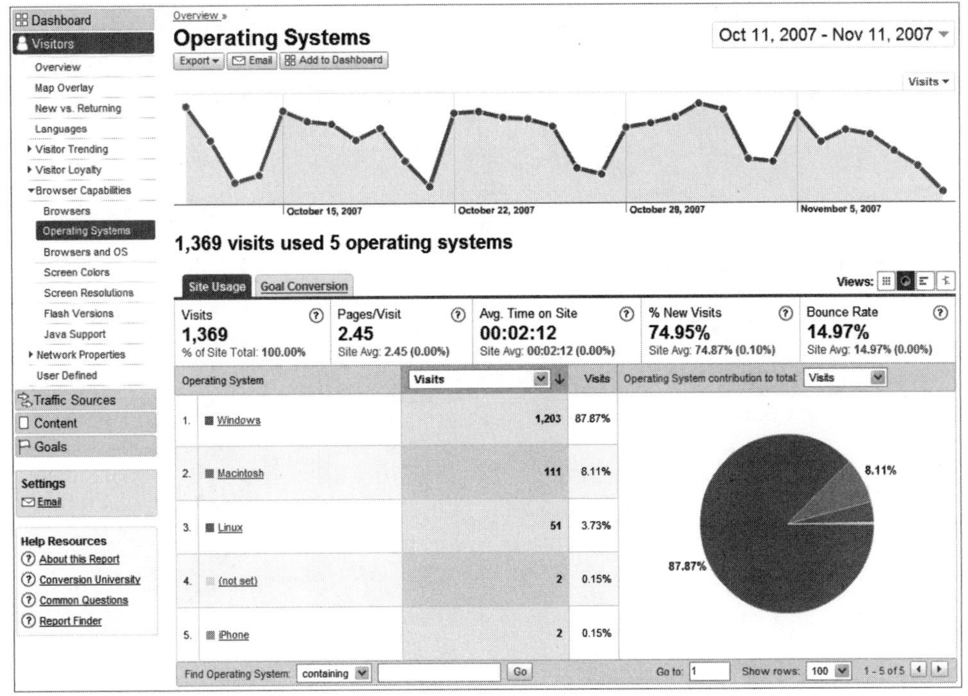

Figure 10.28 Visitor operating systems

Access your visitor screen resolution sizes from the Visitors > Browser Capabilities > Screen Resolutions report, shown in Figure 10.29.

For your KPI reports you should group these into high, medium, and low categories. The boundaries for each group will depend on your business model (do you target mobile visitors)? Generally, for PC users, the following settings are fine:

Low = less than 800 × 600

Medium = 800 × 600

High = greater than 800 × 600

If you see the low segment growing, then it's a safe bet your site is being accessed by mobile devices, and you should cater to that audience accordingly.

Note: Google Analytics can only track mobile devices that support and have JavaScript-enabled browsers and can store cookies. iPhone, BlackBerry, and the latest Nokia phones all have support for JavaScript. However, globally, many cell phones currently do not support JavaScript or cookie storage.

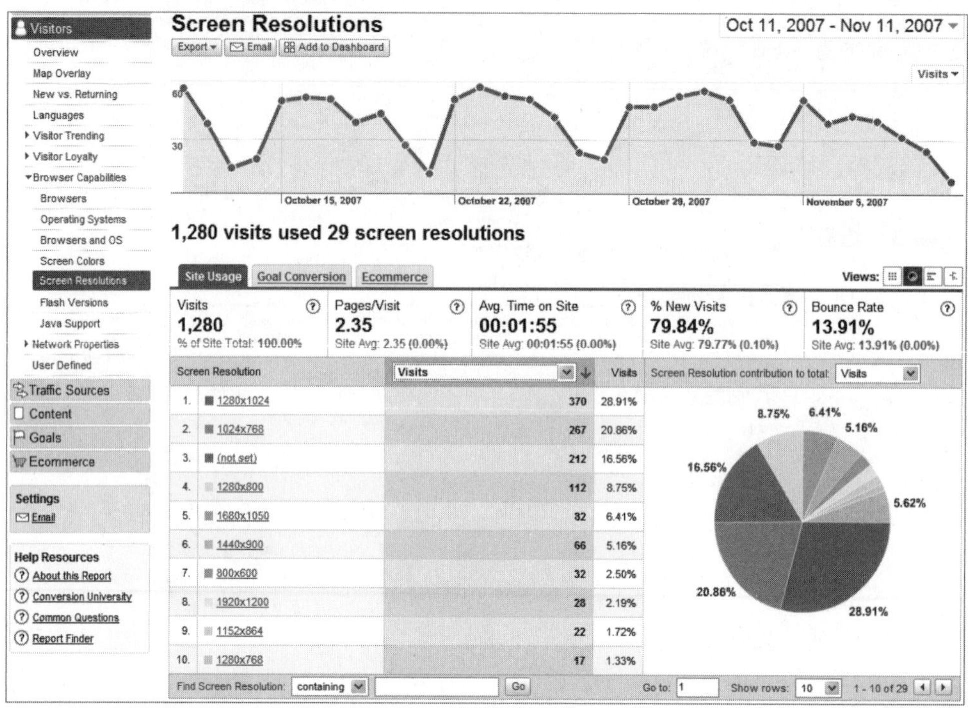

Figure 10.29 Visitor screen resolutions

Percentage Visits with Broadband Connection Speed

The speed at which your visitors access the Internet has obvious implications for webmasters when you are considering adding rich media content to website pages. However, not all parts of the world have broadband access, so even without rich media, slow page download times adversely affect the user experience.

Note: An excellent resource for comparing Internet world statistics is: www.internetworldstats.com. See for example http://www.internetworldstats.com/dsl.htm where of the countries with the highest Internet usage, the top five countries for broadband penetration are Netherlands, South Korea, Sweden, Canada, and United Kingdom respectively. The U.S. is ranked seventh (September 2007).

Regardless of connection speed, a study by Akamai and Jupiter Research identified an acceptable threshold of four seconds for retail web page response times (www.akamai.com/html/about/press/releases/2006/press_110606.html). Also, a September 2006 report by SciVisum (www.scivisum.co.uk/report/malefemale2006/index.htm) showed 78% of online shoppers complained that frustration with website performance has led them at one time or another to stop in mid-transaction. As a rule, I suggest the four second rule is applied to web pages of all industries.

You can view the distribution of visitor connection speeds directly from the Visitors > Network Properties > Connection Speeds report (see Figure 10.30). As you saw when viewing visitors by language setting and screen resolutions, you need to do some grouping here. For example, DSL, Cable, T1, and OC3 are all broadband connection speeds.

Broadband = DSL, Cable, T1, OC3

Dialup = Dialup, ISDN

Table 10.3 Connection Type Acronyms Defined

Term	Description
DSL	Digital Subscriber Line (broadband)
Cable	Similar to DSL (broadband)
T1	Corporate leased line or private wire (fast broadband)
Dialup	Modem (slow-band)
OC3	Optical Carrier 3 (very fast broadband)
ISDN	Integrated Services Digital Network (twice as fast as dialup)

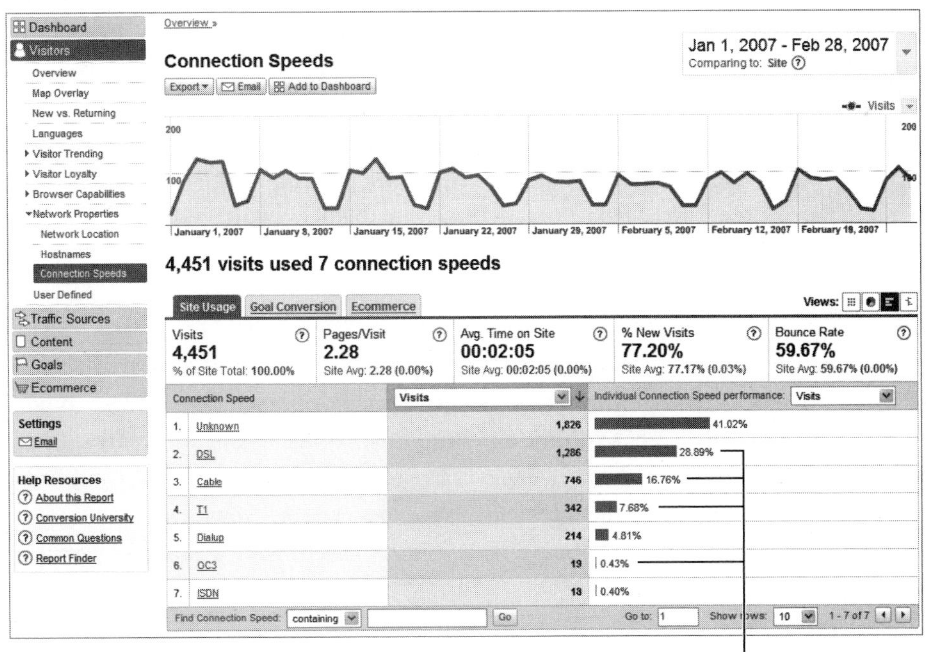

Figure 10.30 Visitor connection speeds

> **How connection speed is determined by Google Analytics**
>
> A visitor's connection speed is determined by their IP address using a third-party database lookup of geo-ip data. The suppliers of this database obtain information from a variety of sources: visitors around the globe who provide details on website location, Internet service providers that allocate IP information, and interpolation and network triangulation of unknown geo-ip addresses from two known ones.
>
> Because of this disparate source of data, there is often a significant percentage of visitors for whom connection speed is unknown. Visitors with unknown connection speeds should always be taken into account.

Percentage of Error Pages Served

This is an obvious metric any webmaster would wish to minimize. It is defined as follows and quoted as a percentage:

$$\text{Percentage error pages served} = \frac{\text{total number of error pages served}}{\text{total number pageviews served}}$$

Tracking error pages was discussed in "Tracking Error Pages and Broken Links," in Chapter 9. You first need to tag your error page template with the GATC. Then, within the Google Analytics configuration settings, apply a filter to place any error page URLs into a virtual subdirectory. This way, you can view the total number of error pages in the Content > Top Content report, as shown in Figure 10.31.

Combing the numbers shown in Figures 10.18 and 10.31 for the example website in question, the percentage of error pages served is 0.27 percent. This is low, but it represents a bad experience for those visitors. A target for this KPI could be to maintain this at less than 0.1 percent.

Internal Search KPIs

Improving the effectiveness of your onsite internal search engine can be critical to your conversion rate and therefore the success of your website. Onsite search is now so important for large websites that it has become an integral part of the navigation system. Even for smaller sites, a good internal search engine can improve the user experience and hence your bottom line, so measuring the internal search experience is a key metric.

Important site search KPIs are available in the Content > Site Search > Overview report, shown in Figure 10.32.

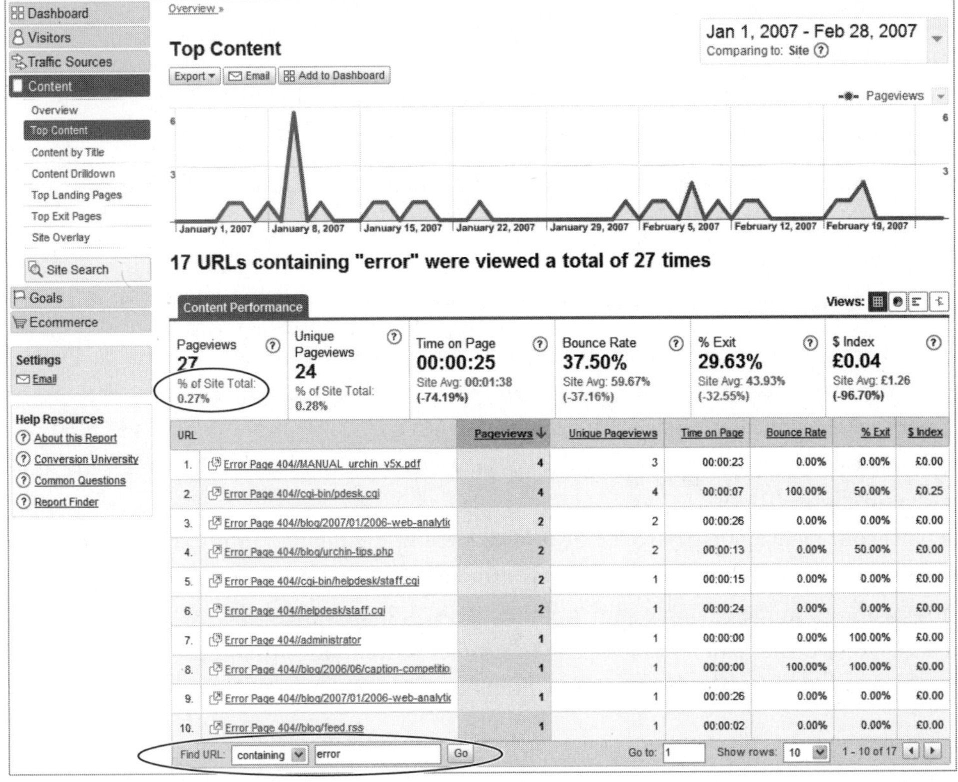

Figure 10.31 Top Content report showing error pages only

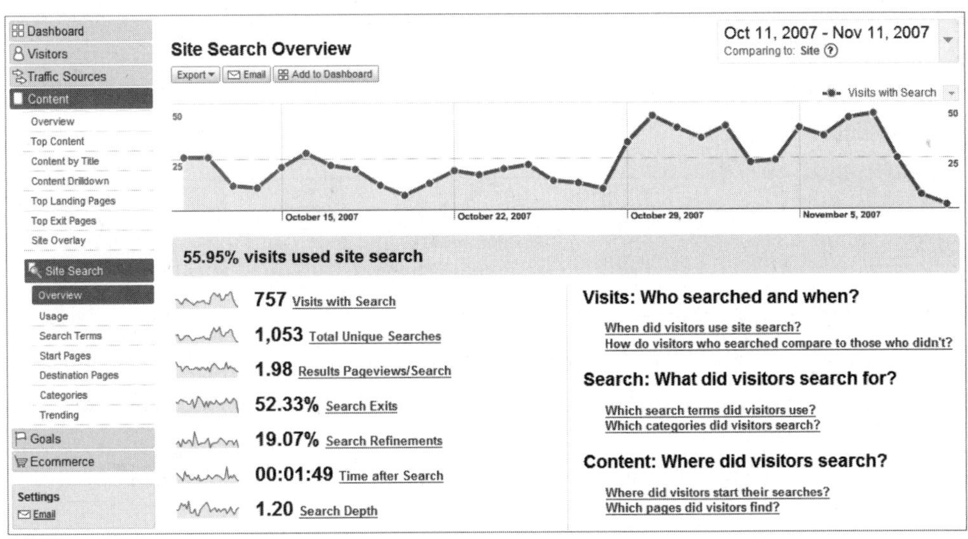

Figure 10.32 Site Search Overview report

Important site search KPIs include the following:

- Percentage of visits that use site search (56.35 percent).
- Average number of search results viewed per search (1.99).
- Percentage of people exiting the site after viewing search results (51.94 percent).
- Percentage of people conducting multiple searches during their visit (19.12 percent). This excludes multiple searches for the same keyword.
- Average time on site for a visit following a search (00:01:49).
- Average number of pages visitors view after performing a search (1.21).

Other important KPIs for site search include how visitors that use this facility compare with those that do not. For example, are site search visitors more likely to convert, spend more money, spend more time on site, view more pages, less likely to bounce, etc.? This can be achieved by viewing the Site Search > Usage report as shown in Figure 10.33.

From Figure 10.33, use the drop-down menu highlighted to select the metric of interest (in this case Revenue). Then divide the metric shown for visits with site search by those without. In the example shown, this is \$1,421.00 / \$392 = 3.63. This means that visits that used site search spent nearly four times more money that those that did not—this is an interesting KPI that indicates how important site search is for this website.

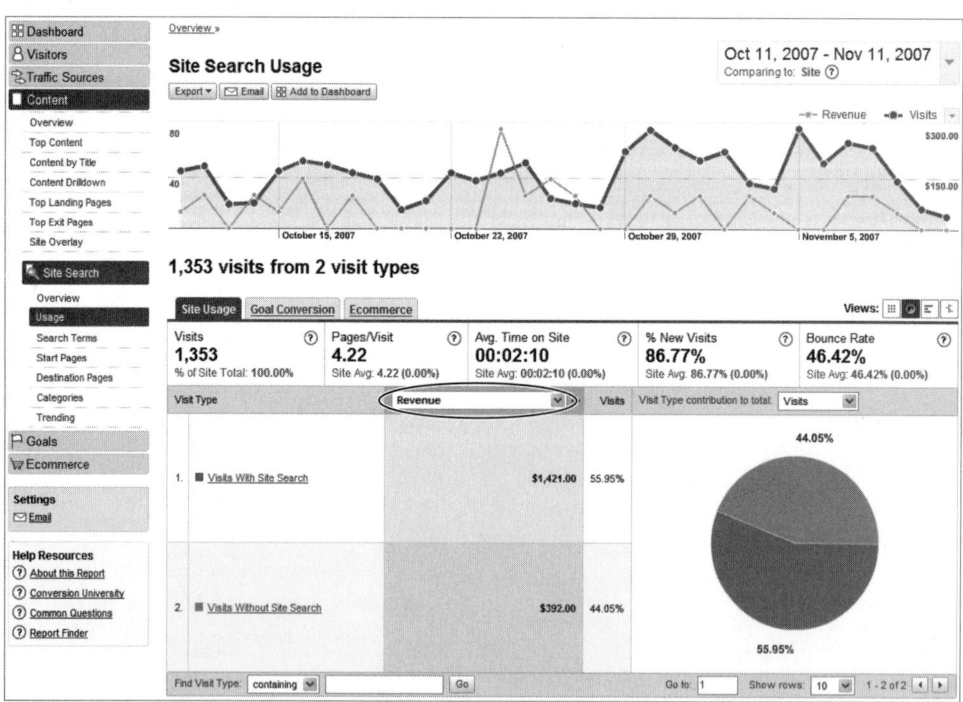

Figure 10.33 Site search usage report

Note: Be aware that when selecting different metrics from the drop-down menu, the row order (and color key) may change depending on which is the highest value. For example, "Visits With Site Search" may be displayed as the first row then revenue is selected, but row number two when bounce rate is selected. This is the same behavior for all reports. That is, the highest value is always displayed first in the data table.

An additional site search KPI to consider is the number of zero result search pages delivered, as shown in Figure 10.33.

To accomplish this you need to ensure that a unique URL is loaded when a page of zero results is shown. In Figure 10.34, this was achieved by setting a virtual pageview within the internal site search engine template page for zero results. The technique uses _trackPageview() as described in Chapter 7.

An interesting observation about Figure 10.34 is that although the time on page is very short for zero result pages (as expected), the percent of website exits is 67 percent lower than the site average. In other words, visitors who receive zero search results appear to try another search, or at least go on to view other pages, on this website.

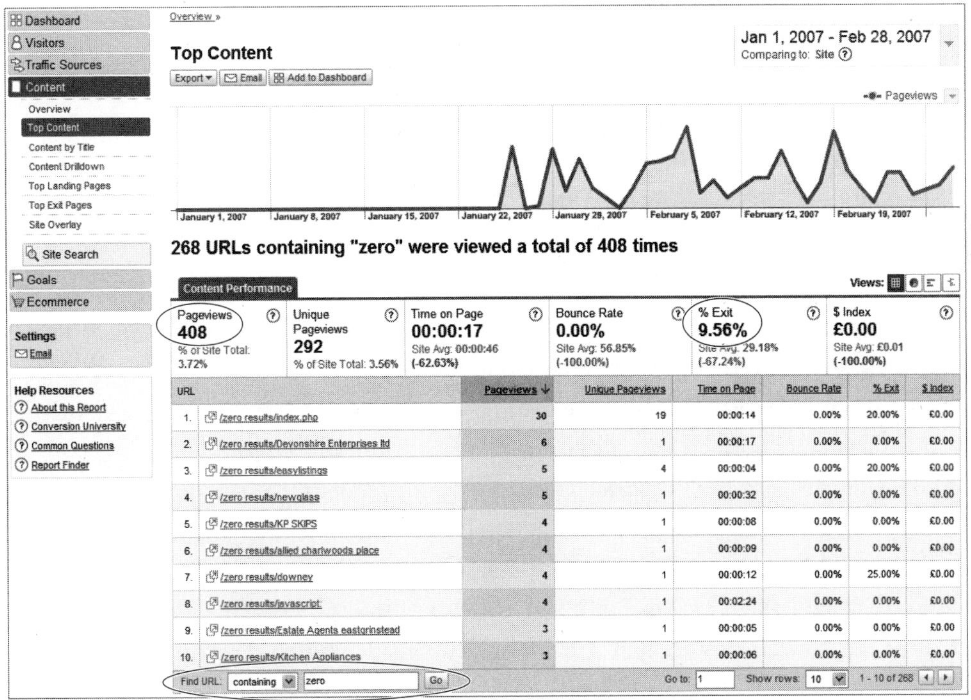

	URL	Pageviews	Unique Pageviews	Time on Page	Bounce Rate	% Exit	$ Index
1.	/zero results/index.php	30	19	00:00:14	0.00%	20.00%	£0.00
2.	/zero results/Devonshire Enterprises ltd	6	1	00:00:17	0.00%	0.00%	£0.00
3.	/zero results/easylistings	5	4	00:00:04	0.00%	20.00%	£0.00
4.	/zero results/newglass	5	1	00:00:32	0.00%	0.00%	£0.00
5.	/zero results/KP SKIPS	4	1	00:00:08	0.00%	0.00%	£0.00
6.	/zero results/allied charlwoods place	4	1	00:00:09	0.00%	0.00%	£0.00
7.	/zero results/downey	4	1	00:00:12	0.00%	25.00%	£0.00
8.	/zero results/javascript:	4	1	00:02:24	0.00%	0.00%	£0.00
9.	/zero results/Estate Agents eastgrinstead	3	1	00:00:05	0.00%	0.00%	£0.00
10.	/zero results/Kitchen Appliances	3	1	00:00:06	0.00%	0.00%	£0.00

Figure 10.34 Percentage of zero search results

Note that the 408 zero result search pages are shown as accounting for 3.72 percent of the total number of pageviews. However, this should not be considered as the KPI. A better metric would be to change the denominator to the total number of search pages viewed, as shown in the following:

$$\text{Percentage zero result pages} = \frac{\text{total number of zero search result pages}}{\text{total number of search pages viewed}}$$

KPI Summary

If you have followed the story in this chapter for the single website used to demonstrate example KPIs, you will recognize the following summary and action items:

1. Blog visitors have different objectives than visitors who are looking to purchase (as discussed for Figure 10.3).

   Action Item: Segment blog visitors into a separate profile so that these may be analyzed in detail. This requires the application of a filter.

2. Forum visitors drive goal conversions (PDF downloads) and are 10 times more likely to do so than any other referrer (as discussed for Table 10.2). However, it is cpc visitors (AdWords) who are driving the transactions (as discussed for Figure 10.4).

   Action Item: Acquire more forum visitors to drive branding, reach, and goal conversions.

   Action Item: Acquire more cpc visitors (AdWords and others) to drive further revenue growth.

   Action Item: Investigate why Google visitors are less likely to convert goals than any other referrer.

3. The example website has excellent trust and design factors that resonate with visitors (as discussed for Figures 10.7, 10.8). That is, new visitors are just as likely to convert as returning visitors (almost).

   Action Item: Reward your web design and development team and ensure that they maintain their current visitor-centric design philosophy.

4. This site has a healthy search engine marketing strategy that is acquiring 77 percent new visitors (as discussed for Figures 10.14, 10.15).

   Action Item: Reward your online marketing team and ensure that they maintain their efforts.

5. Goal conversions are higher for foreign language visitors than for those with English set as their operating system language (as discussed for Figure 10.26).

   Action Item: Investigate the potential for doing business in other languages.

6. Error pages (as discussed for Figure 10.31) are rare but are currently at 0.27 percent of all pageviews.

   Action Item: Aim to reduce error pages by 50 percent to a goal of 0.13 percent for the following month.

7. Internal site search is being used by nearly 56 percent of all site visits (refer to Figure 10.32).

   Action Item: Investigate how to better monetize the site search feature and improve its impact on the user experience.

There are, of course, many other visitor insights provided from working through the KPIs for this example website. However, be careful not to overload your stakeholders in one go. By completing and reviewing the aforementioned seven action items the following month or quarter, you will have built a solid platform from which you can reach the next level of change.

Using KPIs for Web 2.0

Web 2.0 is a phrase attributed to Tim O'Reilly (see www.oreillynet.com/lpt/a/6228 and www.oreillynet.com/pub/a/oreilly/tim/news/2005/09/30/what-is-web-20.html). In effect, Web 2.0 is a buzzword for the next generation of browser applications. According to Wikipedia, "Web 2.0 is a term often applied to a perceived ongoing transition of the World Wide Web from a collection of websites to a full-fledged computing platform serving web applications to end users. Ultimately Web 2.0 services are expected to replace desktop computing applications for many purposes."

The irony is that the technology that drives Web 2.0 is part of today's Web 1.0 technology and has been around for many years—that is, JavaScript and XML. As such, Web 2.0 does not refer to any technical advancements of the Web or the Internet infrastructure it runs on, but to changes in the way the medium is used. That's not to devalue the significance of Web 2.0, as it is this major shift in how users participate and surf the Web that is driving the second generation of interactive web applications.

Web 2.0 applications are usually built using Ajax (asynchronous JavaScript and XML) techniques. Similar to LAMP and DHTML, Ajax is not a technology in itself but a collection of technologies and methodologies combining JavaScript, XML, XHTML, and CSS. Another Web 2.0 technology is Flash. As with Ajax, it has been around for almost 10 years, but has only recently emerged as something more than just cool animation, with its ability to stream video and interact with XML. New up-and-coming technologies include Adobe Flex, Adobe AIR, and Microsoft Silverlight. Collectively, all these technologies are referred to as rich Internet applications (RIAs).

Example Web 2.0 sites

Excellent examples of Web 2.0 websites with RIAs include the following:

Google Maps http://maps.google.com (Ajax)

Google Mail http://mail.google.com (Ajax)

Yahoo Mail http://mail.yahoo.com (Ajax)

Google Docs http://docs.google.com (Ajax)

Tafiti search engine (Microsoft) www.tafiti.com (Silverlight)

YouTube www.youtube.com (Flash and Ajax)

Fox Movies Trailer Library http://silverlight.net/fox/ (Silverlight)

Why the Fuss about Web 2.0?

The techniques employed when developing a website using Web 2.0 technologies separate the components of data, format, style, and function. Instead of a web server loading a discrete page of information combining all those elements, each element is pulled separately. This has tremendous implications when it comes to defining KPIs, as the concept of a pageview all but disappears.

For example, load http://maps.google.com in your browser and navigate to your hometown (usually in the format of "town, country"). Then zoom in and out and pan around by dragging the map around. You can also change to satellite view or a hybrid of both. It is difficult to describe this in words, but if you try it out you very quickly get the idea.

Google Maps is an excellent example of the power and interaction of a Web 2.0 website. When you load the first page there is an initial delay while a JavaScript file is downloaded in the background. This is the controlling file that interacts with your mouse instructions. Note that the page and controlling JavaScript file are only loaded once. Then, as you interact with the map (zoom, pan around, etc.), further data is requested on-the-fly and inserted into the existing page. (The page URL does not change while you do this; the web page itself has become part of the delivery process.) By contrast, a traditional Web 1.0 website would require the reloading of the page to insert each additional map image.

Here we have an example of a visitor requesting one HTML page yet interacting in many different ways—perhaps creating dozens of actions or events (zooming and panning around) and gaining significant benefit from the experience. Clearly, using only pageview data for your KPIs is not going to work if your website contains RIAs.

Note: Tracking Web 2.0 websites is not an issue for Google Analytics. These can even be monetized. See "Event Tracking," in Chapter 7.

Web 2.0 sites are currently rare, but they can have a huge impact. For example, not many people are unaware of Google Maps, Yahoo! mail, or YouTube. The key to their growing success is that the user experience is "cool." Visitors find and interact with content quickly and without waiting for page refreshes. For example, consider the screen shot from YouTube shown in Figure 10.35. All of the areas highlighted are actions or events that the visitor can interact with; that is, they are not pageviews. Essentially, the visitor can multitask with all of these on the same page.

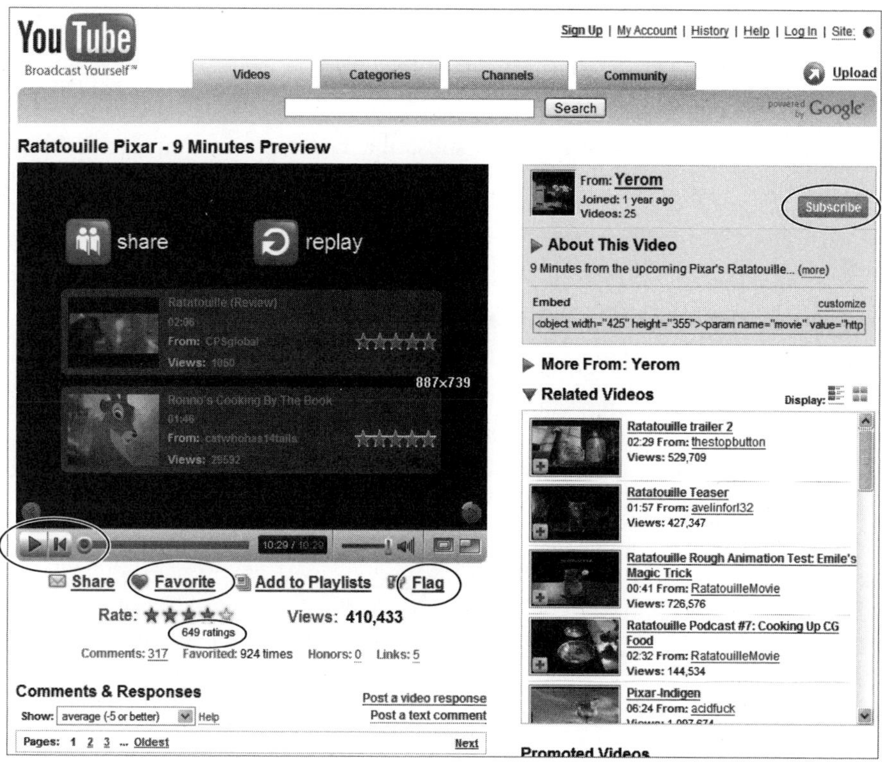

Figure 10.35 Visitor engagements on YouTube

As the number of Web 2.0 RIA sites grows, the requirement to define KPIs for them grows. Rather than think in terms of pageviews, analysts need to think in terms of actions and events that indicate engagement. In other words, what actions do you want your visitors to make in order to classify an engagement?

Engagement was discussed in detail in the section "Content Creator KPI Examples." The principle is the same for RIAs. Without changing your analytical thinking, current KPIs suited to a Web 2.0 environment include the following:

- Percentage new visitors
- Percentage unique visitors
- Average views per visit
- Average visit length
- Average conversion rate

By combining engagement KPIs with event tracking (see Chapter 7), you can then define KPIs such as:

- Percentage of visitors with content views—for example, zoom, pan around, view next message
- Percentage of visitors triggering an event—for example, play, pause, next, rate, advertisement click-through
- Percentage engagement—for example, subscribe, register, comment, rate, add to favorites

Summary

Providing KPIs enables your colleagues to focus on the parts of their online strategy that are most effective at generating more visitors, leads, conversions, and revenue for the business. The key for large organizations is delivering different KPI reports to each stakeholder and ensuring that these are hierarchical.

KPIs for the next generation of websites (Web 2.0) are already starting to emerge but are still in their infancy. Over the next several years, expect to see these focus and expand upon engagement—metrics that are currently used mainly by content managers. Also expect to see a standardization of what *engagement* means for different industries and stakeholder roles.

In Chapter 10, you have learned about the following:

- Setting objectives and key results as an important prerequisite for aligning KPIs with your business
- Selecting and preparing KPIs by translating OKRs into actionable and accountable metrics
- Presenting KPIs in a clear format that business managers recognize and understand
- KPI examples by job role to help you get started with important metrics
- How Web 2.0 and rich Internet applications are changing metrics and KPI definitions

11

Real-World Tasks

By now you may find your eyes glazing over at the scale of the project you have undertaken. However, Google Analytics is one of the easiest web analytics tools to configure, use, and understand. This chapter includes real-world examples of tasks most web analysts regularly need to perform. You will be surprised at how easily you can master the tasks you need to perform to gain insight into the information Google Analytics provides. Even better, you will have a profound impact on the performance of your organization's website.

In Chapter 11, you will learn how to do the following:

Identify poorly performing web pages
Measure the success of internal site search
Optimize your search engine marketing
Monetize a non-e-commerce website
Track offline spending
Use Website Optimizer

Identifying Poorly Performing Pages

With all that visitor data coming in, one thing you will want to do is optimize your pages for the best possible user experience. Often the improvements are straightforward—for example, fixing broken links, changing landing page URLs to match the visitor's intent, or aligning page content with your advertising message. But which pages should you optimize? If your website has more than a handful of pages, where do you start?

Traditionally, for web analytics solutions, identifying pages that underperform has been difficult. However, Google Analytics has several resources and reports to help you. This is not as an exhaustive list, but the following highlights the areas I most commonly turn to:

- $Index values
- Top Landing and Exit Pages report
- Funnel Visualization report

Using $Index Values

The importance of $Index was discussed in Chapter 5 in the section "Content: $Index Explained." In summary, it is a measure of the value of a page and is calculated as follows:

$Index = (goal value + e-commerce revenue) / unique pageviews

Essentially, if page *A* is viewed by visitors who go on to achieve a goal, the value of that goal counts toward the value of page *A*. The more times page *A* is viewed by visitors who achieve goals and the higher the goal value, the greater $Index becomes. This technique is a great way to value pages that are not goals or conversions themselves. Ranking pages by their $Index value enables you to prioritize them for optimization.

Note: It is important to monetize goals in order for the true significance of $Index to be realized. To define a goal value, see "Goals and Funnels," in Chapter 8.

To view the $Index values of your website pages, go to the Content > Top Content report and sort by the $Index column (click the heading). This shows your most valuable pages. However, by default, pages you've defined as your goals are also included here. Clearly, these will always be your highest $Index pages. Therefore, remove these from the list using the inline report filter. The resultant report then reflects your most valuable pages, as shown in Figure 11.1.

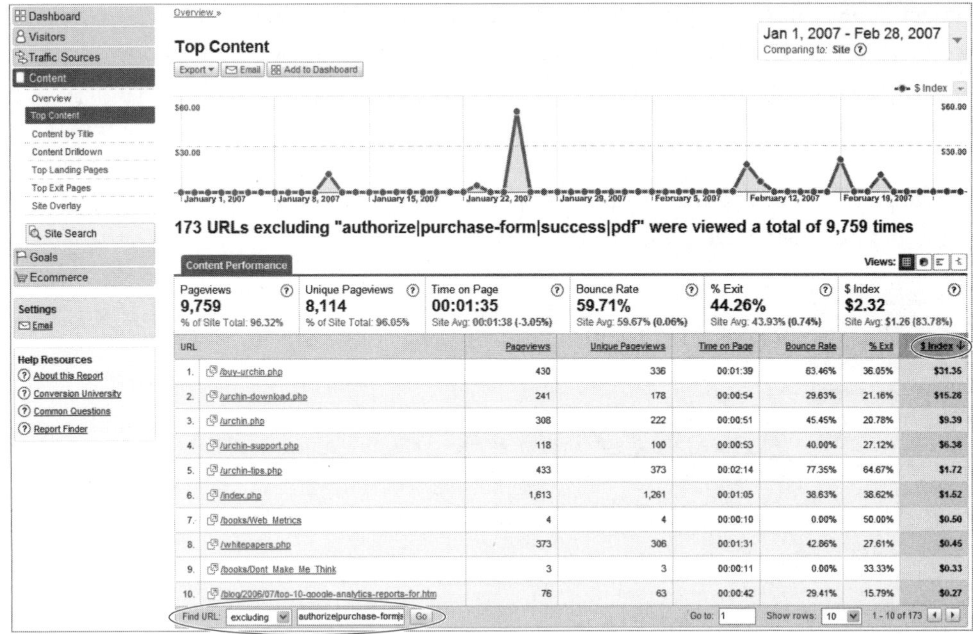

	URL	Pageviews	Unique Pageviews	Time on Page	Bounce Rate	% Exit	$ Index
1.	/buy-urchin.php	430	336	00:01:39	63.46%	36.05%	$31.35
2.	/urchin-download.php	241	178	00:00:54	29.63%	21.16%	$15.28
3.	/urchin.php	308	222	00:00:51	45.45%	20.78%	$9.39
4.	/urchin-support.php	118	100	00:00:53	40.00%	27.12%	$6.38
5.	/urchin-tips.php	433	373	00:02:14	77.35%	64.67%	$1.72
6.	/index.php	1,613	1,261	00:01:05	38.63%	38.62%	$1.52
7.	/books/Web_Metrics	4	4	00:00:10	0.00%	50.00%	$0.50
8.	/whitepapers.php	373	306	00:01:31	42.86%	27.61%	$0.46
9.	/books/Dont_Make_Me_Think	3	3	00:00:11	0.00%	33.33%	$0.33
10.	/blog/2006/07/top-10-google-analytics-reports-for.htm	76	63	00:00:42	29.41%	15.79%	$0.27

Figure 11.1 Viewing high $Index pages with goal pages excluded

Selecting pages for optimization

With your pages ranked in order of their $Index values, it is tempting to simply select the least valuable pages (lowest $Index values) for optimization review. However, it is important to consider your most valuable pages too. For example, you may find that your *payment failure* page has a high $Index, meaning that visitors often see this before finally completing their purchase. Similarly, this could be your *contact page* or *terms and conditions page*, meaning visitors need further information before completing their order.

If so, the report indicates a problem with the conversion process. Perhaps your payment form has an unclear layout or visitors do not have enough assurance to complete their purchase without referring back to other pages. Whatever the reason, pages with both low and high $Index values should be reviewed.

Unexpected pages here—that is, those not obviously related to your goals—indicate an issue with your website structure or its content. Investigate further by clicking on the page link in question within the report table. This takes you to the page shown in Figure 11.2a. From here you can select the Navigational Summary report for that page (see Figure 11.2b), which tells you the visitor's previous and next pages viewed—in other words, how the visitor got to that page and where they went next. From the screen shown in Figure 11.2a, you can also select the Entrance Path report for that page (see Figure 11.2c). This extends the Navigation Path report further by showing visitors who started their visit on the selected page, which pages they viewed next, and on which page they finished their visit.

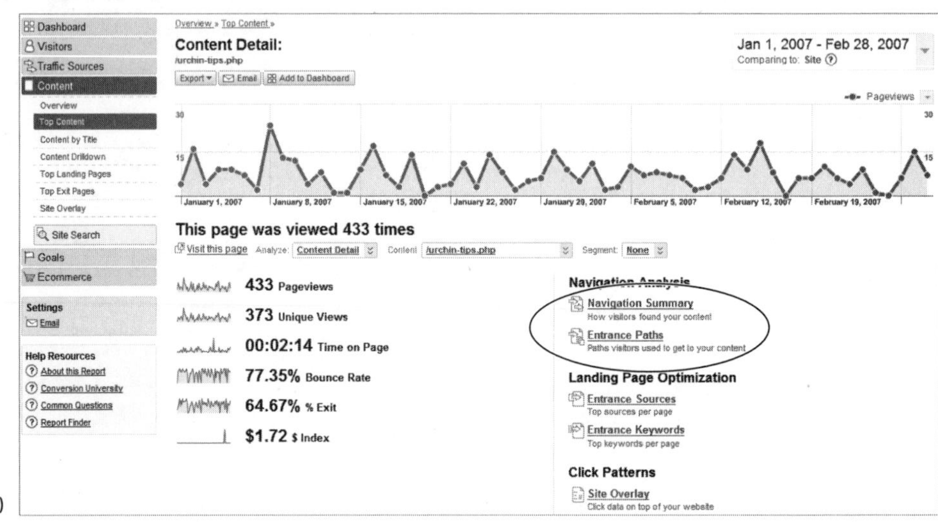

a)

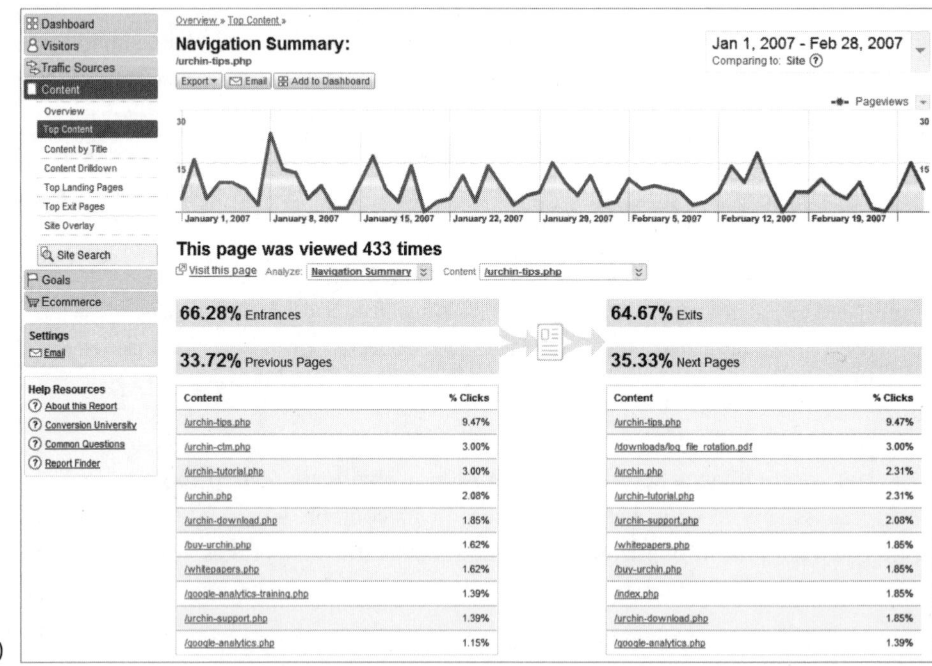

b)

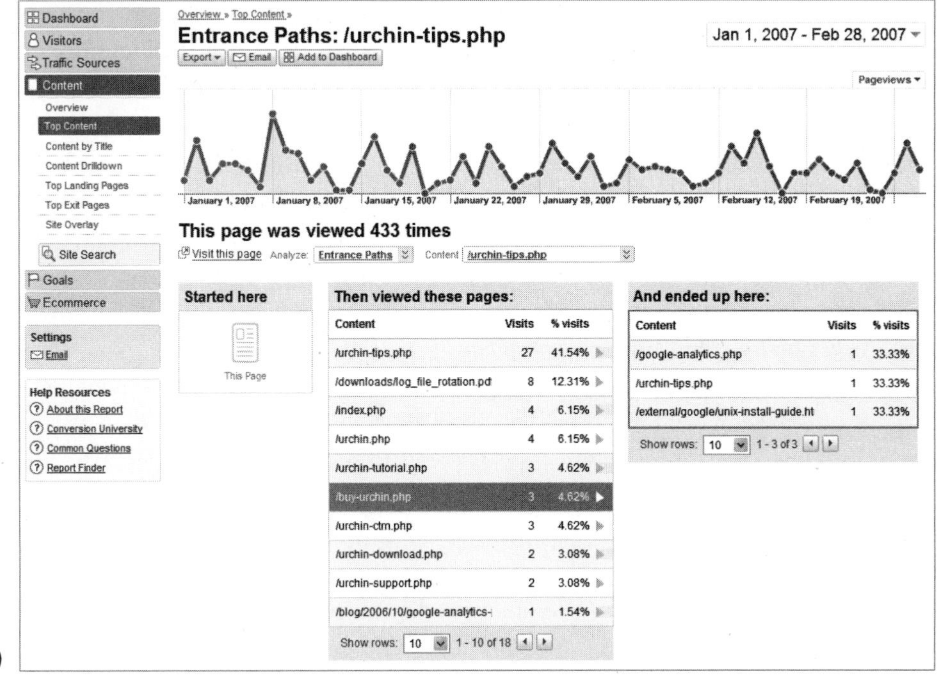

c)

Figure 11.2 (a) The report shown after clicking through on a page link from Figure 11.1; (b) Navigational Summary report; (c) Entrance Path report

Explanation of Figure 11.2c

Visitors who started their visit on the website in question at the page /urchin-tips.php viewed a total of 18 other pages next. Selecting one of these (/buy-urchin.php), shows three visits went on to complete their visit on three different pages: /google-analytics.php, /urchin-tips.php, and /external/google/unix-install-guide.htm.

Once you have a list of the 10 most valuable pages for your website as listed by their $Index values, bring your design or agency team in to discuss improvements. Include a member of your sales team and customer service department in the meeting, and ask them to bring a list of the five most common questions customers ask. Then spend a morning brainstorming.

As an initial exercise, ask the teams to map out what they consider the 10 most important pages for the website and rank accordingly. For each page, solicit a few bullet points explaining why it is important. When these are complete, compare them with your list of the 10 most valuable pages that visitors use—the highest $Index pages. It is hoped that a strong overlap is apparent and you can move on to looking at your least valuable pages. Unfortunately, often there is not; in that case, view the high $Index pages in a browser as a group and try to come up with three reasons why each page is so valuable from a visitor's perspective. Use the Navigational Summary report to assist in this.

The important lesson from this exercise is understanding why visitors value pages that you as a team did not consider valuable. Your next meeting should discuss how can you improve, i.e., increase the value of, those pages the team thought were valuable but visitors did not? Alternatively, is it better to focus resources on the high $Index pages that were missed by your team?

The process just described in an excellent way to get your teams thinking about the value of a page in relation to the end goals for your website, rather than as a page in isolation, which is often the case. There is no such thing as a perfect page that satisfies all visitor needs; there are always areas for improvement. If your team suggests improvements that are not obviously beneficial—for example, "Let's try the sign-up process in Flash,"—consider testing first (see "An Introduction to Website Optimizer," later in the chapter). I recommend that this entire exercise is conducted quarterly.

Having looked at your most valuable pages, it is straightforward to view your least valuable ones. From the Content > Top Content report, reverse the $Index sort order by clicking again on the column heading.

Improving your least valuable pages is an obvious ambition. First check the difference in average $Index values for your least valuable pages compared to your most valuable. Maybe there is little difference, in which case all your pages are valuable!

Tip: As a rule of thumb, in order for a difference in $Index to be significant, consider it being greater than the sum of your average goal value and average transaction value.

With your list of least valuable pages, conduct another meeting with your design or agency teams to discuss improvements. View each page in a browser as a group and consider the page from the visitor's perspective. That is, how is it related to the goals you wish them to complete? It may be that the information contained on those pages isn't relevant and can therefore be removed from your website or combined with another more valuable page. Pruning poor-performing pages in this way helps maintain focus on the remaining website pages.

Note one thing to avoid: Do not combine the assessment of high $Index pages and low $Index pages into one meeting. Although the objectives are the same (page improvement), I have found that mixing these page types into one meeting confuses the issue.

Using the Top Landing Pages Report

As the name suggests, the Content > Top Landing Pages report shows the most popular entrance pages for your visitors (see Figure 11.3).

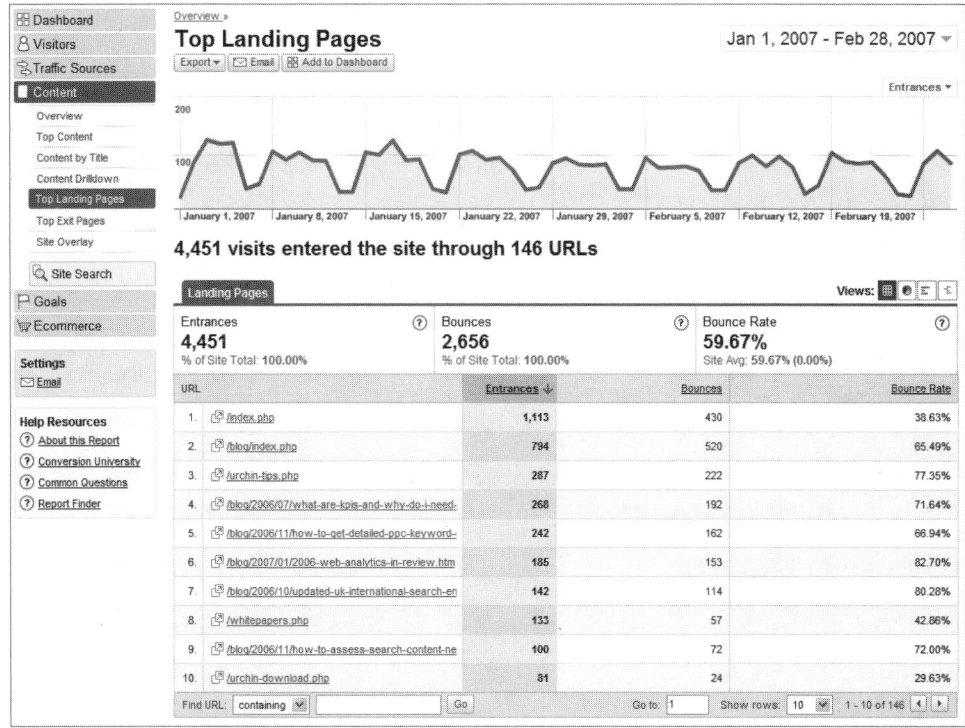

Figure 11.3 Top Landing Pages report

For this report, the bounce rate is the key metric; if visitors are arriving at the landing page and then leaving the site after viewing that one page, that is generally considered poor engagement. If a landing page has a high bounce rate, then it means that the content of that page did not meet the visitor's expectations. Hence, you need to look at what the visitor's expectations were, which means looking at the referral details.

Note: The assumption here is that you are not attempting to write the perfect blog or news article, for which it might be expected that visitors would be happy reading just a single page.

For each page URL with a high bounce rate (as a rule of thumb I define "high" as a bounce rate of greater than 50 percent), click on its link in the report. This takes you to the same Content Detail report shown in Figure 11.2a.

For assessing bounce rates, the Navigational Analysis report of Figure 11.2b is not required, as the entry point and exit point are the same page for those visits that bounce. Similarly, click patterns are not relevant for bounce visits. The key reports to view are therefore within the Landing Page Optimization section—namely, entrance sources and entrance keywords—as these refer to your visitor's expectations before arriving on your website.

Assessing Entrance Sources

As the term suggests, *entrance sources* are the referring websites that lead visitors to your site—for example, search engines, paid advertising, and e-mail links. An example report is shown in Figure 11.4.

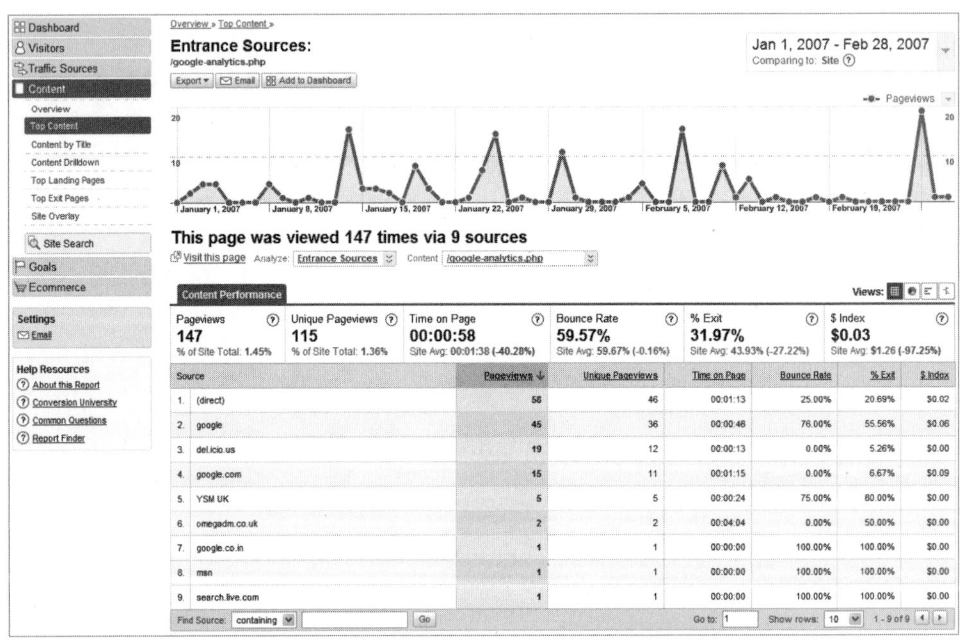

Figure 11.4 Entrance Sources report

Discuss this report with your marketing team by considering the following three perspectives:

- Offline marketing initiatives
- Paid search campaigns
- Search engine optimization (SEO)

In the report shown in Figure 11.4, the source labeled "direct" could be the result of offline marketing efforts whereby people have seen your ad and remembered your web address. If you observe a high bounce rate from this source, then look at how you are targeting visitors by offline methods. A common mistake is to send visitors for a specific campaign to a generic home page, leading to poor traction with the visitor. Later in this chapter I discuss how to overcome this (see "Tracking Offline Marketing").

Note: The label "direct" will also be applied to visitors who bookmark your website (add to favorites) and any non-web referral link that has not been tagged correctly, such as e-mail links and embedded links within digital collateral. To ensure that these are tracked, refer to "Online Campaign Tracking," in Chapter 7.

From Figure 11.4, identify any paid search campaigns. Row 5, "YSM UK," is one such example from the Yahoo! Search Marketing pay-per-click network. Pay-per-click advertising is an excellent way to target search engine visitors with a specific message (ad creative) and specific content (landing page URL). Any high bounce rates observed from these sources should be investigated immediately, as they reflect poor targeting or a misaligned message. For example, a common mistake is using time- or price-sensitive information in your ad creatives that is outdated when the visitor clicks through. Another area to look at is the geo-targeting of visitors; for example, do your pricing and delivery options match the expectations of visitors from different locations?

From a SEO perspective, think in terms of the visitor experience, as ultimately this is what search engines are trying to emulate with their ranking algorithms. For example, do your page title tag and description metatag match the rest of your page content? Do you have pages on your site that are irrelevant to searchers and should therefore not be listed on the search engines? This could be, for example, your privacy policy or your mission statement to be carbon neutral this year. Do your pages have sufficient content written in good grammar and without spelling mistakes? These commonly overlooked errors often ruin the user experience.

Also consider link referrals from other websites. A visitor that follows a link from another web site that turns out to be out of context is obviously a poor experience and waste of time for the visitor (it can also have a negative impact on your SEO rankings). If you find referral links with high bounce rates, use the Traffic Sources > Referring Sites report to investigate further. From here you can identify the referring site and view the exact page that visitors clicked through, to arrive on your website (see Figure 11.5). Sometimes a simple, polite e-mail to the webmaster of the referring site can pay you dividends. Specify that you want to ensure that links are in context and point to a relevant, specific landing page on your website. Provide any necessary details in your e-mail.

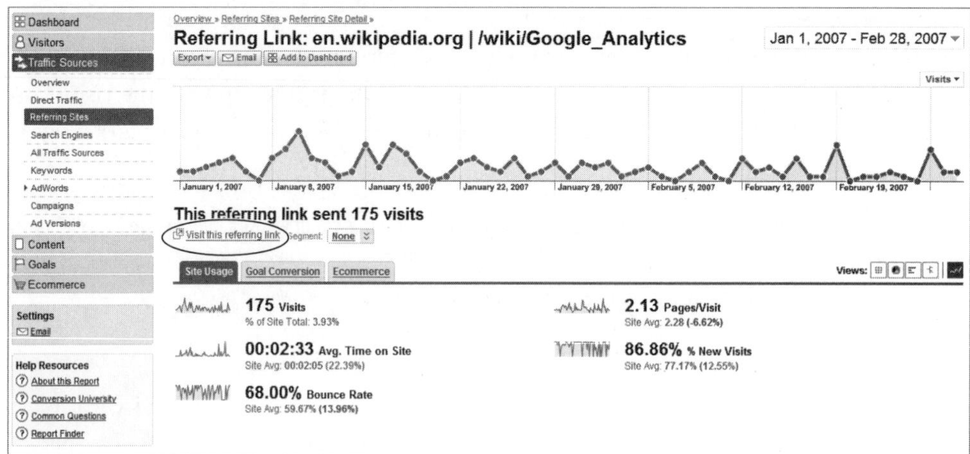

Figure 11.5 Entrance Sources report

Assessing Entrance Keywords

The Entrance Keywords report focuses on those visitors who have used search engines to arrive on your website—both paid and non-paid (organic) search engines. In effect, this report is direct market research—visitors are informing you exactly what content they expect to see on the page they arrive at on your site, as shown in Figure 11.6.

As with the Entrance Sources report, high bounce rates here (greater than 50 percent) is an indicator that something may be amiss with your online marketing. Assuming your web server performance is not an issue, look at your visitor targeting, message alignment, and page relevancy as described in the previous section.

Following this, consider the Entrance Keywords report as an opportunity to build page content around the listed keywords. For example, in Figure 11.6, row 3 shows a search term of "google analytics tutorial," yet the site in question, which contains a great deal of Google Analytics information, does not have such a tutorial. Perhaps they should, and, more to the point, perhaps they should monetize this process.

Using Funnel Visualization

As shown in Chapter 5, funnel analysis is an important process that helps you recognize barriers to conversion on your website, including the checkout process. I have often seen how understanding the visitor's journey within a website, followed by subsequent changes to improve the process, can lead to dramatic improvements in conversion rates and therefore the bottom line. In the following example, I saw a tenfold increase!

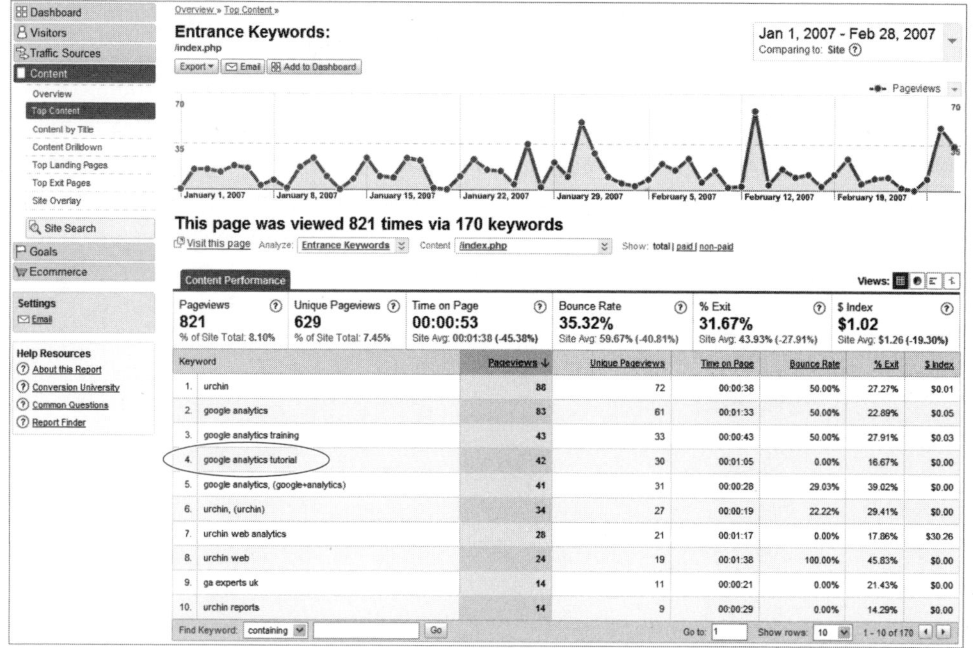

Figure 11.6 An Entrance Keywords report

Note: According to data shown in the 2007 Online Retail Checkout report from e-consultancy (www.e-consultancy.com/publications/online-retail-checkout-2007), the average abandonment rate for visitors that enter a shopping cart is around 60 percent. Of this, 12 percent are abandoned before the final checkout, that is, during the funnel process. This leaves 48 percent as the average checkout abandonment rate. In order words, the transaction revenue obtained by site owners is just under half of the revenue that customers are in the process of spending.

An ideal funnel process would schematically look like Figure 11.7, whereby there is a gradual decrease in visitors (width of funnel) due to self qualification pageviews (height of funnel) by, for example, price, feature list, delivery location, stock availability, and so on.

Figure 11.7
An ideal schematic wine goblet funnel shape

Figure 11.8 shows a real-world schematic example of a poorly performing checkout process for a travel website. Please note, I am quite biased when it comes to travel websites. On the whole, they tend to be poorly built from a user's viewpoint. They are pretty, with a lot of colorful images and inspiring photographs, but I never seem to have a good experience when it comes to actually booking my travel plans, let alone a great one. (An exception is www.sterling.dk, a Danish airline with a class-leading travel website from a user experience point of view.) However, as a wise person once said to me, "Your biggest problem is also your greatest opportunity."

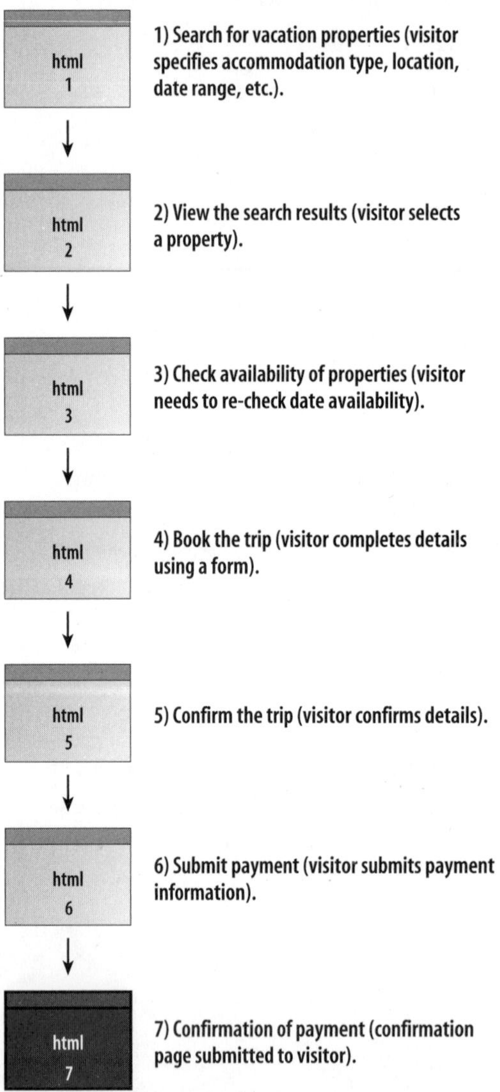

Figure 11.8 Schematic funnel process for a travel website

Figure 11.9 is the same funnel process as reported by Google Analytics.

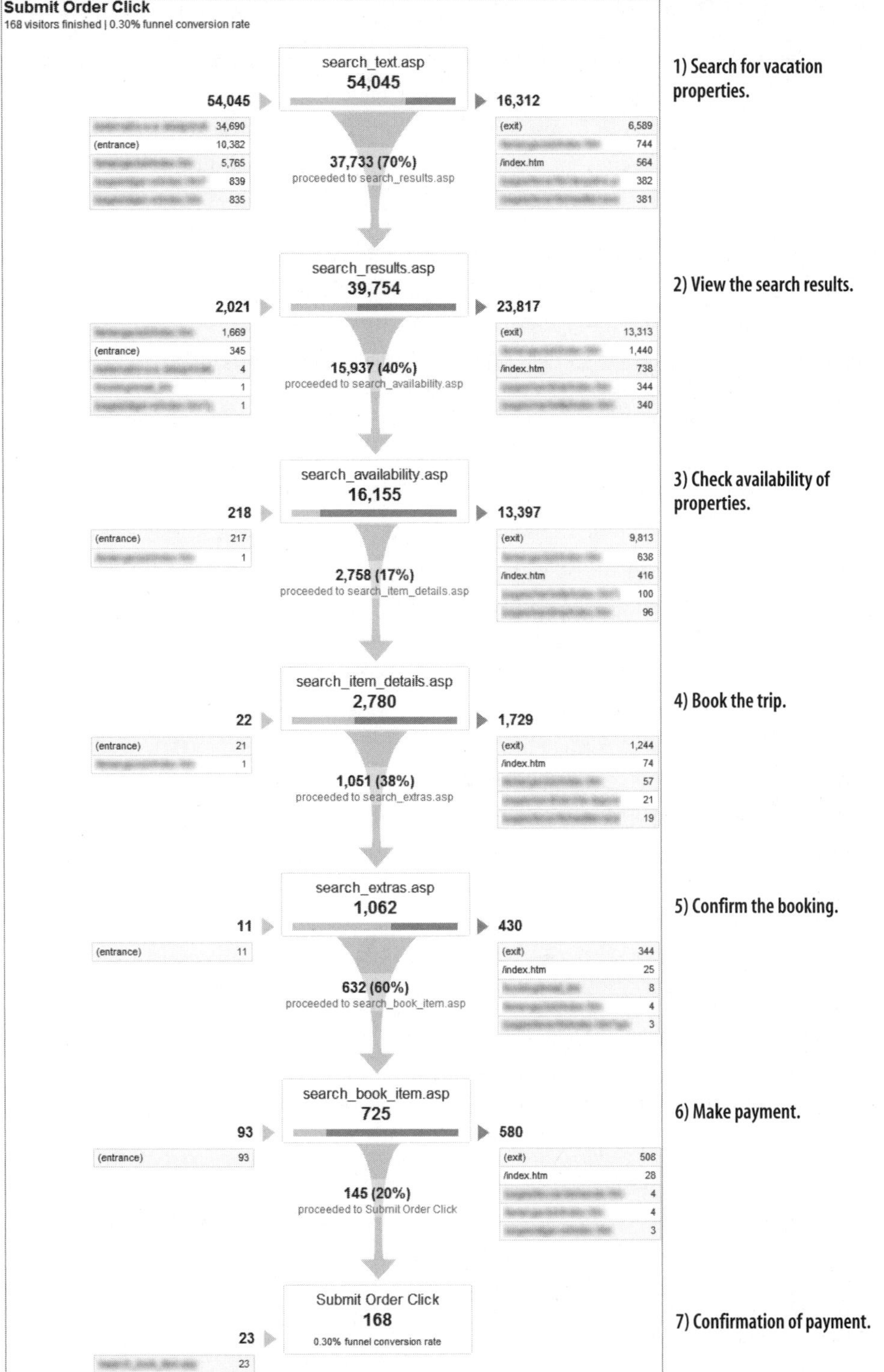

Figure 11.9 Funnel Visualization report (page names obfuscated for anonymity)

Issues with the Funnel Presented

Issue 1 The most obvious metric that stands out in Figure 11.9 is the end conversion rate—a woefully poor 0.30 percent. Put another way, 99.70 percent of all visitors abandon the booking process. Considering the cost of acquiring those visitors by both paid and non-paid search, that means a very, very negative return on investment.

Note: Although this funnel example is an extreme case, it never ceases to amaze me that online purchase rates can be so low and are accepted as such. For example, a July 2007 Forrester Research report showed U.S. retail websites convert an average of 2-3 percent of their site visitors into buyers. Surely we can do better than have 97-plus percent of visitors leave a website without conversion? I hope that having read this far, you will agree that it is laudable and entirely possible to improve this significantly.

Issue 2 Looking at the entire booking process, the length of the funnel, at seven steps, appears overly long. From user-experience experiments, it is widely known that users do not like long checkout processes. That's obvious to anyone who uses the Web! The most effective method to reduce cart abandonment is to streamline the number of steps in the process, and this can be applied here. For example, step 5 (Confirm Booking) is superfluous because all booking details are displayed at each preceding step.

Issue 3 The first step in the process begins with the search_text.asp page. This is the page where visitors search for their holiday property (hotel, villa, apartment). From this page, 30 percent drop out of the funnel.

Issue 4 Following step 1, the search results page (step 2) loses 60 percent of remaining visitors; over half of these (13,313) exit the site completely. This is clearly a pain point and should be red-flagged as a problem page.

Issue 5 Looking at the Check Availability page (step 3), 83 percent of remaining visitors drop out of the funnel; again, the vast majority are site exits (60 percent).

Issue 6 The next steps in the system have similar problems, but the killer is step 6, which is when payment details from the visitor are requested. Out of the 725 visitors that have had the stamina and persistence to get through what is obviously a difficult process, 80 percent of them (580) abandon at this last step; the vast majority leave the website completely.

Representing the result of these issues schematically, rather than the ideal funnel shape of Figure 11.7, this website owner has a funnel shape more like what is shown in Figure 11.10, with two clear pain points in the process that lead to large-scale abandonment.

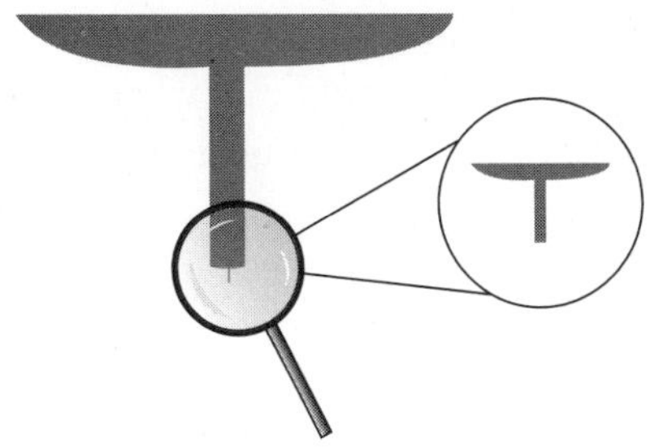

Figure 11.10
Stacked champagne glass schematic funnel shape

Action Points from the Funnel Visualization

Understanding the real-world funnel process of Figure 11.9 and its problems took me less than one hour because the data is so clearly presented. Of course, correcting such issues obviously takes longer; you need to understand why this happened. This is something that web analytics tools cannot do; they cannot tell you why visitors are abandoning your booking process.

To address this, you could deploy a survey system—a survey that pops up when a visitor abandons the booking process or leaves your website. However, if your visitors are leaving because of a bad experience, they usually won't want to spend further time on your site explaining what went wrong. Often, though, a little lateral thought on your part and visiting your own website as if you were a potential customer can go a long way. For example, in this scenario I focused on steps 3 and 6, where the vast majority of visitors were abandoning the booking process. That led to the development of several solutions:

1. Improve the availability checker page.

   Step 3 (the availability checker) indicates either a total lack of accommodation availability, in which case the website owners should turn down the visitor acquisition "tap" and save marketing budget, or a malfunction in the process of selecting available dates.

   Lack of availability was not an issue. By checking the website manually, no errors were found with the availability checker, but the process was quite clunky and difficult to interpret. That is, dates were nonclickable with selection controls located below the fold of the page—that is, not visible without scrolling.

2. Correct the layout of the payment form.

   Step 6 (the payment form) required some additional thought. Although the form was considered to be overly long, it did not make sense that such persistent visitors would bail out at the penultimate step (visitors were aware of their progress by the numbering of the steps – for example, with the heading "Step X of Y"). To test for problems, I tried the process of booking a holiday.

What I immediately discovered when clicking to submit my dummy payment details was an error page. In addition, the error page did not indicate what caused the error. Using the back button, I checked all the required fields and tried again—same error page, no message indicating what the error was. This process was repeated many times with no further insight. It really did appear to be a mystery as to why I could not complete my payment.

In fact, the problem was staring me in the face. The credit card type (Amex, Visa, MasterCard) was pre-selected as Amex by default. However, this HTML drop down menu was not aligned with the other form fields—it was to the extreme right of the page when everything else was left justified; I simply didn't see it. I was filling in all my details correctly and hadn't noticed the default setting for the credit card type. In fact, I hadn't noticed the card type drop down at all.

Determining credit card type

The initial digits of a card number can be used to identify the card type, as shown in the following table:

Card Types	Prefix	Width
American Express	34, 37	15
Diners Club	300 to 305, 36	14
Carte Blanche	38	14
Discover	6011	16
EnRoute	2014, 2149	15
JCB	3	16
JCB	2131, 1800	15
Master Card	51 to 55	16
Visa	4	13, 16

A form selector may offer the benefit of redundancy to check for user error, but skipping this step can streamline the payment process.

Now the explanation of step 6 abandonment is clear. Visitors were receiving the error page, which was probably the straw that broke the camel's back after such a difficult and tortuous booking process, and so they abandoned the site.

3. Track all error pages to understand what your visitors are experiencing.

Part of the difficulty in identifying the problem visitors were experiencing in step 6 was due to the fact that the subsequent error page was not being tracked. Had it been, using the methods described in Chapter 9, the investigation could have taken place much more quickly.

4. Show clear instructions in your error pages.

   Even if an investigation into the low conversion rate had not been undertaken, visitors could have corrected the payment problem themselves—that is, if they were told what the problem was. Clearly, this is not a solution to the problem, but it is certainly better than slamming the door in their face with nothing more informative than "Error—please try again."

Summary of Funnel Visualization

Presenting these findings to the client was groundbreaking. They had hired and fired several search engine marketing agencies in the belief that they were receiving poorly qualified leads, resulting in such a low (0.3 percent) conversion rate. In fact, the problem was entirely on their site—that is, a poor user experience.

Funnel analysis shows both the power and the weakness of web analytics as a technique for understanding visitor behavior on your website. The power is in identifying the problem areas during a typical path visitors take; for that, your web analytics is capable of telling you what happened and when. That in turn enables you to focus your efforts at improving the particular problem page. The weakness of web analytics is that it does not tell you why visitors made the choices they did. To understand why visitors behave in a non-anticipated way, you need to investigate—either directly yourself (try a checkout or booking on your own website) or by conducting a survey or usability experiment.

Tip: If visitor usability is a new term for you, check out the excellent book by Steve Krug, *Don't Make Me Think*, as a background read before contacting a specialist agency.

Measuring the Success of Site Search

Site search is the internal search engine of your website that visitors often substitute for a menu navigation system. For large websites with hundreds or thousands of content pages (sometimes hundreds of thousands), internal search is a critical component for website visitors to enable them to find what they are looking for quickly. Internal search engines generally use the same architecture as an external search engine, such as Google. In fact, companies such as Google and Yahoo! sell their search solutions to organizations so they can provide their own site search engine.

Important site search KPIs were discussed in the section "Webmaster KPI Examples," in Chapter 10. In addition to the Site Search Overview report (refer to Figure 10.32), one of the things you will want to know is what keywords visitors are typing once they arrive on your website. The idea is that once you know these keywords, you include them (or exclude as negative keywords if they are not relevant to you) in your paid and organic campaigns, as well as ensure that landing pages are optimized for them. This is discussed in the next section, "Optimizing Your Search Engine Marketing." An example Site Search Terms report is shown in Figure 11.11.

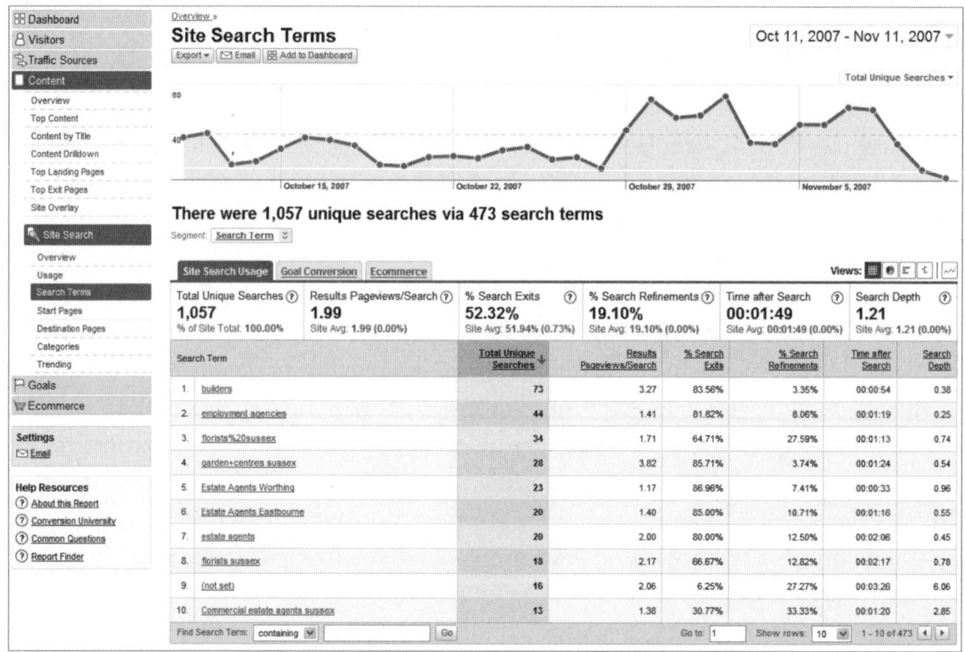

	Search Term	Total Unique Searches	Results Pageviews/Search	% Search Exits	% Search Refinements	Time after Search	Search Depth
1.	builders	73	3.27	83.56%	3.35%	00:00:54	0.38
2.	employment agencies	44	1.41	81.82%	8.06%	00:01:19	0.25
3.	florists%20sussex	34	1.71	64.71%	27.59%	00:01:13	0.74
4.	garden+centres sussex	28	3.82	85.71%	3.74%	00:01:24	0.54
5.	Estate Agents Worthing	23	1.17	86.96%	7.41%	00:00:33	0.96
6.	Estate Agents Eastbourne	20	1.40	85.00%	10.71%	00:01:16	0.55
7.	estate agents	20	2.00	80.00%	12.50%	00:02:06	0.45
8.	florists sussex	18	2.17	66.67%	12.82%	00:02:17	0.78
9.	(not set)	16	2.06	6.25%	27.27%	00:03:26	6.06
10.	Commercial estate agents sussex	13	1.38	30.77%	33.33%	00:01:20	2.85

Figure 11.11 Site Search Terms report showing keywords used

Note: The value of the Site Search Terms report shown in Figure 11.11 should not be underestimated. Visitors are actually telling you what they would like to see on your website, in their own language, using their own terminology. Perhaps you assumed "widgets" was the commonly known name of your product but you find out that people are searching for "gadgets," or people are looking for "widgets with feature X," which your manufacturing team hadn't thought of. Site search is the most direct visitor feedback you can obtain without infringing on visitor privacy.

In addition to viewing what search terms are used on your website, you should track how these convert by viewing the Goal Conversion and Ecommerce tabs. A useful metric for this is the Per Search Value, as shown in Figure 11.12.

This is similar in principal to $Index, described in Chapter 5. $Index measures the value of a page according to whether that page is used by visitors who go on to complete monetized goals or e-commerce transactions. Analogous to this, the Per Search

Value is a measure of the value of a site search term. That is, did visitors who used a particular site search term go on to complete a transaction or monetized goal? The higher the Per Search Value, the greater value that term is to the success of your website. Therefore, make use of the Per Search Value when prioritizing which search terms to overlap with your website marketing.

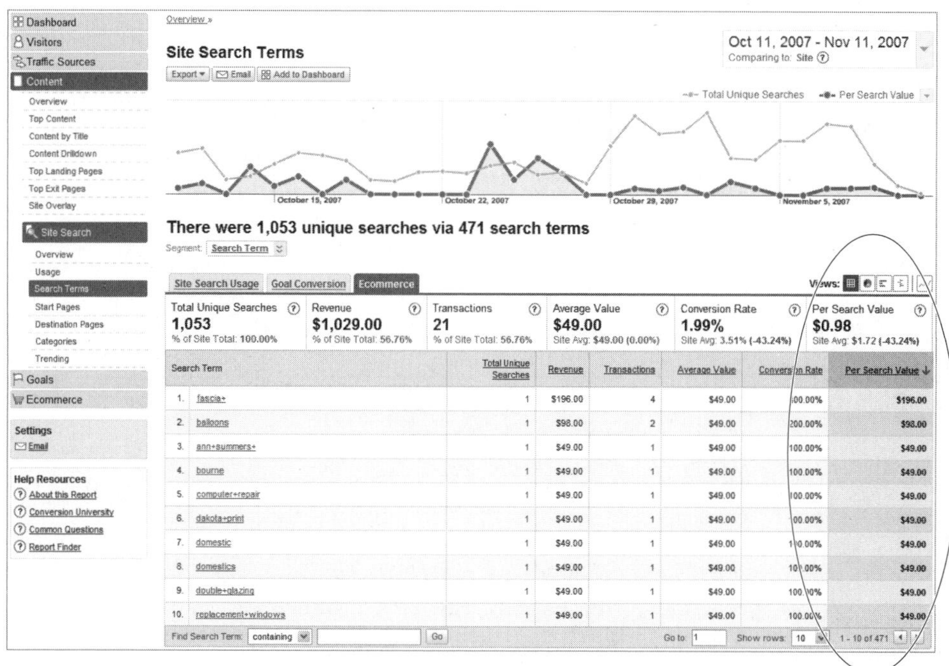

Figure 11.12 The value of site search terms

Beyond looking at site search terms used, how do visitors who use your site search facility compare to those who do not? I illustrate this with a series of screenshots taken from the Content > Site Search > Usage reports.

From Figure 11.13, you can see that the percentage of visits resulting in a visitor using the site search facility is the same as those who did not (50.7 percent and 49.27 percent respectively). However, the bounce rate for those who used site search is much lower (37.67 percent), hence a better user experience is inferred for those visitors.

Note: Having a bounce rate reported for site search visitors may sound contradictory. How can a visitor who conducts a search bounce if they landed on a page, conducted a site search, and viewed the results—that is, viewed a minimum of two pages? For this example website, site search result pages are also directly listed on referrer sites and in search engine result pages such as Google. Therefore, the landing page is a site search result in itself.

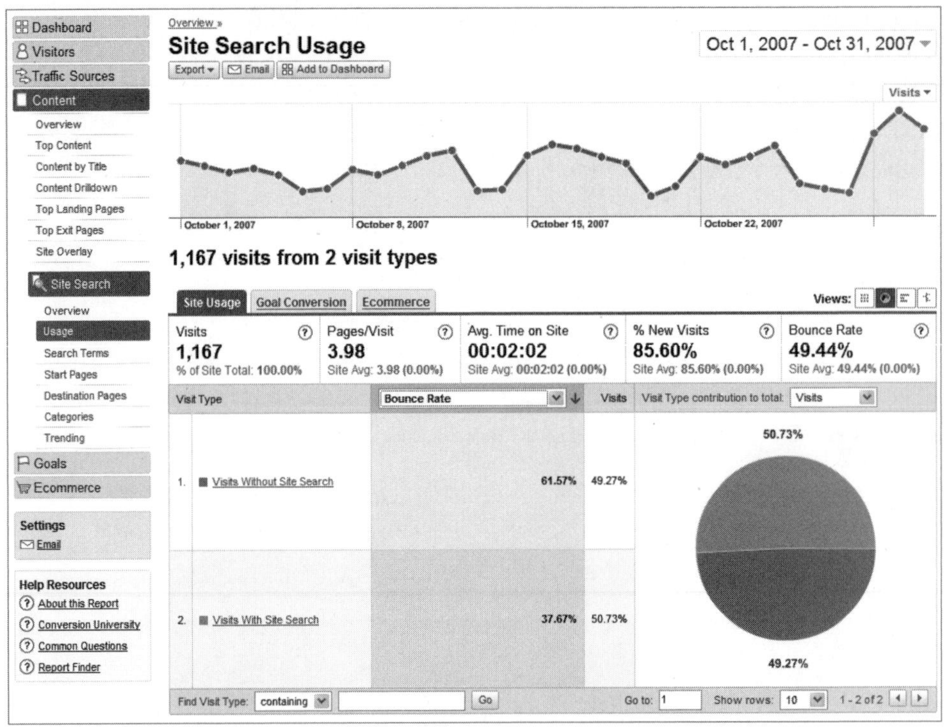

Figure 11.13 Bounce rate comparison of visitors who use site search and visitors who do not

Select other key metrics from the drop-down list highlighted in Figure 11.14. The ones I focus on in addition to bounce rates are as follows:

- Goal conversion rate

$$\text{Goal Conversion Rate} = \frac{\text{Number of Conversions}}{\text{Number of Visits}} \times 100$$

- Revenue

$$\text{Revenue} = \text{Goal Value} + \text{E-commerce Value}$$

- Average value

$$\text{Average Value} = \frac{\text{Goal Value} + \text{E-commerce Value}}{\text{Number of Conversions} + \text{Number of Transactions}}$$

- E-commerce conversion rate

$$\text{E-commerce Conversion Rate} = \frac{\text{Number of Transactions}}{\text{Number of Visits}} \times 100$$

- Per Visit Value

$$\text{Per Visit Value} = \frac{\text{Goal Value} + \text{E-commerce Value}}{\text{Number of Visits}}$$

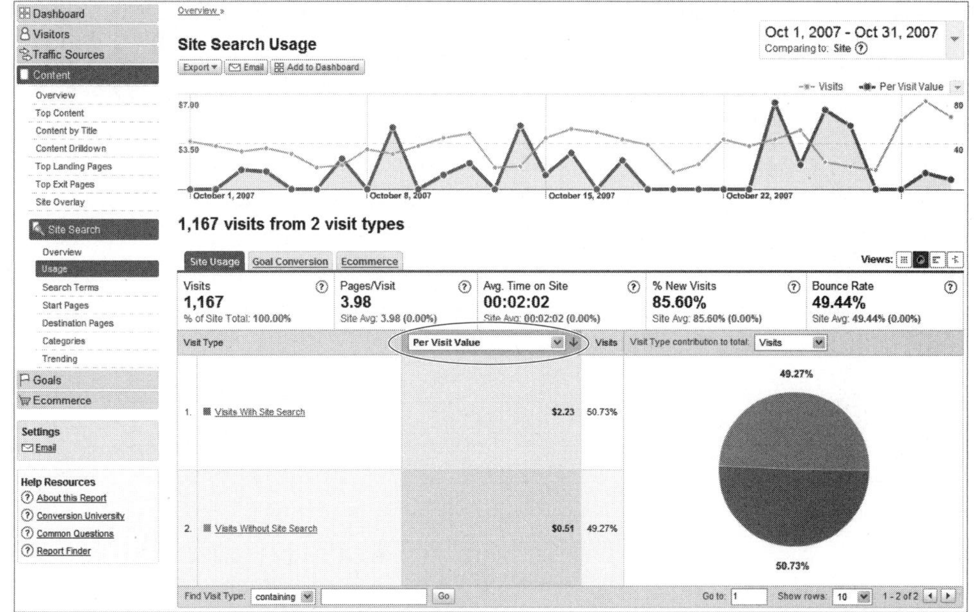

Figure 11.14 The Per Visit Value difference from using site search

A particular favorite of mine is the Per Visit Value. As shown in Figure 11.14, a visitor who uses site search is more than four times as valuable as a visitor who does not. Armed with that information, meet with your web development team (those responsible for your internal site search engine) and discuss with them what plans they have for developing and growing the site search service. Before doing so, use the following formula to calculate the revenue impact that site search is having on your website:

$$\text{Revenue Impact of Site Search} = \left(\text{Per Visit Value with Site Search} - \text{Per Visit Value without Site Search}\right) \times \text{Number of Visits with Site Search}$$

From Figures 11.14 and 10.32, for the example website shown:

$$\text{Revenue Impact of Site Search} = (2.23 - 0.51) \times 759$$

$$= \$1305.48 \text{ per month}$$

This metric puts you in a great position to help your development team budget for further investment in site search. To put this value into context, this represents over 80 percent of the monthly revenue for this website.

What if the metrics are reversed—that is, visits that used site search have a lower Per Visit Value than those that don't? This would result in a negative revenue impact of site search—that is, its use is costing you money. It is possible that such a result could be valid. For example, in some scenarios, finding information can best be served by a directory type structure of navigation, rather than a search engine. Yahoo! built its business on this principal. Generic keywords, that is those with multiple contexts (for example, golf car versus golf club), and location-specific keywords are a few examples where navigating a directory structure may serve the visitor better than using your site search. However, a negative revenue impact of site search usually indicates an issue with the quality of the results returned to the visitor.

So far, we have assumed that your internal site search engine is working well—producing accurate and informative results regarding visitors' searches. To get a handle on whether this is valid, look at the zero results produced by your site search engine, as discussed in the section "Webmaster KPI Examples," in Chapter 10. The Top Content report is shown here again in Figure 11.15. The zero result pages are denoted by the virtual path, of the form "/zero/*keyword_used*."

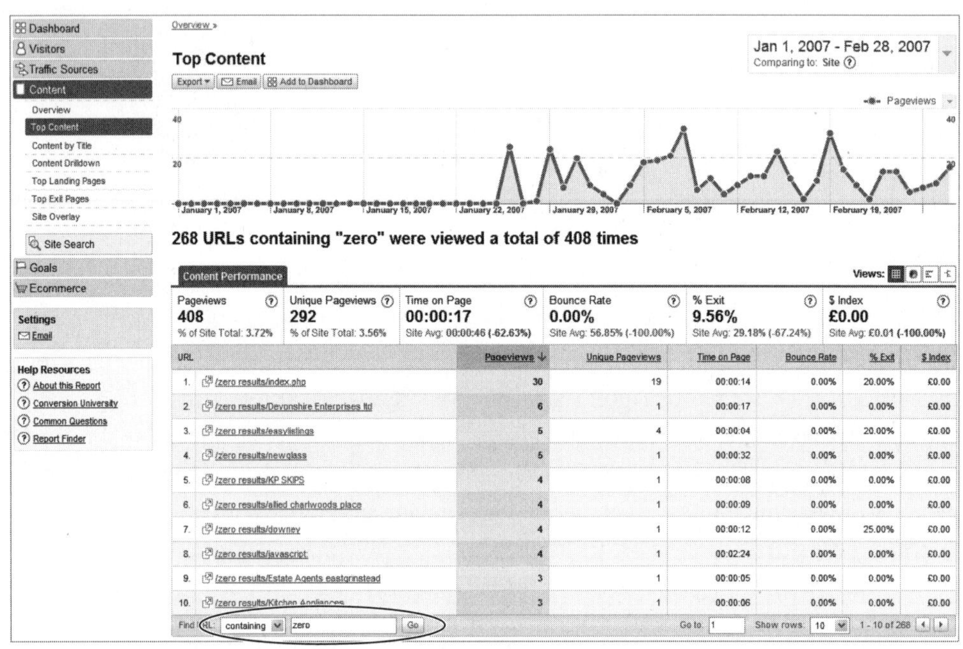

Figure 11.15 Site search zero results and keywords

Export this list into Excel and highlight the ones that are directly related your website content. Meet with your web development team to ascertain why such relevant terms produce zero results. Maybe you have overlooked misspellings, regional differences (think "holiday" versus "vacation"), or visitors using terminology they are not familiar with that needs to be considered. However, it may be that there is a problem with how your site search engine works or is configured. Is it picking up newly created or modified pages? Can it index PDF files? How is it ranking results?

Site search engines are often configured once and then forgotten about in this respect. That's a mistake. Websites evolve rapidly, including new content and new technologies. If site search users have a lower revenue impact without good reason, then it is costing you money. Present this figure to the head of your web team and schedule a meeting to discuss improvements. Showing a dollar amount is a much better motivator than saying, "Our site search is not working as effectively as our navigation system."

With your export list of zero result site search terms, highlight the keywords visitors used that are not relevant to your organization but are related to the business sector you are in. If the number of these is significant (more than a few percentage points of the total number of unique searches), then meet with your product or service team to discuss their meaning. Perhaps the product team never thought people would want to search for feature *X* combined with product *Y*. Your site search data could provide valuable insight into this. For example, an action item may be to build a specific landing page for product *XY* to gain further feedback from those visitors.

Optimizing Your Search Engine Marketing

If you own a commercial website, then you want to drive as much qualified traffic to it as possible. Online marketing options include search engine optimization (non-paid search, also known as organic search), paid search advertising (also referred to as pay-per-click or cost-per-click), e-mail marketing, banner displays, and social networks (comments and links left on myspace.com, forums, blogs, etc.).

All of these visitor acquisition methods have a cost—either direct, with the media owner or indirect, in management fees—though there is nothing stopping you as a do-it-yourself enthusiast. Optimizing your marketing campaigns using Google Analytics data can achieve cost savings and expose significant opportunities for your business. This section focuses on the essential steps for optimizing your search engine marketing (SEM), both paid and non-paid, including the following:

- Keyword discovery (paid and non-paid search)
- Campaign optimization (paid search)
- Landing page optimization and SEO (paid and non-paid search)
- AdWords ad positioning optimization (paid search)
- AdWords day parting optimization (paid search)
- AdWords ad version optimization (paid search)

Keyword Discovery

When optimizing for SEM, one of the things you will constantly be on the lookout for is ideas for adding new, relevant keywords to your campaigns. These can be broad (for example, "shoes"), bringing in low-qualified visitors in the hope they will bookmark your page or remember your brand and website address for later use, or very specific (for example "blue suede shoes"), which are highly targeted to one of your products and could lead to an immediate conversion on a visitor's first visit.

Several tools are on the market, both free and commercial, to help you conduct keyword research—for example, the Google AdWords Keyword Tool (https://adwords.google.com/select/KeywordToolExternal), the Yahoo! Keyword Assistant Tool (http://searchmarketing.yahoo.com/en_GB/rc/srch/intro101.php), and Microsoft's Adlab (http://adlab.microsoft.com). These enable you to discover what people are searching for that may be related to your products or services and in what numbers. The tools help you determine which search keywords are most frequently used by search engine visitors and then help you identify related keywords, synonyms, and misspellings that could also be useful to your marketing campaigns. Clearly, being language- and region-specific is important; for example, "tap" and "holiday" are terms used in the U.K. that in the US are more commonly known as "faucet" and "vacation," respectively.

In addition to such "offsite" tools, your Google Analytics reports contain a wealth of information that can help you hunt for additional suitable keywords. There are two areas to look at: search terms used by visitors to find your website from organic search engines, and internal site search queries—that is, those used by visitors after they are on your website.

Farming from Organic Visitors

The Traffic Sources > Keywords report is dedicated to referral keywords—keywords used by visitors that come from all search engines, both paid and non-paid search (see Figure 11.16). As an initial exercise, click the "non-paid" link from the Show menu, and then export all of your non-paid keywords. Compare these with those targeted by your paid campaigns. Organic terms that are not in your paid campaigns are excellent candidates to be added to your pay-per-click account. After all, you will wish to maximize your exposure to relevant search terms.

Note that when adding keywords used by organic search visitors to your pay-per-click campaigns, you should consider your current organic search rankings for those terms. For example, if you are number one for your brand or product name in the organic results, should you also add this to your paid campaigns? If you do, you are likely to cannibalize your own free organic traffic. On the other hand, you would also remove a competitor from the search engine paid results; that is, you would occupy more "shelf space" on the results page. Generally, I advise not adding organic keywords to your pay-per-click campaigns when you already occupy any of the top three organic positions for that particular search engine.

From the screen shown in Figure 11.16, click the search term itself for further details. Then cross-segment by source to display which search engines referred that keyword. An alternative way to see which search engine is associated with which keywords is to view the Traffic Sources > Search Engines report. Clicking the search engine listed will show the keywords used by your visitors on that search engine.

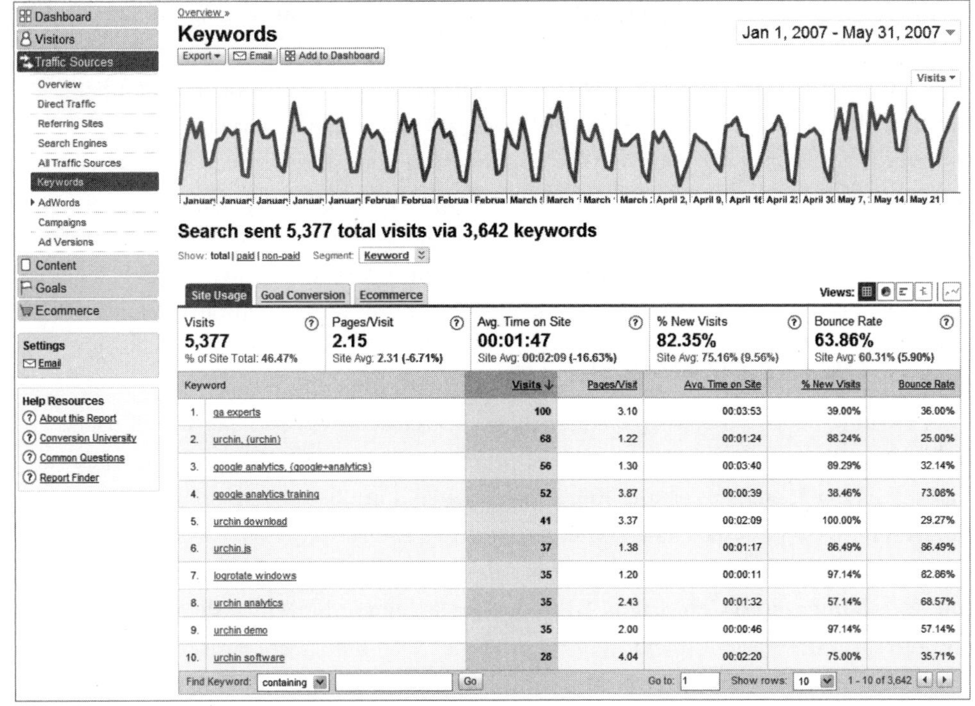

Figure 11.16 Keyword research

Farming from Site Search Visitors

If your site has an internal search engine to help visitors find what they are looking for, then this is an excellent feedback mechanism for your marketing department—that is, visitors telling you exactly what they want to see on your website. We looked at measuring the success of site search in the preceding section and also in Chapter 10, in "Webmasters KPI Examples."

From within your Google Analytics account, export your site search keywords and compare them with those in your paid search accounts (pay-per-click). Site search keywords not in your pay-per-click accounts are strong candidates to be added. You can prioritize these by considering not only their prevalence in site usage reports, but also whether they produce goal conversions and e-commerce transactions. The use of the Per Visit Value for this was discussed earlier in the section "Measuring the Success of Site Search."

When selecting new keywords from your Site Search reports, if you have a good landing page ranked in one of the top three organic search engine positions for a particular search engine, I suggest that you do not add that term to your paid campaigns for that search engine. As mentioned in the previous section, this is likely to cannibalize your own free organic traffic.

In addition to comparing keywords from site search with your paid campaigns, you should also compare them with your non-paid search terms. Perhaps there are variations in usage or spelling you can take account of in your page content. Perhaps visitors are using relevant keywords after they are on your site that you are not aware of. For example, visitors looking for books may also use keywords such as "how-to guides," "manuals," "whitepapers," and "tech sheets" on your internal site search. This is a perfect opportunity to build and optimize your website content for those additional, related terms.

Campaign Optimization (Paid Search)

After farming for new keywords from organic search referrers and site search users, and adding these to your paid campaigns (if applicable) and to the content of relevant pages, the next stage is to ensure that these keywords are optimized—that is, that they give you the best possible chance of conversion.

Within the Traffic Sources report is a dedicated section for AdWords. This enables you to drill down into Campaign, Ad Group, and Keyword level for details of conversion rates, return on investment, margin, and more. As a business entity, you want to invest more in campaigns that produce more conversions and leads for you than in those that merely create visibility for your brand. However, care must be taken here because Google Analytics gives credit for a conversion to the last referrer. In other words, spending more on campaigns that are reported as generating conversions and culling those that don't may result in you chopping off the head that feeds the tail.

What about other PPC networks?

Currently, within Google Analytics you can track visitors from any search engine, indeed any referral—not only can you track which search engine visitors came from but also their paths and conversion rates, down to campaign and keyword level. However, at present, cost data can only be imported from AdWords. That is, ROI and margin data can only be calculated for AdWords visitors.

For those keywords that convert, you should optimize your investment by setting the maximum cost-per-click (cpc) you can afford within AdWords. The caveat is that your return on investment should be positive—that is, revenue generated should be greater than your costs. Google Analytics has no idea what margins you operate under, so you need to factor that in as well. The following is an example:

$$\text{ROI} = \frac{\text{Revenue} - \text{Cost}}{\text{Cost}} \times 100\%$$

If your ROI for a keyword is 500 percent, this means you are receiving a $5 return for every $1 spent on AdWords. Assuming revenue of $600 from $100 spent, this is calculated as follows:

$$\begin{aligned} ROI &= \frac{600 - 100}{100} \times 100\% \\ &= 500\% \end{aligned}$$

You need to take into account your operating profit margin here to get the true figure, so if your operating profit margin (excluding marketing costs) is 50 percent, your true gross profit ROI is 200 percent:

$$ROI_{gross\ profit} = \frac{Revenue \times \frac{Profit\ Margin}{100} - Cost}{Cost} \times 100\%$$

This means you can spend 200 percent more money (three times as much) on this keyword in AdWords without producing a negative ROI. Clearly, you do not want to reach this maximum; otherwise, what's the point of being in business? Therefore, select a level that you are comfortable with—that is, one that drives more traffic to your website while still being profitable.

Note: At the beginning of a campaign launch, your ROI may be negative as you build up brand awareness and visibility for your website. Visitors to a new website (new to them) usually require multiple visits before they convert. However, a negative ROI should only be acceptable for a short period of time—on the order of weeks, depending on your circumstances.

To calculate the maximum amount to spend on customer acquisition—the maximum cost per acquisition (cpa_{max})—use the following procedure:

$$cpa_{max} = \frac{Revenue \times \frac{Profit\ Margin}{100}}{\frac{ROI_{gross\ profit}}{100} + 1}$$

For this calculation, I have assumed a target $ROI_{gross\ profit}$ of 150 percent and a profit margin of 33 percent. The revenue used is taken as the Average Value (average order value) from Figure 11.17.

$$cpa_{max} = \frac{\$49.00 \times 0.33}{1.5 + 1}$$

$$cpa_{max} = \$6.47$$

This is the total cost you are willing to pay in order to achieve an average order of $49.00.

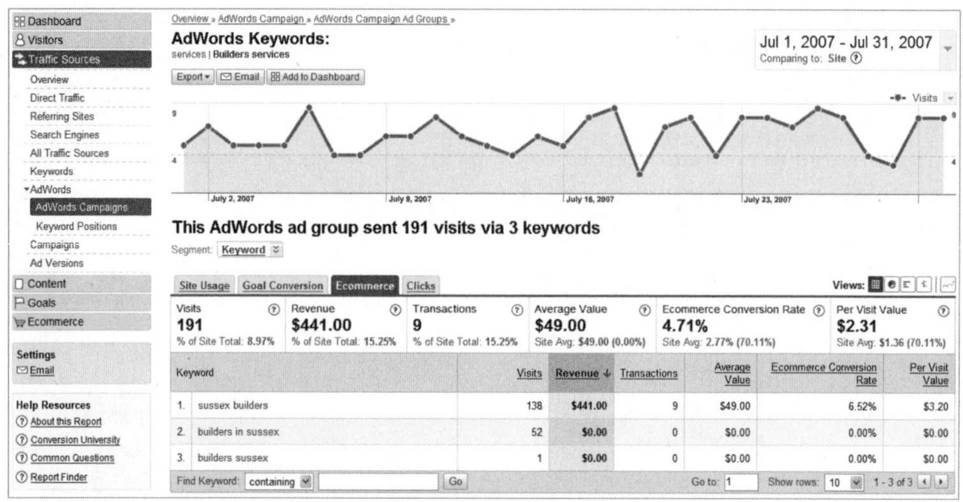

Figure 11.17 AdWords Campaigns report

Knowing your conversion rate for each keyword, you can now calculate your maximum cost per click (cpc_{max}) allowed for that keyword. The conversion rate for purchases is taken from the Google Analytics reports for the keyword "sussex builders" (see Figure 11.17).

$$cpc_{max} = cpa_{max} \times (\text{Ecommerce Conversion Rate} / 100)$$

$$cpc_{max} = \$6.47 \times (6.52 / 100)$$

$$cpc_{max} = \$0.42$$

For this example keyword, "sussex builders," you could bid up to $0.42 in AdWords to generate as much traffic as possible and be assured that you will make a gross profit of $1.50 for every $1 spent. Of course, the actual bid you pay in AdWords is determined by the market—that is, how many competitors are also bidding on the same keyword and how effective their ads are at gaining click-throughs. This is the basis of the AdWords Quality Score system.

The result of the described technique is that you will never overbid for your AdWords keywords. Even if you reach your cpc_{max} within your AdWords account, you will still maintain a 150 percent $ROI_{gross\ profit}$. In Google Analytics this would be reported as an ROI of 657 percent, as follows:

$$ROI = \frac{49 - 6.47}{6.47} \times 100\%$$

Simplifying the Task

The calculations of cpc_{max} appear cumbersome when written on paper, but with a spreadsheet they are actually quite simple, as shown in Figure 11.18. First, within the Traffic Sources > AdWords > AdWords Campaigns report, drill down to the keyword level by clicking through on the report table. Export your AdWords data from Google Analytics to a CSV file (or schedule a report e-mail on a regular basis). Note that when you do this, all AdWords data is exported together—that is, the data contained in all the tabs of Figure 11.17b (Site Usage, Goal Conversion, Ecommerce, Clicks) will be included in the CSV file. Open the file in Excel. From this spreadsheet, you only require three columns of data: the keyword, the average value, and the e-commerce conversion rate; the rest can be discarded. From the screen shown in Figure 11.18, inputting your profit margin (cell E2) and desired ROI (cell E3) will display the cpc_{max} (column F).

	A	B	C	D	E	F
1	# ------------------------------					
2	www.mysite.com			**Profit Margin %**	33	
3	AdWords Keywords:			**ROI (gross profit) %**	150	
4	July 1, 2007	July 31, 2007				
5	# ------------------------------					
6						
7	# ------------------------------					
8	# Table					
9	# ------------------------------					
10	**Keyword**	**Visits**	**Average Value ($)**	**Ecommerce Conversion Rate**	**CPAmax ($)**	**CPCmax ($)**
11	sussex builders	138	49	0.065217391	6.468	0.421826083
12	builders in sussex	52	55	0.01	7.26	0.0726
13	builders sussex	1	34	0.05	4.488	0.2244
14	# --					
15						

Figure 11.18 Excel spreadsheet to calculate per keyword cpc_{max}

Note: You can download this Excel template from www.advanced-web-metrics.com/scripts.

As you can see, the cpc_{max} calculation is at the keyword level throughout—for example, the average ROI, average value, and e-commerce conversion rate. It is more likely, however, that you will be bidding a single cpc amount for groups of keywords—that is, ad groups. In that case, the more focused your ad groups are, the more accurate the cpc_{max} calculation will be. Consider the following ad group examples:

Single AdWords Ad group for a mix of generic and specific keywords:

Ad	Keywords targeted:	Bid terms:
Blue suede shoes Go out on the town in style. High quality at online prices. mysite.com/shoes	(general shoes)	shoes, fun shoes, blue suede shoes, turquoise suede shoes, fancy dress shoes, stylish suede, shoes, stylish shoes

Clearly, the average ROI, average value, and average conversion rate for this group will have a large variance, as a broad spread of keywords are targeted. To provide better targeting and receive improved metrics, this should be split into two focused ad groups—for example, a specific shoes type ad group (suede shoes) and a less specific ad group (general shoes):

Two AdWords Ad groups for generic keywords and more specific keywords:

Ad	Keywords targeted:	Bid terms:
Blue suede shoes Go out on the town in style. High quality at online prices. mysite.com/shoes	(suede shoes)	blue suede shoes, turquoise suede shoes, stylish suede shoes
Stylish suede shoes Go out on the town in style. High quality at online prices. mysite.com/shoes	Keywords targeted: (general shoes)	Bid terms: shoes, fun shoes, fancy dress shoes, stylish shoes

By dividing your AdWords keyword list into more focused keyword themes, you will get a much better handle on which keywords and campaigns are working for you, and therefore better metrics (conversion rate, average value) to optimize your cpc_{max} values. Of course, your landing pages for each ad group should also be optimized for those keywords, and that is discussed next.

Landing Page Optimization and SEO (Paid and Non-paid Search)

A landing page is the page your visitors land on (arrive at) when they click through from a search engine results page. Landing pages should be focused around the keywords your visitors have used—that is, keywords relevant to what they are looking for—and be as close to the conversion point as possible. That way, you give yourself the best possible chance of converting your visitors into customers.

For paid search, controlling which landing page a visitor arrives at is straightforward: you enter the URL in your pay-per-click campaigns. For example, in AdWords, each ad group should have its own unique landing page relevant to the displayed advertisement. For all paid search campaigns, you need to append tracking parameters to your URLs. This is done automatically for you in AdWords, but you must apply this manually for other paid networks (see "Online Campaign Tracking" in Chapter 7).

For non-paid search, controlling landing pages is much harder to achieve because the search engines consider all pages on your website when deciding which are most relevant to a visitor's search query. If you describe a product on multiple pages, then any or all of these may appear in the search engine results. However, the highest ranked page may not be your best converting page. By optimizing the content of your best converting page, you can influence its position within the search engine results, thereby gaining a higher position than other related pages from your site. Landing page optimization is therefore a subset of search engine optimization (SEO).

Robots.txt

The file robots.txt can be used to stop search engines from indexing certain pages on your website. For example, create a text file in the root of your web space named robots.txt with the following contents:

```
User-agent: *
Disallow: /offer_codeX
Disallow: /offer_codeY
```

This file tells all search engines that obey the robots exclusion standard to not index the files specified. For more information on the robots exclusion standard, see www.robotstxt.org.

Principles of SEO and Landing Page Optimization

For both paid and non-paid search visitors, you want to ensure that the landing page is as effective as possible—that is, optimized for conversion—once a visitor arrives. That does not mean the visitor's next step is necessarily to convert from this initial landing page; the landing page could be the beginning of the relationship, with the conversion happening much later, or on a subsequent visit, for example.

By optimizing the content of your landing pages for a better user experience, you not only increase conversions for all visitor types, but also improve your organic search engine rankings. Often the effects of this optimization process can be dramatic.

A key part of the optimizing process is understanding why visitors landed on a particular page of your website in the first place. The keywords they used on the referring search engine will tell you this. Within Google Analytics you can view keywords for your top landing pages in one of several ways:

- From the Traffic Sources > Keywords report, click a keyword, and then cross-segment by landing page.
- From the Content > Top Landing Pages report, click a landing page and select Entrance Keywords from the Landing Page Optimization section.

Generally I prefer the latter—that is, focusing on a landing page and viewing which search keywords led visitors to it. This method is referrer agnostic, meaning you cannot tell whether your visitors arrived on a particular landing page by clicking on an organic listing or on a paid ad. This difference is not important; a visitor arriving on your website by a well-targeted link (paid or non-paid) should be just as likely to convert regardless of the referrer used.

For the optimal user experience, focus your landing pages around a particular keyword theme, such as a specific product or service. The exception to this would be your home page, which shouldn't be used as a landing page except for your company or brand-name keywords.

Note: Your home page is generally poor as a landing page for anything other than your company name. This is simply because by its very nature your home page is a generalist page that focuses on creating the right image, branding, and mission statements. Usually you will notice low conversion rates, low $Index values, and high bounce rates for this page, which is expected. Therefore focus your efforts on your content pages.

A *keyword theme* is a term used in search engine marketing to describe a collection of keywords that accurately describe the content of a page. For example, if you sell classic model cars, keyword themes would center on particular makes and models, such as the following:

"classic alpha romeo model car"

"replica model alpha romeo"

"classic alpha romeo toy car"

Less product-specific pages—for example, a category page—would use a less specific keyword theme:

"model cars for purchase"

"classic toy cars for sale"

"scale model cars to buy"

As a rule of thumb, themes generally consist of 5–10 phrases per page that overlap in keywords (the preceding examples list three such phrases for each page). More than 10 overlapping phrases dilutes the impact and effectiveness of the page, from the perspective of both the user experience and search engine ranking. If you already have a page that targets more than 10 keyword phrases, consider creating a separate page to cater to the additional keywords.

As an initial exercise, view your top 10 landing pages from your Content > Top Landing Pages report. For each page listed in the report, click through on its link to its Content Detail report and then again on its Entrance Keywords report. Print out the top 10 entrance keywords and repeat this process for each of your landing pages. Then visit your website and print out the content for each of your top 10 landing pages.

Note: If your Entrance Keywords report for each landing page contains hundreds of table items, it may be because it is a poorly focused or targeted page. Check the landing page URLs specified in your paid campaigns. Are they pointing to the most appropriate page? If not, change them accordingly.

For each landing page URL, view the two corresponding printouts. Is the page content tightly focused around its listed Entrance Keywords report? This is quite a subjective process, though as a guide if you read the first three paragraphs (or approximately the first 200 words) of your landing page and you don't come across every one of your top 10 entrance keywords, then the page can be said to be unfocused. The extent of this is relative to the percentage of missing entrance keywords from those first paragraphs.

If you determine that a landing page is unfocused, revise its content, ensuring that all 10 of your top target keywords are placed within the first 200 human-readable words (that is, not HTML code). Pay particular attention to placing keywords in your paragraph headings—for example, assuming a target keyword of "blue widget," you could use a heading of `<h1>Our blue widget selection</h1>`

Text versus images

Machine-readable text is text that can be selected within your browser, and copied and pasted into another document or other application such as Word or TextPad. If you cannot do that, then the text is likely to be a rastered image (GIF, JPG, PNG, etc.) or another embedded format such as Flash. Both methods can convey an effective message to a visitor, and in some cases brand managers prefer the image format when referring to a product or company name. However, it is doubtful this has any impact on conversions, so long as these are referenced as images or logos elsewhere in the page design. In addition, it's common for headings to be provided as images so that nonstandard fonts and smoothing or special effects can be used.

For SEO rankings, however, machine-readable text is king. The inappropriate use of images or other embedded content as headings will be detrimental to your SEO efforts. Search engines ignore images for ranking purposes and embedded objects such as Flash can only be partially indexed. Although it is a good practice to include the alt tag (alternative text attribute) for each image to improve the display and usefulness of your document for people who use text-only browsers or have reading disabilities, it has very little positive impact on search engine ranking. Therefore, where possible, use HTML and CSS to style your text, rather than images or Flash.

Other prominent areas that are not visible on the page where you should place your target keywords include the title tag and description metatag. Using the same keyword examples, these could be written as follows:

```
<title>Purchase blue widgets from ACME Corp</title>
<meta name="description" content=" ACME Corp, the blue widgets division of
BigCorp, is a US sales and support channel for the industry leading blue
widget package." />
```

Page title tags are visible by reading the text in the title bar at the top of your browser (blue in Windows, silver on a Mac), but visitors generally do not read this on your page, as it is located separately from your content, above the browser menu and navigation buttons. However, the title tag is the same text that is listed as the clickable link on search engine result pages and is very important for SEO ranking purposes.

A best practice tip is to also include your target keywords within call to action statements and make these a hyperlink to the beginning of a goal process—the Add to Cart page, for example. This is illustrated with the following text examples:

Example 1 To purchase and get a free gift click here.

Example 2 Purchase blue widgets and get a free gift with your first order.

Example 2 contains three important elements that have proven to be many times more effective than example 1 (see "An Introduction to Website Optimizer," later in the chapter, for ways to test this hypothesis):

- The call to action statement contains the target keywords.
- The call to action keywords are highlighted as a hyperlink.
- The hyperlink takes the visitor to the start of the goal conversion process.

The techniques described here for optimizing and focusing your landing pages will undoubtedly increase your conversion rates and decrease page bounce rates regardless of visitor referral source. In addition, as a consequence of improving the user experience, such changes also have a significant and positive impact on your search engine rankings. Therefore, once you have optimized the top 10 landing pages, move on to the next 10.

From a paid search point of view, you need to ensure that campaigns point to one of these optimized landing pages—or create new ones. The worst possible thing you can do is use your home page as the landing page. If you take away only one lesson from this section, let it be this: avoid this mistake!

A note on SEO ethics

When optimizing your landing pages to place keyword phrases in more prominent positions, always consider the user experience. Overly repeating keywords or attempting to hide them (using CSS or matching against the background color, for example), though not illegal, will inevitably result in your entire website being penalized in ranking and possibly removed from search engine indexes altogether—and this can happen at any time without warning, even years later.

Although it is possible to get back into the search engine indexes once you have removed the offending code, this can be a long, drawn-out process that damages your reputation. Essentially, spamming the search engines is not going to win you any friends, either from your visitors or the search engines themselves, so it's best to avoid it.

Optimizing landing pages for better performance is a complicated business; indeed, it's a specialized branch of marketing. However, here is a 10-point summary for you to follow that will give you a solid start:

1. Always put your visitors and customers first; design for them.
2. Use dedicated landing pages for your campaigns, for both paid and non-paid visitors.
3. Ensure that landing pages are close to the call to action.
4. Structure your website content around keyword themes of 5–10 overlapping keywords and phrases.
5. Place your keyword-rich content near the top of the page. Think like a journalist writing for a newspaper, with structured titles, headings, and subheadings that contain keywords.
6. Use keywords in your HTML <title> tags.
7. Use keywords in your anchor links—that is, HTML <a> tags.
8. Avoid placing text in images or Flash or other embedded content.
9. Use a robots.txt file to control what pages are indexed by search engines.
10. Never "keyword stuff" or attempt to spam the search engines; it's not worth it and you can achieve better results by legitimate means.

Note: It's important to recognize that I have only attempted to cover the principals of SEO. Many factors affect your search engine rankings. The more important ones are page content (keyword density, keyword prominence), site architecture, internal link structure, and the number and quality of incoming links from other websites. For further in-depth reading on the subject, see *Search Engine Marketing Inc.*, by Bill Hunt and Mike Moran (IBM, 2005).

AdWords Ad Position Optimization

As shown in Chapter 5, Google Analytics contains an AdWords report called Keyword Positions (see Figure 11.19). This report provides metrics for your AdWords keyword performance on a per-position basis—in other words, the number of visits you received while in ad position 1, 2, 3, and so on. In fact, it's not just visitor numbers you can view on a per-position basis; 15 metrics are available. These include the following:

- **Visits**
- Pages/Visit
- Average Time on Site
- **% New Visits**
- Bounce Rate
- Goal 1 Conversion Rate [conversion rates for goals 2–4]
- Overall Goal Conversion Rate
- [Revenue, Transactions, Average Value, Ecommerce Conversion rate, **Per Visit Value**]
- **Per Visit Goal Value**

Values in square brackets [ ] may not be present in the report, as they depend on your specific configuration—for example, whether e-commerce is enabled or you have goals defined.

The three metrics I suggest you focus on for ad position optimization are highlighted in bold: visits, percent new visits, and either the per visit goal value or per visit value, if you have an e-commerce website. That is not to say that the others are not useful, but in an analyst's world of information overflow, I like to keep things as simple as possible, and these are my favorites.

For example, I deliberately avoid pages per visit and average time on site, as these metrics can be misleading without further detailed investigation. On the one hand, a high value for these metrics could indicate a visitor is engaging with you. On the other hand, it could mean that visitors are lost in your navigation or confused by your content. Bounce rates and individual goal conversion rates are best viewed in other reports such as the Traffic Sources > AdWords > AdWords Campaigns report.

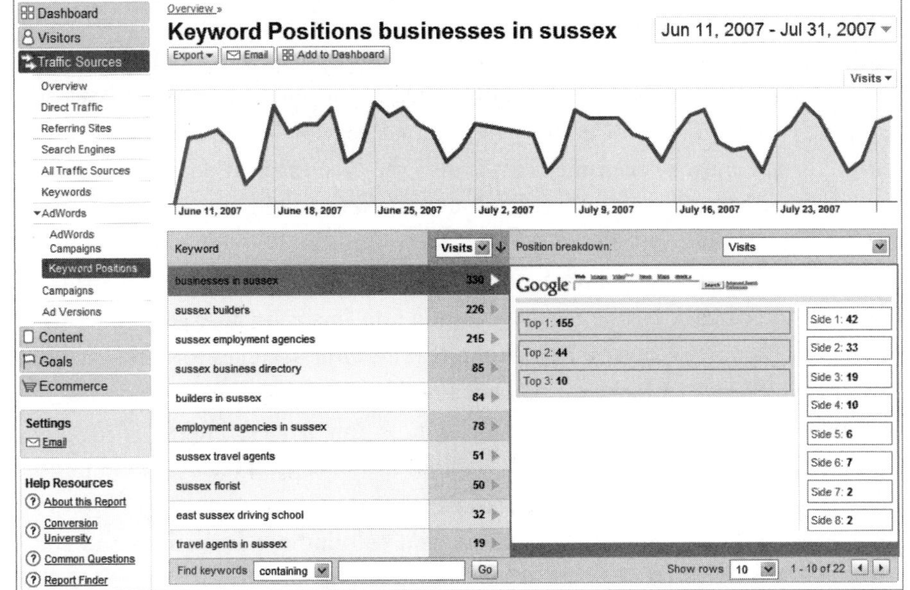

Figure 11.19 AdWords Keyword Position report

For the remaining metrics, I use the per visit value (or per visit goal value if you do not have e-commerce reporting) as an excellent proxy for revenue, transactions, average value, and conversion rate metrics. If the per visit value is healthy, then so are all the others.

With an understanding of how your ads perform by position, you can set position preference within your AdWords account, as shown in Figure 11.20 and discussed next.

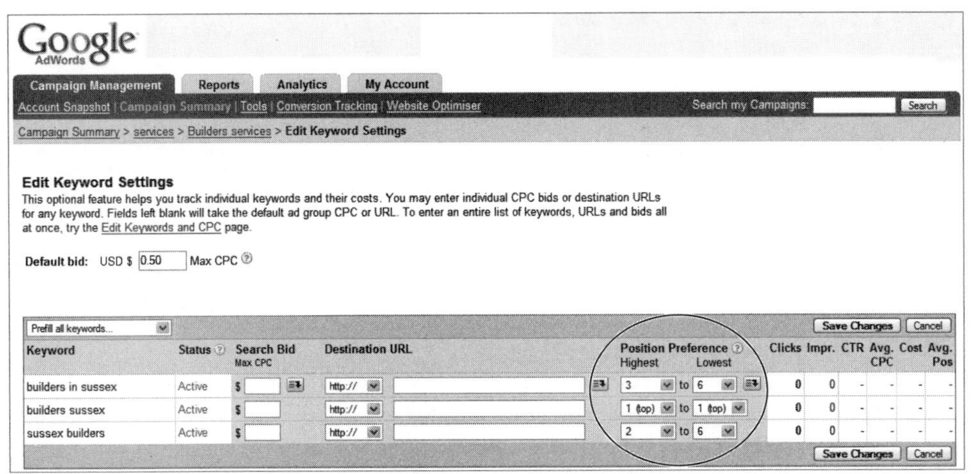

Figure 11.20 Setting your position preference in AdWords

Optimizing Positions by Visits

Acquiring the most traffic for the least cost is an obvious ambition of all marketers. A common misconception with pay-per-click advertising is that the higher the ad position, the more traffic you will receive. Certainly for focused keyword phrases with no ambiguity, that holds true most of the time. This is also the case for bids on brand terms, with the caveat that it is not necessary to bid on brand terms (and not advisable) if you already have a high organic placement—that is, a top-three organic ranking.

For more generic terms, visitors tend not to follow this pattern of behavior; that is, most visitors do not click on the highest position ad. This is because generic terms can have different visitor intentions. For example, if a visitor searches for "blue suede shoes," are they interested in footwear or Elvis Presley? Similarly, searches for "golf driving" could be looking for a car, golf equipment, or golf driving ranges.

Because of this ambiguity (use of less-focused keywords), advertisers tend to use broad match in their ad campaigns in order to capture as many visitors as possible who *may* be interested in their product. For example, you will find car dealerships and golf equipment suppliers advertising alongside search results for "golf driving." The same is true for "blue suede shoes"—footwear suppliers advertising alongside music download sites. The result is a blurring of the click-through distribution by position (see Figure 11.21).

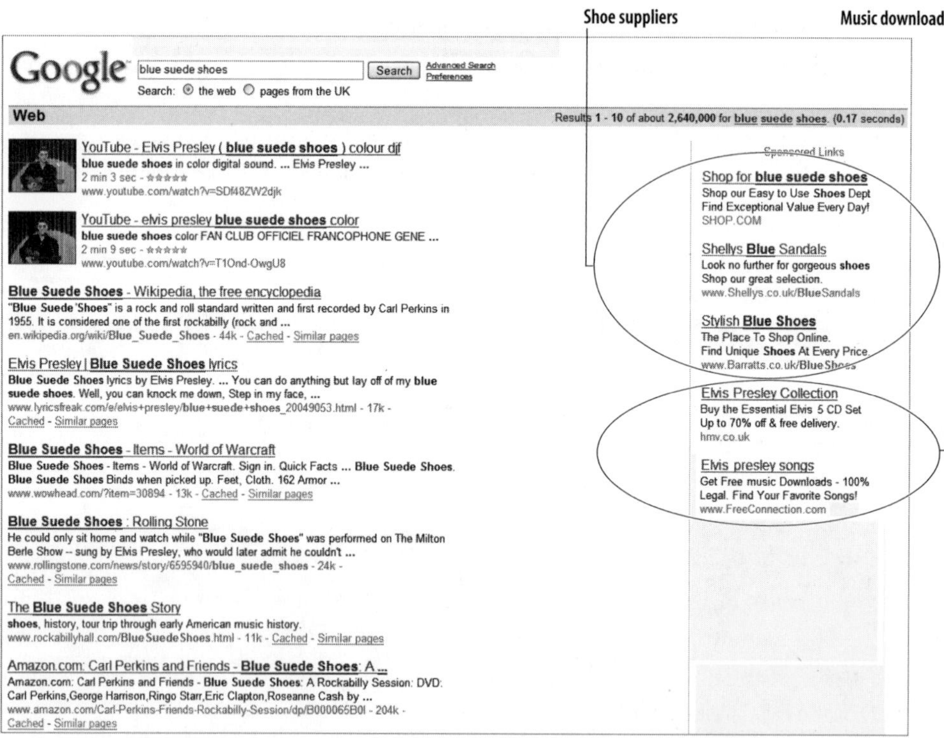

Figure 11.21 Ad performance (click-through rate) can vary significantly for ambiguous search terms

You can take advantage of this blurring by viewing your Google Analytics Keyword Position reports, and adjusting your bids within AdWords Position Preference to be placed in the most effective position for your target audience. For example, if you are a music retail site and the top three AdWords positions for the bid term "blue suede shoes" are for shoe suppliers, then you only need to bid to position four for your ad to be in the number one position for your sector. There is no point in being number one overall, as you will be paying a premium to be placed higher than non-relevant competitors.

To determine whether most people searching for "blue suede shoes" are referring to Elvis or footwear, you can use the data available at Google Trends: http://trends.google.com. As indicated in Figure 11.22, currently and historically there is a considerably greater volume of search queries for footwear than music downloads. For a music retailer, bidding to position three would avoid the expense of acquiring potentially irrelevant traffic from such an ambiguous search term.

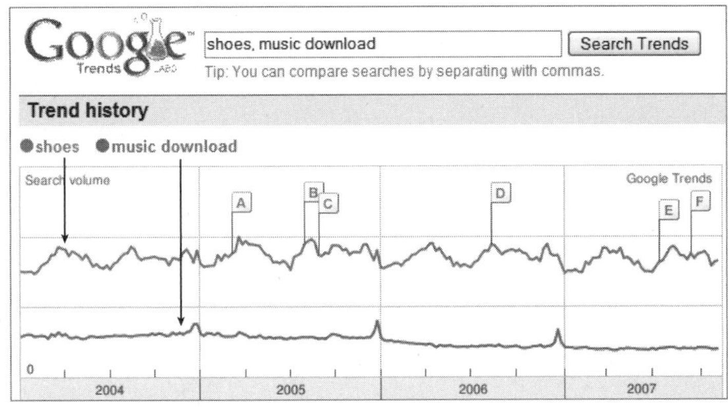

Figure 11.22
Google Trends data comparing query volumes on the Google search engine

Optimizing Positions by Percent New Visits

Percent new visits is an interesting metric to view by ad position. When running a paid campaign, you hope that the visitors you are acquiring are almost all new visitors—people coming to your site for the first time. If instead significant proportions are repeat visitors (I consider this as greater than 25–30 percent), then you need to look at your visitor acquisition strategy. If people are not ready to convert on their first visit, then you want them to either bookmark your website or at least remember your company or product name. That way, a follow-up search by visitors for your brand keywords should bring you to the top of the organic (free) results, saving you the cost of a repeat pay-per-click visitor.

Sometimes the top three ad positions on Google—those ad positions that occur at the top of a search results page (as apposed to the right-hand side)—can attract significant numbers of repeat visitors. This is probably because they are at the top of the search results page and in the direct line of sight for the searcher, just below the search box. Because of this, those positions can lead to significant click-throughs without the visitor bothering to view the rest of the results page—that is, without seeing your top organic position. If you find this is the case for your repeat visitors, consider not advertising in these positions using the Position Preference settings of your AdWords campaigns.

Optimizing Positions by Per Visit Value

Per visit value is probably the most important metric to be viewed on a per-position basis. The per visit value is calculated as follows:

$$\text{Per Visit Value} = \frac{\text{Goal Value} + \text{E-commerce Value}}{\text{Number of ad click-throughs while in Position X}}$$

This metric tells you how valuable an ad position is to your business on a per-visit basis, and it can vary wildly by position. For example, Figure 11.23 shows that the highest value positions for the selected keyword ("sussex builders") occur in positions 4–6. It therefore makes sense to focus the AdWords position preference for this keyword in those positions. However, the caveat here is that these positions will receive less traffic than positions 1–3, so compare this metric with the revenue by position metric (select from the Position breakdown drop-down menu) to see the total value of these positions to your business.

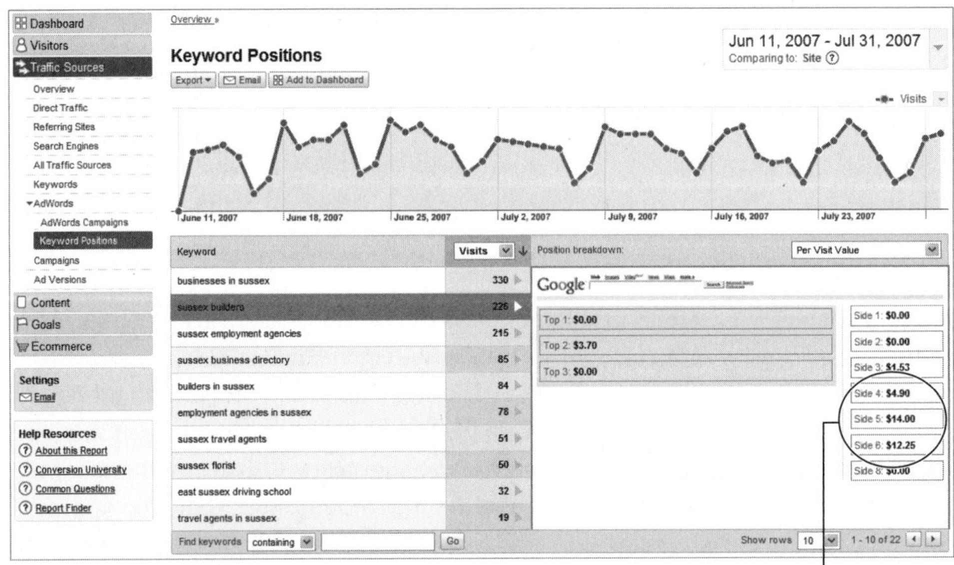

Figure 11.23 AdWords per visit value by position

AdWords Day Parting Optimization

By knowing at what time of day your website is being accessed by visitors, you can better tailor your advertising campaigns to match. For example, if you are a business-to-business website, then most of your visits will probably occur during normal working hours. Rather than display your ads in equal distribution throughout the day, it would make sense to run and maximize your pay-per-click campaigns at around the same time visitors will be looking on the Web. Of course, time zones should be taken into consideration.

Other examples include targeting a younger audience during after-school hours; targeting magazine readers who are likely to be online in the early evenings; targeting social networking sites whose potential audience is most likely to be online from 5:00 p.m. to 1:00 a.m.; and coinciding with radio advertisements, where remembering a website URL can be difficult and the interested audience may conduct a search to find your site based on the terms used in the radio ad.

By viewing hourly reports, you can view the distribution of your visitors throughout the day. Hourly visitor reports are available in the Visitors > Visitor Trending section (see Figure 11.24).

Generally, I advise against looking at short time frames such as a single day, as visitors over short periods can vary significantly and randomly, making reports difficult (if not impossible) to interpret. Instead, select a longer period and ensure that the date range includes relevant days of the week for you. For a business-to-business website, for example, select Monday to Friday, or use Friday to Sunday if your target audience is more likely to be looking for your products or services in their leisure time. In addition, try to choose a discrete day range—one that does not overlap with national holidays if that would affect your visitor numbers. Of course, whatever business you are in, compare different segments, such as weekend visitors to weekday visitors.

From Figure 11.24, which has had no day parting optimization, you can see that there are fewer visitors in the early morning (midnight until 7:00 a.m.), significant numbers from then on, with a clear peak in the afternoons (2:00–5:00 p.m.). If you have e-commerce reporting enabled, compare your day parting visitor information with when transactions take place: Go to the E-Commerce > Total Revenue report.

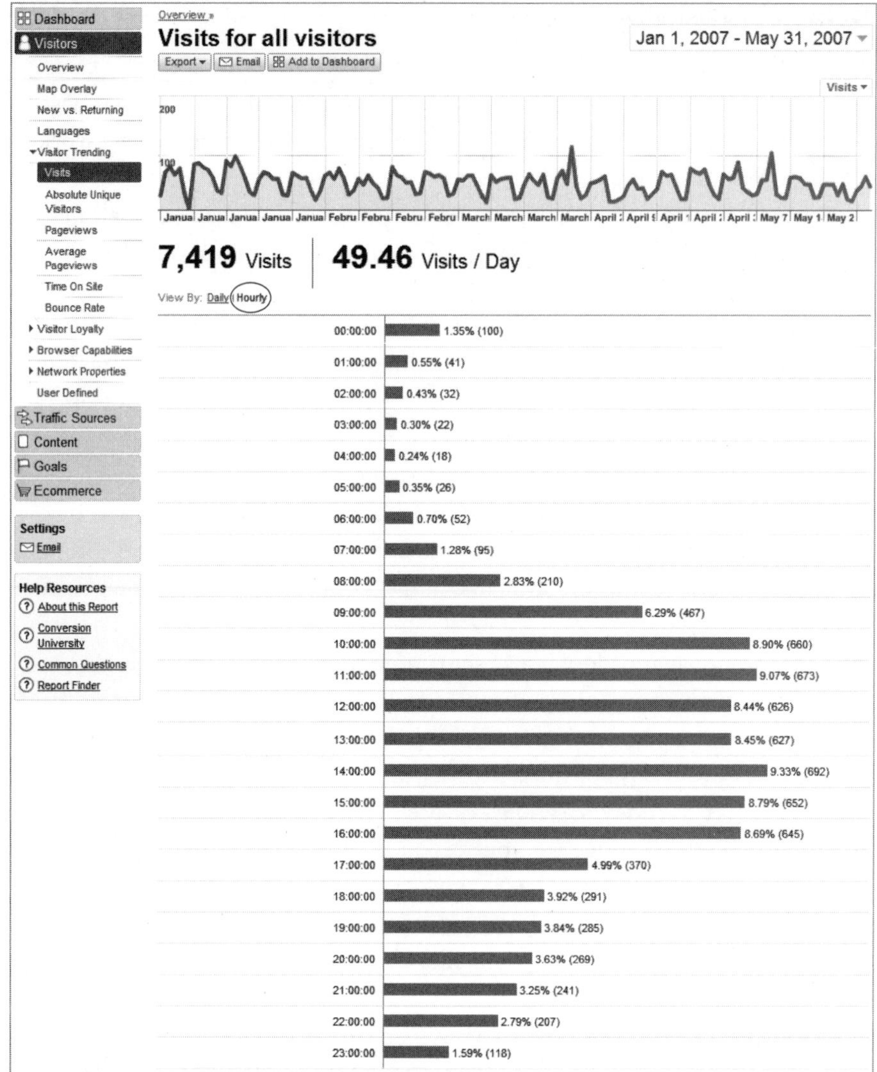

Figure 11.24 Viewing hourly reports

Use this information to optimize your paid campaigns by setting ads to display on or around these time frames, both when visitors are in a research frame of mind (just visiting) and when they are ready to purchase. Figure 11.25 shows you how to achieve this within the AdWords Ad Scheduling page. Not only can you schedule when your ads are displayed, you can also vary your bids for ads on a given time or day. For example, if your default bid is $1.00, you can set a custom percent of bid entry for Tuesdays as 120 percent—that is, your bid for Tuesdays only would be $1.20. By this method, you would be spending money on acquiring paid visitors at periods when they are most likely to be looking and purchasing, and at a price that is most advantageous to you.

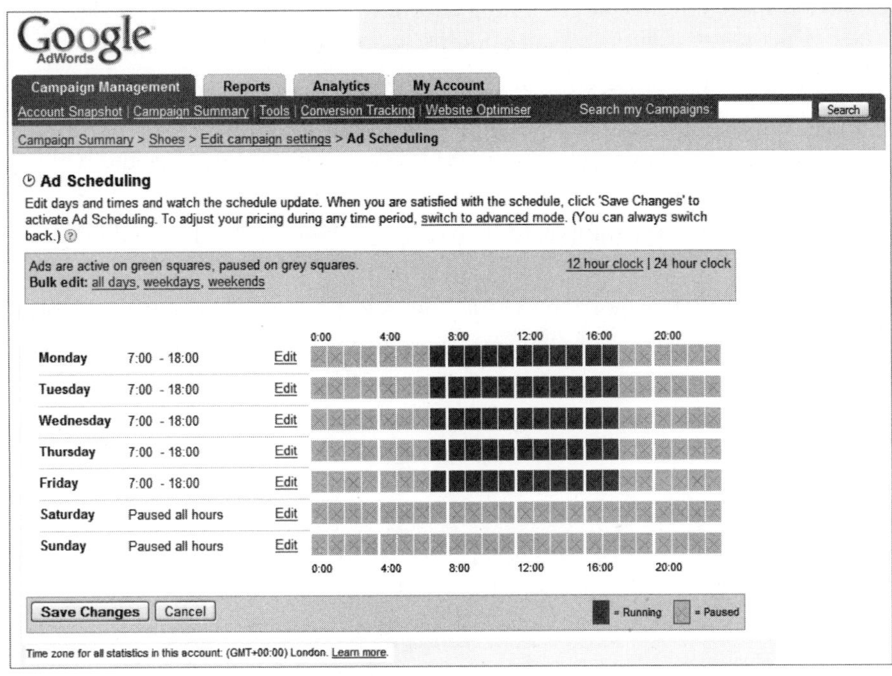

Figure 11.25 Ad scheduling within Google AdWords

Time zone considerations

To take advantage of day parting reports, ensure that your paid campaigns are specific to a particular time zone. For example, don't mix your paid campaigns by displaying the same ad to both a U.S. and a U.K. audience. Time zone settings for AdWords are on a per-account basis. If you have different time zone audiences, then create separate AdWords accounts for them.

Time zone settings for Google Analytics can be set on a per-profile basis. *However*, if you link your Google Analytics account to your AdWords account as described in Chapter 6, then your AdWords time zone and country settings take precedence and you cannot realign them within Google Analytics.

If time zone and other regional specifics (language, currency) are important for you, the best practice advice is to use a one-to-one relationship of Google Analytics and AdWords accounts. You can run an aggregate Google Analytics account by adding an additional GATC to your pages (see "Collecting Data into Multiple Analytics Accounts" in Chapter 6).

AdWords Ad Version Optimization

When creating your pay-per-click campaigns in AdWords, how do you know whether one ad creative is more effective at generating click-throughs than another, similar ad? For example, is the headline "Blue suede shoes" better for you than "Turquoise suede shoes" or "Unique suede shoes"? Of course, you don't know the answer to this, and that's the point: no one does. It's up to your audience to decide. Even after you know the answer, it's like the English summer weather: It can still change quickly and without warning. To determine which ad performs best, use *ad version testing*.

Ad version testing is a method used by pay-per-click networks that enables you to display different ad versions for the same target keywords. With Google AdWords, ads can be rotated in equal proportion to a random selection of visitors—for example, five ads each showing 20 percent of your total impressions. You can maintain this and view results in your Google Analytics reports. Alternatively, you can allow AdWords to optimize the display of your ads, favoring the better-performing ones by showing more impressions of the ad that receives more click-throughs.

Note: This is a simplified description of how ad version testing works within AdWords. Optimized ad serving actually favors ads with higher historic click-through rates and quality scores. For more information on AdWords quality scores, see http://adwords.google.com/support/bin/answer.py?answer=21388.

Figure 11.26 shows four different ads for the same target keywords.

Ad	
Stylish suede shoes Go out on the town in style. High quality at online prices. mysite.com/shoes	**Ad version 1**
Suede shoes Go out on the town in style. High quality at online prices. mysite.com/shoes	**Ad version 2**
Turquoise suede shoes Go out on the town in style. High quality at online prices. mysite.com/shoes	**Ad version 3**
Blue suede shoes Go out on the town in style. High quality at online prices. mysite.com/shoes	**Ad version 4**

Figure 11.26 Four different AdWords ad versions targeting the same keywords

Google Analytics tracks different AdWords ad versions with no additional configuration changes required. Ad version results appear automatically in your reports as long as you have the Google Analytics auto-tagging box ticked within your AdWords account (see Chapter 6).

To track ad versions for other paid referral sources, such as Yahoo! Search Marketing and Microsoft adCenter, you need to add tracking codes to your landing page URLs as discussed in Chapter 7 in the section "Online Campaign Tracking." Specifically, the utm_content parameter is required to differentiate ad versions.

As you can see in Figure 11.27, "Stylish suede shoes" is receiving the vast majority of click-throughs from AdWords (no AdWords impression optimization applied). From the drop-down menu item shown, you can also view other visit metrics for each ad version. In addition to the Site Usage report, you should view the ad versions data in the Goal Conversion and Ecommerce reports.

Figure 11.27 Ad version testing results

Tip: Each of the AdWords ad variations shown in Figure 11.26 has a unique headline. It is these headlines that appear in your Google Analytics Ad Versions report. It is not yet possible to report on ad variations that use the same headline, differing only in body text. Note that turning off auto-tagging in AdWords and attempting to use manual tracking parameters as an alternative will not work.

Check the Goal Conversion and Ecommerce reports to confirm that "Stylish suede shoes" is performing better from a conversion and revenue point of view (click the tabs within the Ad Versions report). For example, it may be that the attractive headline of "Stylish suede shoes" is better for visitor acquisition (click-throughs), but when it comes to visitors interacting with your website, perhaps "Blue suede shoes" converts better and generates more revenue. If that is the case, then take advantage of this discrepancy and create separate ad groups for each, so you can run separate bidding strategies.

Assuming the Goal Conversion and Ecommerce reports show a similar trend as in Figure 11.27, you can then either enable Google's ad serving optimization, which will favor "Stylish suede shoes," or disable (pause) the remaining ad versions and focus all your pay-per-click efforts on "Stylish suede shoes."

As an aside, you can also use ad version testing for non-pay-per-click campaigns by using the utm_content tracking parameter. For example, within your e-mail marketing campaigns, you could test the effectiveness of HTML e-mail versus plain-text messages by appending utm_content=plain-text or utm_content=html to website links within your e-mail. If you use the utm_campaign tracking parameter in this way, then you should also take advantage of using the other campaign tracking parameters available to you (see "Online Campaign Tracking," in Chapter 7).

Monetizing a Non-E-Commerce Website

For non-e-commerce websites, understanding and communicating website value throughout your organization is key to obtaining buy-in from senior management. After all, you want to make changes to improve your bottom line; but without an associated dollar value, that can be difficult to achieve. By gaining executive support, you will be able to procure investment for content, infrastructure, and online marketing. The problem is that many executives' eyes glaze over when they see yet another set of charts on visitor metrics. "Our site doesn't sell anything, so who cares?" is a common response. Identifying the monetary value of your visitor sessions is a proven way to get executive attention and it can help keep the company website from becoming just someone's pet project.

Google Analytics provides two mechanisms for demonstrating website monetary value:

- Assigning goal values
- Enabling e-commerce reporting for your non-e-commerce site

The key to both approaches lies in knowing the worth of goal conversions to your business. For example, if a PDF brochure is downloaded 1,000 times and you estimate that one of these downloads results in a customer with an average order value of \$250, then each download is worth \$0.25 (\$250/1,000). If 1 in 100 downloads converts into a customer, then each PDF download is worth \$2.50 to you and so on.

For each goal, you need to ask yourself how many of these goal conversions are required to create a customer and what is the average value of that customer. The Google Analytics Goals > Overview report shows how many conversions you get to each of your site goals. Initially, you'll need to estimate the percentage of goal conversions that result in paying customers, but you'll be able to fine-tune your estimates later as you collect more information. Once you can estimate the value of each of your site goals, it is straightforward to monetize your website.

Tip: If you are struggling to estimate goal values, start the process off by first evaluating your least significant goal. Give this a value of 1 (as with assigning all goals in Google Analytics, the actual amount is unitless—the symbols $, £, €, etc., are labels). For more valuable goals, use a multiple of the least valuable one. For example, if your least valuable goal is a PDF download and your next more valuable goal is a subscription request that is five times more valuable to you, then assign goal values of 1 and 5, respectively.

Approach 1: Assign Values to Your Goals

Every site has at least one goal; quite often it has several. Non-e-commerce sites have PDFs and other document files to download, product demonstrations, brochure requests, quote requests, subscription signups, registrations, account logins, blog comments, content ratings, printouts—even the humble mailto:link (e-mail address link) can be considered a goal and tracked with Google Analytics (see "Event Tracking," in Chapter 7).

Assigning a goal value is straightforward, and is described in "Goals and Funnels" in Chapter 8. Adding values to goals enables you to gain additional metrics in your Google Analytics reports, such as the average per visit goal value ($/Visit) as shown in the Traffic Sources section, and average page value ($Index) as shown in the Content report section. In addition, you can view individual and total goal values in the Goals > Goal Value report. However, you can obtain far more detailed reporting by using the technique outlined in the following second approach.

Approach 2: Enable E-commerce Reporting

By setting up your non-ecommerce site as an e-commerce website in Google Analytics, you'll be able to do the following:

- Have an unlimited set of goals (standard goal reporting is currently fixed at four goals per profile)
- See the amount of time and number of visits it takes for visitors to convert
- View a breakdown of how much each "product" (goal) contributes to your website revenue
- Group goals into categories
- List specific "transactions" (individual goals)

Here is an example to illustrate the last bullet point and the capability the expanded reports will give you. Imagine you are a publisher of content with hundreds of PDF files available for download (probably behind a registration system). Perhaps you also have abstracts available for free. Using a wildcard such as *.pdf in your goal configuration setup will tell you how many goal conversions you receive for PDF files. However, it does not tell you how many PDF files were downloaded, because visitors can only convert once during a session for a particular goal, even though they may have downloaded several PDF files.

To ascertain the total number of downloads, you need to view the Goals > Goal Verification report; if you wish to see the different types of PDF downloads (for example, specification, help, brochure, price guide), apply inline filters. Clearly, this is not scalable.

The solution is to enable e-commerce reporting so that every individual PDF download can be tracked as a product, grouped into categories, and monetized, as shown in Figure 11.28.

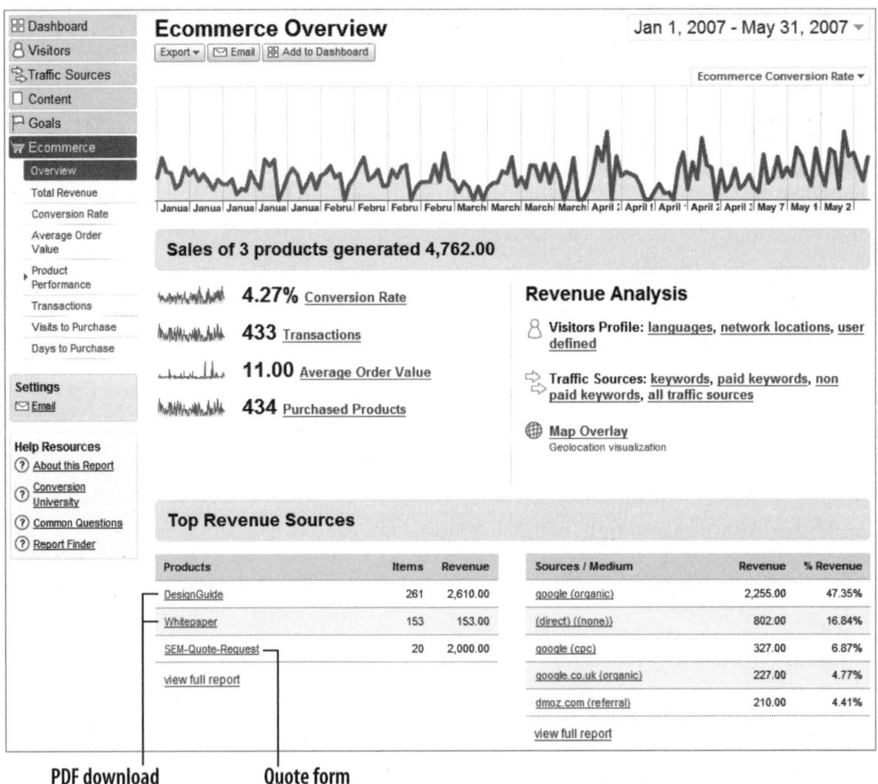

Figure 11.28 An e-commerce report for non-e-commerce goals

Note: Goal conversions measure whether a visitor has converted, and this can only happen once during a visitor session. For example, if your goal is to download any PDF file and a visitor downloads five PDF files in one session, that is recorded as one goal conversion with one goal value.

These are just a few examples of what you will see. Using this approach, you gain additional aggregate information as well as more specific goal and goal conversion information. How this is achieved is discussed next.

Tracking a Non-E-commerce Site As Though It Were an E-commerce Site

First, you need to tag each goal page with e-commerce tracking information (see "E-Commerce Tracking," in Chapter 7). You will be able to leave some of the e-commerce fields blank. For example, assume that one of your goals is for a visitor to click a mailto: link. Visitors who click this cannot leave their name or address by this action, so you will not be able to collect this information. It is also against the Google Analytics Terms of Service to track personally identifiable information (see www.google.com/analytics/tos.html). The following are example goals tracked with e-commerce fields.

Generating unique order IDs

In all of the examples given, it is important that you assign your own unique order ID to each transaction. To do this, add the following code just above the </head> HTML tag of each page that you are tracking with e-commerce fields:

```
<script type="text/javascript">
function getOrderID(){
   // generate a random order id
   var randomnumber=Math.floor(Math.random()*1000);
   var currentTime = new Date();
   var month = currentTime.getMonth()+1
   var timeStamp = "" +currentTime.getFullYear() + month +
currentTime.getDate() + "-" +currentTime.getHours() +
currentTime.getMinutes() + currentTime.getSeconds() +"-" + randomnumber;
   return(timeStamp);
}
</script>
```

Continues

Generating unique order IDs *(Continued)*

When the page is loaded, this generates a unique order ID of the form YYYYMMDD-hhmmss-*XXX*, where *XXX* is a random number between 0 and 999. This enables tracking of up to 1,000 orders per second, and enables you to keep order IDs in a logical structure that can be searched for within the reports later, if required.

With the script in place, generate an order ID by calling the JavaScript function getOrderID() as shown in the examples.

1. Pseudo e-commerce fields for a mailto: link goal

   Add the following e-commerce fields to the page with the mailto: link to be tracked, anywhere after your GATC:

```
<script type="text/javascript">
   orderNum = getOrderID();
   pageTracker._addTrans(
     orderNum,                  // order ID - required
     "",                        // affiliation or store name
     "1",                       // total - required
     "",                        // tax
     "",                        // shipping
     "",                        // city
     "",                        // state or province
     ""                         // country
   );
   pageTracker._addItem(
     orderNum,                      // order ID - required
     "brian@mysite.com",            // SKU (stock keeping unit)
     "Email link",                  // product name
     "Contact",                     // category or variation
     "1",                           // unit price - required
     "1"                            // quantity - required
   );
</script>
```

The preceding code are arrays and can therefore be collapsed into single lines. From now on I use the abbreviated form of assigning e-commerce values as follows:

```
<script type="text/javascript">
   orderNum = getOrderID();
   pageTracker._addTrans(orderNum, "", "1", "", "", "", ➥
   "", "");
   pageTracker._addItem(orderNum, "brian@~CAmysite.com", ➥
   "Email link", "Contact", "1", "1");
</script>
```

As you can see, most of the e-commerce fields are blank, as you cannot know the shipping address of someone who simply clicks your e-mail link. A value of $1 and a quantity of 1 have been assigned.

2. Pseudo e-commerce fields for a PDF download goal

Add the following e-commerce fields to the page with the PDF link to be tracked, anywhere after your GATC:

```
<script type="text/javascript">
   orderNum = getOrderID();
   pageTracker._addTrans(orderNum, "", "10", "", "", "", ➥
   "", "");
   pageTracker._addItem(orderNum, "", "PDF Brochure", ➥
   "Download", "10", "1");
</script>
```

Here, a PDF download has been categorized as “Download” and given a value of $10; the quantity remains 1. If you have multiple PDF files on the same page, then you could categorize them and value each differently, perhaps by language or by content. This is shown in example 4.

3. Pseudo e-commerce fields for a contact form goal

Add the following e-commerce fields to the page with the contact form submission to be tracked, anywhere after your GATC:

```
<script type="text/javascript">
   orderNum = getOrderID();
   pageTracker._addTrans(orderNum, "", "50", "", "", "", ➥
   "", "");
```

```
    pageTracker._addItem(orderNum, "", "Form submission", ➡
    "Contact", "50", "1");
</script>
```

This example assumes a value of $50 per form submission.

With your e-commerce fields in place on the pages that contain goals, next you need to decide how to call these values into Google Analytics. This is done using the JavaScript call to the pageTracker._trackTrans() function. For the preceding three examples you can use the following:

1. <a href = "mailto:email@address.com" onClick = "pageTracker._trackTrans();">
2. <a href = "file.pdf" onClick = "pageTracker._trackPageview('/downloads/file.pdf');pageTracker._trackTrans();">
3. <form action = "formhandler.cgi" onSubmit = "pageTracker._trackTrans();">

Note the use of trackPageview for example 2. This is not directly related to what we wish to achieve but it should be used as a best practice technique—that is, capturing the PDF download as a virtual pageview. For more details on virtual pageviews, see "trackPageview( ): the Google Analytics Workhorse," in Chapter 7.

4. Pseudo e-commerce fields for multiple PDF goal downloads: a special case

   This is a special case in which multiple PDF downloads appear on the same page. For this scenario, the e-commerce event handler needs to be called for each PDF download link. That way, they receive a different transaction ID. This is required as you cannot have multiple item transactions with this pseudo e-commerce method. Use the following format for each:

```
<a href = "file1.pdf" onClick =
"pageTracker._trackPageview('/downloads/file1.pdf');
orderNum=getOrderID();pageTracker._addTrans(orderNum,
"", "10", "", "", "", "", ""); pageTracker._addItem(orderNum, "", "PDF
Brochure", "Download", "10", "1");pageTracker._trackTrans();">
<br>
<a href = "file2.pdf" onClick =
"pageTracker._trackPageview('/downloads/file2.pdf');
orderNum=getOrderID();pageTracker._addTrans(orderNum,
"", "5", "", "", "", "", ""); pageTracker._addItem(orderNum, "", "PDF
Specification Sheet", "Download", "5", "1");pageTracker._trackTrans();">
```

   Here, two PDF downloads have been categorized and given values of $10 and $5, respectively. If a visitor clicks on both of these files (or repeatedly clicks on the same file), then each is tracked as a separate transaction, as the function getOrderID() is called on each occasion. Assuming there is a minimal delay in loading the HTML page in question into the visitor's browser, the transaction IDs for these two files will be very similar—for example, varying only in the ss-XXX part of the string YYYYMMDD-hhmmss-XXX.

Approach 2 provides significant benefits

By enabling e-commerce reporting on your non-e-commerce website, you can see at a glance the referring sources that lead to goal conversion, time to purchase, visits to purchase, average order value, which keywords convert best, and more. If you were to use the first approach, you would need to navigate to each goal page and determine the information separately—and that can be quite tricky with 500 PDF whitepapers, 10 application downloads, three mailing list subscriptions, two quote request forms, and a contact us form!

Tracking Offline Marketing

Having a unified metrics system that can report on key performance indicators from Web, print, display, radio, and TV— all in one place—and one that can track the correlation between all visitors who start in one channel and cross over into others before converting, has been a long-sought analytics nirvana for many a marketer.

Some vendors have attempted to achieve such a system, with varying degrees of success. The barriers of technical difficulty (bringing information from disparate systems together) and issues with data alignment (e.g., how do you compare a web visitor who has specifically searched for information to a passive TV viewer?) means to date, such a high cost and resource investment is one that few organizations have availed themselves with.

Despite such difficulties, many inroads are being made to overcome them. The open-source nature of Google's application programming interface (API) model for making data accessible goes some way towards making this happen. Current Google APIs include AdWords, Google Maps, and Google Earth, though a public Google Analytics API is not yet in place. If and when such an API becomes available, customers would be able to stream their Google Analytics data directly out and into their own applications, and potentially import data back into Google Analytics. This could be as simple as real-time updates to KPI tables in Excel, or the merging of web data with CRM data.

So far, no nirvana product that is easy to adopt with a price tag that does not hinder adoption yet exists. However, Google Analytics can still provide you with a great deal of insight in terms of measuring your offline marketing campaigns. Consider the chart shown in Figure 11.29. This chart measures the uplift in web visitor numbers while running a print advertising campaign.

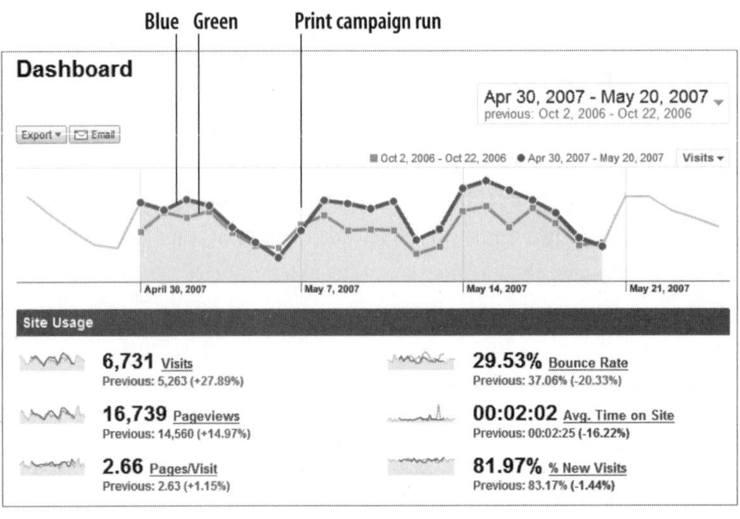

Figure 11.29 Observed uplift in visitors from offline marketing initiatives

Both lines in Figure 11.29 represent a three-week time frame; green is data for autumn, blue is data for the following spring. A magazine ad ran for the last two weeks in spring (May 7 to May 20). As you can see, the uplift over the entire three-week period in visit and pageview numbers is significant, at plus 28 percent and plus 15 percent, respectively. Page bounce rate is also reduced, at minus 20 percent. Any seasonality is taken into account by displaying data one week prior to the print ad campaign—that is, the visit numbers closely align between the two time periods (including pageview data, though this is not shown). The hypothesis is therefore that the print campaign drove the uplift.

Tip: Take care when comparing date ranges as it is important to align with the days of the week; that is, compare Monday with Monday, and so on. Seasonality also needs to be considered; otherwise, you may be giving undue credit to an offline campaign. If possible, try to normalize your numbers by taking into account the background growth in visitor traffic you may receive between the time periods considered.

The strong correlation observed in Figure 11.29 does not equal 100 percent causality. A better solution to gain more certainty is to combine offline campaigns with unique landing page URLs that these visitors use, and there are a number of ways to do this.

Track offline visitors using the following:

1. **Vanity URLs** Recommended when you have strong product brand awareness, with all your web content hosted on a single central domain. Examples include ThinkPad, iPod, Castrol, Gillette, Colgate, Aquafresh, Big Mac, Fanta, Snickers, and so on.
2. **Coded URLs** Recommended when you have a strong company brand or when your products already have separate websites. Examples include IBM, Microsoft, Google, Kellogg's, Kodak, BMW, and any product that relies on model numbers for identification, such as cell phones, cars, printers, cameras etc.
3. **Combining with Search** Recommended when your brand values are less significant than your product or service values or your target audience is more price oriented than brand oriented. Examples include the vast majority of small- to medium-size businesses, the travel industry, the insurance sector, utilities, groceries, and office supplies.

Using any or all of these methods is acceptable, and you can likely think of others.

Note: The example names given for tracking offline visitors are for brand recognition only. They do not reflect the actual website architecture or strategies of the sites in question.

Using Vanity URLs to Track Offline Visitors

If your website content is held at www.mysite.com and you have a strong product brand that has greater awareness than your company brand, consider using a vanity URL of www.myproduct.com for your offline campaigns such as television, radio, and print. Use your website (www.mysite.com) only to promote to an online audience.

Clearly, you don't want to build two separate websites to promote to offline and online audiences. Their needs are the same; the only difference is how they find your website. Apart from the resource overhead, you should not build duplicate pages, as this will get you penalized by the search engines.

To avoid duplicate content, apply permanent redirects to your vanity URLs. Redirects on your web server capture the different URLs of your offline visitors, append tracking parameters, and then automatically forward them through to your main content website. The process takes a small fraction of a second to perform and shows no visible difference to your offline visitors. They type in a vanity URL (www.myproduct.com) and arrive on your official website (www.mysite.com) with tracking parameters appended. In effect, you are pretending to have product specific websites to your offline visitors, using this to differentiate, then redirecting them to your actual content.

With a redirect in place, you can view offline visitors by identifying the campaign variables used. In Figure 11.30, the offline ad is identified in the reports by the medium "Print."

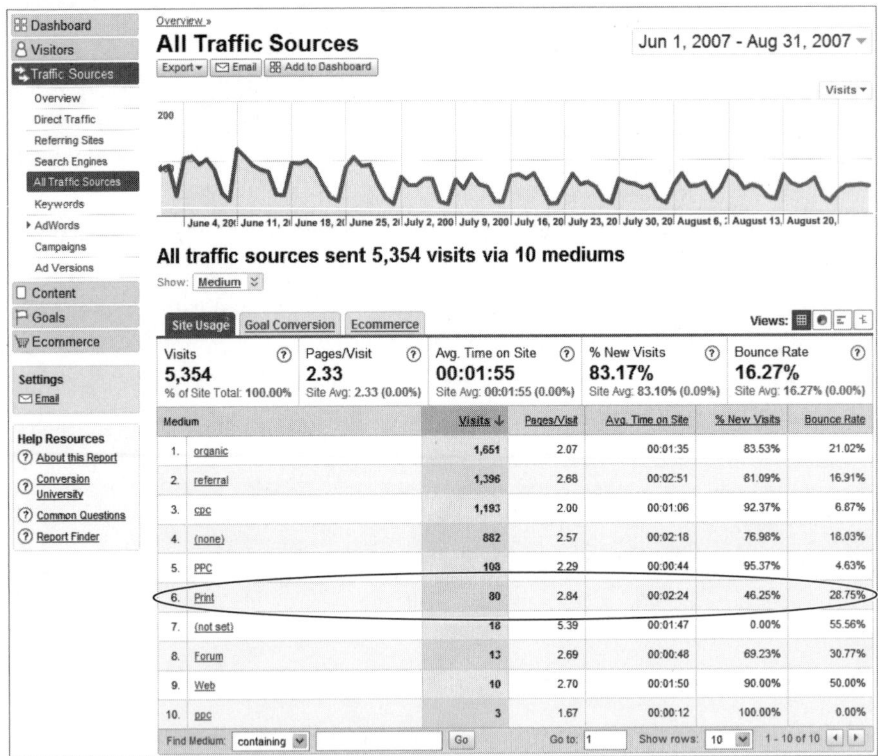

	Medium	Visits	Pages/Visit	Avg. Time on Site	% New Visits	Bounce Rate
1.	organic	1,651	2.07	00:01:35	83.53%	21.02%
2.	referral	1,396	2.68	00:02:51	81.09%	16.91%
3.	cpc	1,193	2.00	00:01:06	92.37%	6.87%
4.	(none)	882	2.57	00:02:18	76.98%	18.03%
5.	PPC	108	2.29	00:00:44	95.37%	4.63%
6.	Print	80	2.84	00:02:24	46.25%	28.75%
7.	(not set)	18	5.39	00:01:47	0.00%	55.56%
8.	Forum	13	2.69	00:00:48	69.23%	30.77%
9.	Web	10	2.70	00:01:50	90.00%	50.00%
10.	ppc	3	1.67	00:00:12	100.00%	0.00%

Figure 11.30 Viewing visit details from an offline (print) campaign

Using vanity URLs for managing offline campaigns is very effective, assuming you have multiple domains to use and the product you are selling is not trademarked or protected by someone else, preventing you from using it as part of a domain. In addition, this won't work if you already have your products hosted on separate websites (see the following section on using coded URLs).

Using Coded URLs to Track Offline Visitors

If your company brand has greater awareness than your products, then consider using coded URLs within your offline campaigns. These are of the following form:

```
www.mysite.com/offer_codeX
www.mysite.com/offer_codeY
```

Coded URLs are unique to your offline campaigns; they are not displayed anywhere on your website and are not visible to the search engines. That is, your content should be visible to the search engines, but this will be via a different online-only URL such as www.mysite.com/productX.

Using redirects effectively

Redirects are an important aspect of using vanity URLs, as they avoid any duplicate content issues (bad for SEO) and allow campaign variables to be appended to the final URL destination.

Two types of redirects are possible: permanent (status code = 301) and temporary (status code = 302). From a search engine optimization point of view, it is important to apply permanent redirects so that the final destination URL is the one that is indexed by the search engines; otherwise, the search engines ignore the content.

The following is an Apache example of redirecting `www.myproduct.com`, used only for offline campaigns, to `www.mysite.com`, your official online web address containing your content. In this Apache server example, the rewrite code is placed in the virtual host configuration section for `www.myproduct.com` in the `httpd.conf` file. Other web servers use a similar method:

```
<VirtualHost>
   ServerName www.myproduct.com
   RewriteEngine on
   RewriteCond %{HTTP_USER_AGENT} .*
   RewriteRule .*
http://www.mysite.com/?utm_source=magazineX&utm_medium=print&utm_➥
campaign=March%20print%20ad [R=301,QSA]
</VirtualHost>
```

The rewrite code requires the `mod_rewrite` module to be installed. Most Apache servers have this by default (see `http://httpd.apache.org/docs/mod/mod_rewrite.html`). Ensure that the `RewriteRule` is contained on one line within your configuration file (up to and including `QSA]`); and if spaces are required, use character encoding (`%20`).

In this example, Google Analytics campaign variables are used so that you can uniquely identify the offline campaign, as described in the section "Online Campaign Tracking," in Chapter 7. These are then permanently passed onto the official website using the Apache `mod_rewrite` option *query string append* (QSA) to ensure that any other query parameters are added. After a redirect takes place, you should see your campaign variables in the address bar of your browser. If not, the QSA has not worked and this will need to be resolved.

For the example redirect given, the offline visitor can be identified in your Google Analytics reports anywhere the source, medium, and campaign variables are displayed. In this case, the source is "magazine," the medium is "print," and the campaign is "March print ad." This works when the only offline campaign running is a print ad, as in Figure 11.30. Had this vanity URL been required for other offline campaigns running at the same time, then both the `utm_source` and `utm_campaign` tracking variables would need to be changed to "offline." You cannot differentiate simultaneous offline campaigns using this method.

By using coded URLs in your offline marketing, you will know that visitors to the subdirectory `/offer_codeX` must have come from your offline ad; there is nowhere else to find it. Of course, there is always the possibility that the visitor will only remember your domain (mysite.com) and not the specific landing page (offer_codeX) required to distinguish them from direct visitors; this is common for strong brands. It is therefore important that your offline campaign provides a compelling reason for the visitor to remember your specific URL. This can be the promotion of special offer bundles, voucher codes, reduced pricing, free gifts, competitions, unique or personalized products, and so on, that are only available by using the specific URL you display in your offline campaigns.

A useful tip when employing this technique is to use a landing page URL that can be remembered easily, tying it in with your message and the medium. This sounds like common sense, but you would be surprised what a little thought can achieve for you. For example, for a TV campaign you could consider the following:

```
www.mysite.com/10percent
www.mysite.com/getonefree
www.mysite.com/twofourone (or /2for1, /241)
www.mysite.com/xmas
www.mysite.com/sale
```

Identifying with your TV branding slogan or campaign message can be a very effective way at keeping your full URL in the viewer's mind, as this associates your website with their viewing activity.

As with the use of vanity URLs, redirecting visitors is required. This enables you to avoid producing duplicate content and appends tracking parameters to the landing page. The only difference here is that the redirection is applied to a subdirectory, not the entire domain. This is desirable if your products are already hosted as separate websites.

Even without redirection, as long as the URLs remain unique to your offline campaigns and are neither shown as links within your website nor indexed by the search engines, you will still be able to measure the number of offline visitors to these specific pages. Redirection is there to help you compare campaigns within your Google Analytics reports.

Combining with Search to Track Offline Visitors

When your brand values are less significant than your product or service values or your target audience is more price orientated than brand oriented, remembering a URL can be difficult for your potential visitors—your brand is simply not strong enough to gain tractions. An alternative technique is to use search as part of your offline message, such as running a radio ad that uses something like "Find our ad on Google by searching for the word *productpromo* and receive 10 percent off your first order."

Redirecting a subdirectory

This example uses the Apache mod_rewrite module, which most Apache servers have installed by default. See http://httpd.apache.org/docs/mod/mod_rewrite.html.

```
<VirtualHost>
   ServerName www.myproduct.com
   RewriteEngine on
   RewriteCond %{HTTP_USER_AGENT} .*
   RewriteRule /xmas.* /productX/?utm_source=channel-➥
   123&utm_medium=tv&utm_campaign=March%20tv%20ad [R=301,QSA]
</VirtualHost>
```

Ensure that the RewriteRule is contained on one line within your configuration file (up to and including QSA]); and if spaces are required, use character encoding (%20).

By creating an AdWords ad just for this campaign, targeting a unique word or phrase that is unlikely to be used by non-related searchers, you not only provide a strong incentive for visitors, but also directly assign these visitors to a specific offline marketing effort.

This extra step of asking your potential audience to first go elsewhere has a small drawback: You pay for the click-through on your AdWords ad. However, using a unique search phrase means you should be the only bidder and hence would pay as little a one cent per click-through. The upside is considerable: You have full control of the ad wording and landing page URL. That means each campaign (print, TV, display, radio) can have a separate landing page and hence is completely traceable, without the need for redirection.

Example keywords to use in your AdWords campaign include the following:

- 10percent
- productX101
- whyCompanyName
- 1-800-123-BIKE—your toll free number (U.S.)
- 207-123-4567—your telephone number
- Signal House, Station Road—the first line of your address

Tip: Check your AdWords listing regularly, as competitors may pick up your campaigns and start to bid on the same keywords!

An Introduction to Website Optimizer

Website Optimizer is a web page testing tool that enables you to compare pages (A/B testing) and test elements within a page (multivariate testing). Marketers will be familiar with A/B testing - a binary test to compare the effectiveness (usually a conversion rate) of a statistical element, such as one product image versus another. For example, page A is shown to 50 percent of new visitors selected at random, while page B is shown to the remaining 50 percent of visitors. If page A is better at generating conversions than page B, then page A is declared the winner and subsequently shown to all visitors. Another page, or page section, can then be tested, such as product title A versus product title B.

Multivariate testing is used to evaluate images, headlines, descriptions, colors, fonts, content, and so on *within* a page in order to understand which combinations provide better conversions. According to Wikipedia (http://en.wikipedia.org/wiki/Multivariate), multivariate statistical analysis describes "a collection of procedures which involve observation and analysis of more than one statistical variable at a time." The key phrase here is "more than one statistical variable at a time."

Note: Most multivariate and A/B testing tools enable you to vary the percentage of visitors shown test alternatives. For example, show page A to 10 percent of visitors selected at random and page B to a further 10 percent, leaving the remaining 80 percent to see the original content.

The great advantage of A/B testing is that it is simple to set up, obtain results, and make a change. It can be used to test design layout—for example, should the menu navigation system be at the top or left side of the page, is black text on a white background better than white text on a dark background? The iterative nature of A/B testing enables you to gain results quickly and is particularly useful when answers to macro-questions are required—is version A better than version B or not?

However, A/B testing is limited when you want to try numerous elements—that is, test versions A to Z or even hundreds of combinations. With multiple page elements—for example, multiple product images, titles, and descriptions on the same page—A/B testing can be laborious and time consuming to implement and obtain results. Another caveat is that A/B testing cannot tell you whether one page element affects the conversion rate of another; for example, what if the product title affects how visitors perceive the product image?

Multivariate testing is the solution to this; it enables you to test multiple page elements simultaneously and determine what, if any, correlations exist between elements. The purpose is to evaluate the best combination of all page elements to create a winning page—that is, generate more conversions.

AMAT: Where Does Testing Fit?

You have set up your website, initiated marketing to bring relevant traffic, viewed your visitor reports, and notice a page is underperforming. You've identified the problem, and various teams have come up with suggestions. Which one do you pick as the replacement? This common problem can sometimes end the entire process; people just don't know what to do next. This is precisely where testing fits.

Multivariate and A/B testing is a crucial element that dovetails into the web marketing life cycle (AMAT):

- Acquire visitors.
- Measure interactions.
- Analyze results.
- Test improvements.

As Figure 11.31 shows, AMAT allows for a continuous cycle of improvement, providing a measurable process by which you can increase the conversion rates for your website, right down to a page-by-page basis if required.

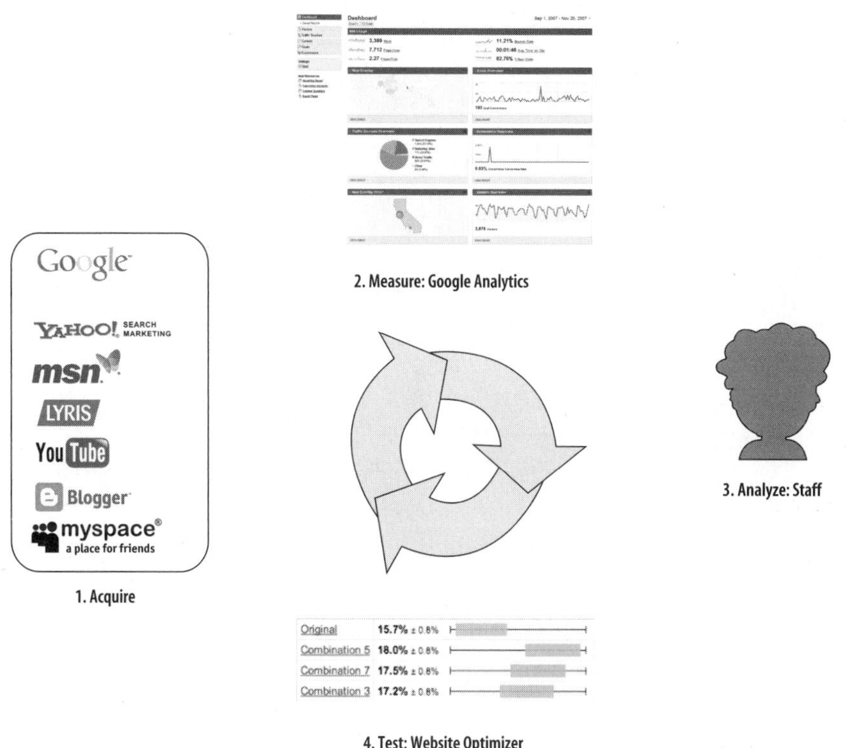

Figure 11.31 The web marketing life cycle (AMAT)

Getting Started: Implementing a Multivariate Test

There is a close relationship between Website Optimizer and Google Analytics—the conversion data used in Website Optimizer reports comes from the same database system. In addition, implementing a test requires adding page tags, similar to the process for the GATC. The following section outlines the principles of a Website Optimizer implementation. A fuller description is available from http://services.google.com/websiteoptimizer.

Similar to Google Analytics, Website Optimizer is integrated with AdWords and is accessed from within your AdWords account, as shown in Figure 11.32.

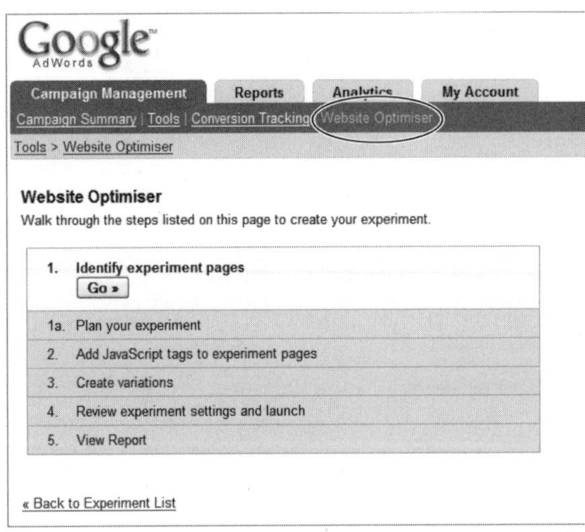

Figure 11.32 Accessing Website Optimizer from AdWords

From the initial screen, there are four steps to create a test:

1. Identify a test page and conversion goal.
2. Tag both pages.
3. Create new content variations to test.
4. Review and launch.

Step 1: Pick a Test Page and Conversion Goal

From the myriad of pages on your website, how do you choose which pages to test? Sometimes you will have a clear idea. Perhaps it is a product feature page where <title> tags, headlines, image selection, and description text have all received input—that is, requests for change—from various parts of your organization or customer feedback. If you haven't already identified a page to test, view your Content > Top Content report in Google Analytics. Pages with a high bounce rate, high exit rate, or high $Index value should be considered suitable candidates for testing (see Figure 11.33). As an additional resource, also consider the steps discussed earlier in the section "Identifying Poorly Performing Pages."

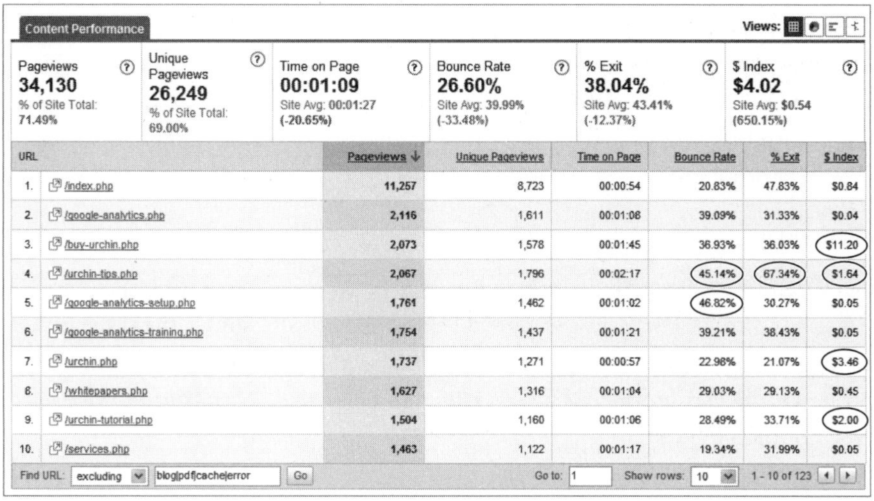

Content Performance — Views:

Pageviews	Unique Pageviews	Time on Page	Bounce Rate	% Exit	$ Index
34,130	26,249	00:01:09	26.60%	38.04%	$4.02
% of Site Total: 71.49%	% of Site Total: 69.00%	Site Avg: 00:01:27 (-20.65%)	Site Avg: 39.99% (-33.48%)	Site Avg: 43.41% (-12.37%)	Site Avg: $0.54 (650.15%)

	URL	Pageviews ↓	Unique Pageviews	Time on Page	Bounce Rate	% Exit	$ Index
1.	/index.php	11,257	8,723	00:00:54	20.83%	47.83%	$0.84
2.	/google-analytics.php	2,116	1,611	00:01:08	39.09%	31.33%	$0.04
3.	/buy-urchin.php	2,073	1,578	00:01:45	36.93%	36.03%	$11.20
4.	/urchin-tips.php	2,067	1,796	00:02:17	45.14%	67.34%	$1.64
5.	/google-analytics-setup.php	1,761	1,462	00:01:02	46.82%	30.27%	$0.05
6.	/google-analytics-training.php	1,754	1,437	00:01:21	39.21%	38.43%	$0.05
7.	/urchin.php	1,737	1,271	00:00:57	22.98%	21.07%	$3.46
8.	/whitepapers.php	1,627	1,316	00:01:04	29.03%	29.13%	$0.45
9.	/urchin-tutorial.php	1,504	1,160	00:01:06	28.49%	33.71%	$2.00
10.	/services.php	1,463	1,122	00:01:17	19.34%	31.99%	$0.05

Find URL: excluding blog|pdf|cache|error Go — Go to: 1 — Show rows: 10 — 1 - 10 of 123

Figure 11.33 Selecting testing pages from Google Analytics' Top Content report

Selecting pages with a high bounce rate Pages with a high bounce rate (single-page visits) are obvious targets for testing, as they are poorly performing pages for you. Segmentation is important here; rather than view bounce rates for all visitors, drill down into the Top Content reports and view particular segments such as source, medium, and campaign names. For example, it may be that your paid campaigns have low bounce rates while your organic search visitors have high bounce rates. If that is the case, then it is your differences in visitor acquisition that require investigation, not the landing page. Assuming your online marketing campaigns have been optimized, consider selecting landing pages with a bounce rate exceeding 50 percent as a starting point.

Selecting pages with a high exit rate Pages that show a high exit rate are the places from which visitors leave your website. That is not necessarily a bad thing. However, if the exit page is not a goal page (or your contact details), then it is likely you wanted that person to stay on your site. As with bounce rates, consider initial testing for pages whose exit rate is higher than 50 percent.

Selecting pages with a high or low $Index value $Index is a measure of the value of a particular page (see "Content Reports: $Index Explained," in Chapter 5). It is calculated by determining whether that page occurs in the path of those visitors who convert and weighting it by the goal value. Excluding the goal page itself, pages with a high $Index will have a high influence on goal conversions and so are ideal candidates for testing. Similarly, important pages that are receiving a low $Index should also be considered for testing.

For your test goal conversion page, you can use the same goal URLs as those defined in your Google Analytics configuration, or you can select others. As with Google Analytics, you can use virtual pageviews and wildcards as the conversion goal, so /download/.* and /cgi-bin/*.pl can be defined as goal pages as long as such files are being tracked with Google Analytics. Pages that cannot receive a tracking code, such as PDF files, cannot be used.

Step 2: Tag Both Pages

With your test and goal page URLs selected, you need to insert page tags to control the experiment and track the results. Figure 11.34 schematically shows the three different tags required for this. These tags are snippets of JavaScript code that are provided in the Website Optimizer interface. The tracking and conversion scripts are simple modifications of the GATC.

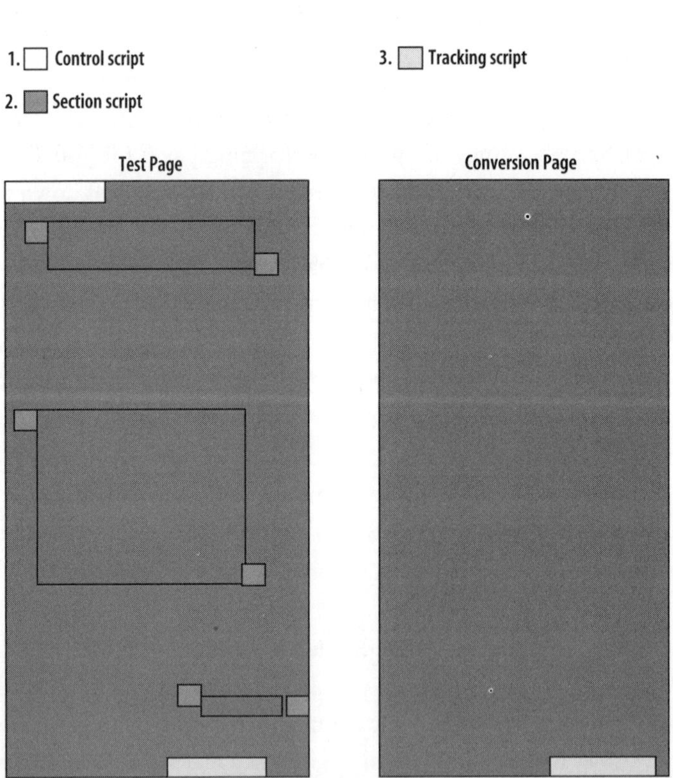

Figure 11.34 Schematic tagging of pages for an experiment

The three different page tags required are as follows:

1. Control script

   The control script governs the progress of the experiment. It contacts Google servers to retrieve appropriate content variations (the actual variations are maintained on Google servers). The control script also ensures that a user's view of the page remains consistent and that multiple views of the same page by the same user do not affect the experiment statistics.

   The control script must be placed before any section scripts and before all displayable content. In most cases, this will be in the HTML `<head>` section of the test page.

2. Section scripts

   Section scripts are used to define sections of page content that will be varied in the experiment. Most things can be included within a section—for example, text, script, graphics, and so on, or all of these in one contiguous block. The limit is 10 kilobytes of HTML (about 10,000 characters).

   If you are testing more than one section, then each section requires a unique name. Section names are case sensitive and can be up to 20 characters long. Try to use meaningful names—for example, "headline" or "product_photo" to make it easier to interpret your reports.

3. Tracker scripts (two)

   These scripts trigger Google Analytics data collection and ensure that page refreshes are counted properly. The tracker script is added to both the test page and the conversion page and should immediately follow your GATC—that is, placed after all displayable content in each page, just above the `</body>` tag.

   A generic example illustrating the positioning of the scripts is shown here:

```
<html>
 <head>
   ...
   <script><!- Control script
      ...
   //-><\script>
</head>
 <body>
     ...
     <displayable content>
     ...
     <script><!- Page section script
       ...
   //-><\script>
     <displayable content>
```

```
   ...
</body>
   ...
   <script><!-Your regular GATC
     ...
   //-><\script>
   <script><!- Optimizer tracking script
     ...
   //-><\script>
   ...
</html>
```

Custom variables

If you have the following custom variables in your GATC, then you need to create a new script, setting the customized variables to the same values set in your GATC:

```
pageTracker.setDomainName
pageTracker.setAllowHash
pageTracker.setSessionTimeout
pageTracker.setCookiePath
```

This new script should be in its own set of <script> tags and placed immediately *above* the Website Optimizer control script, in the header area of your page. It is important to declare the pageTracker object first, exactly as it appears in your GATC, as shown here:

```
<html>
<head>
  ...
  <script>
  var pageTracker = _gat._getTracker("UA-12345-1");
  pageTracker.setDomainName("none")
  pageTracker.setAllowHash(false)
  pageTracker.setSessionTimeout("3600")
  pageTracker.setCookiePath("/path/of/cookie")
  </script>

  <script><!- Control script
     ...
  //-><\script>
  ...
</head>
```

Once you have installed all the tags, validate them within Website Optimizer. If errors are detected, fix these before continuing. Website Optimizer will not let you proceed to the next step without validation. There are two methods of doing this:

- Provide the URLs for your test and conversion pages. Website Optimizer will access them and validate.
- If your test pages are not externally visible—for example, if they are part of a purchase process, behind a login area, or inaccessible for some other reason—you can upload the HTML source files.

If you are unable to validate the tags using either method—for example, if you are using a dynamic web page that Website Optimizer cannot view—opt out of the validation step and take responsibility yourself.

Step 3: Create New Content Variations to Test

At this step, variations of section content are added within the user interface by simply pasting plain text or HTML content into the box provided, as shown in Figure 11.35. This is required for each variation. Once completed, you can then preview each combination that your visitors might see.

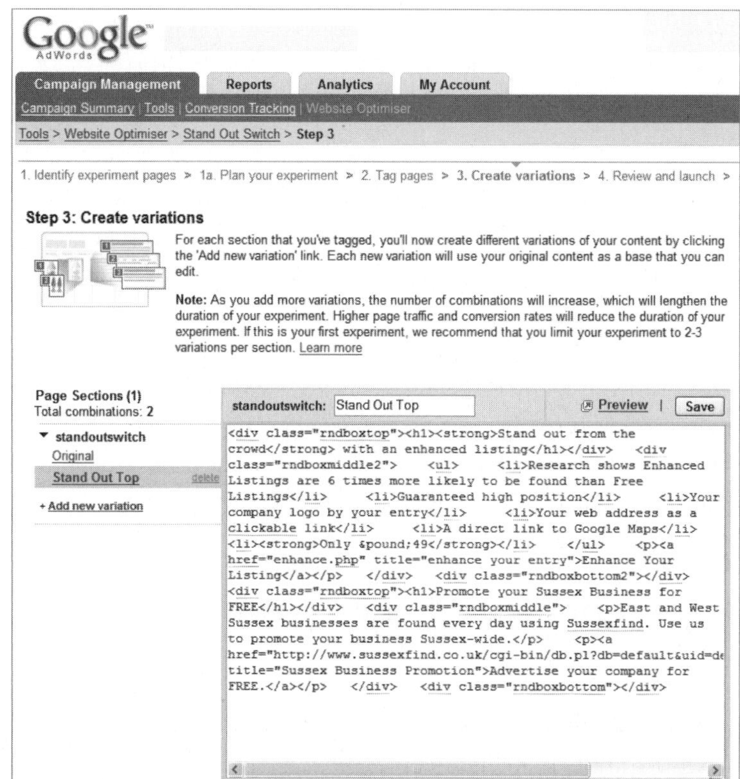

Figure 11.35 Adding variations for your test page

Note that the content variations used for testing are hosted on Google servers; the original content remains hosted by yourself or your hosting provider. Each time a visitor views your test page, Google servers insert your variations randomly. Once a visitor has received a particular combination, that combination remains fixed for the duration of that visitor's session.

The number of combinations is important. When your test page is displayed during an experiment, Website Optimizer is testing the performance of not only individual variations, but also the combined effect of all page sections on the page. For example, in an experiment with two page sections—headline and image with two and three variations, respectively—the following six combinations will be tested (2 × 3 combinations):

1. original headline + original image
2. original headline + new image
3. original headline + new image2
4. new headline + original image
5. new headline + new image
6. new headline + new image2

Extending this to three page sections with three variations for some description text, you will have 18 combinations (2 × 3 × 3). This has obvious implications regarding the length of time the experiment needs to run in order to produce meaningful results (see below).

Step 4: Launch

This is where you enter the percentage of traffic to include in the experiment (1–100%); the more traffic included, the faster the experiment will run. Before launching, it is worthwhile to make a final check of your experiment settings. Once you start the experiment, you will not be able to change the parameters; a new experiment must be created.

After you click Start, you will return to the experiment workflow page, which has an additional section describing the progress of this experiment and the number of impressions and conversions tracked so far. Your test page will start showing different combinations immediately, but there is a delay of about an hour before reports begin displaying data. Figure 11.36 is a schematic representation of how Website Optimizer works.

How Long Will an Experiment Take?

The progress of the experiment and the estimated duration depend entirely on the amount of traffic seen on your test and conversion pages. As a guide, when selecting test pages choose pages that receive thousands of pageviews and are part of a conversion process that results in hundreds of goal conversions. The period it takes to achieve this in your Google Analytics reports is a good guide to how long it will take for your experiment to run for each variation.

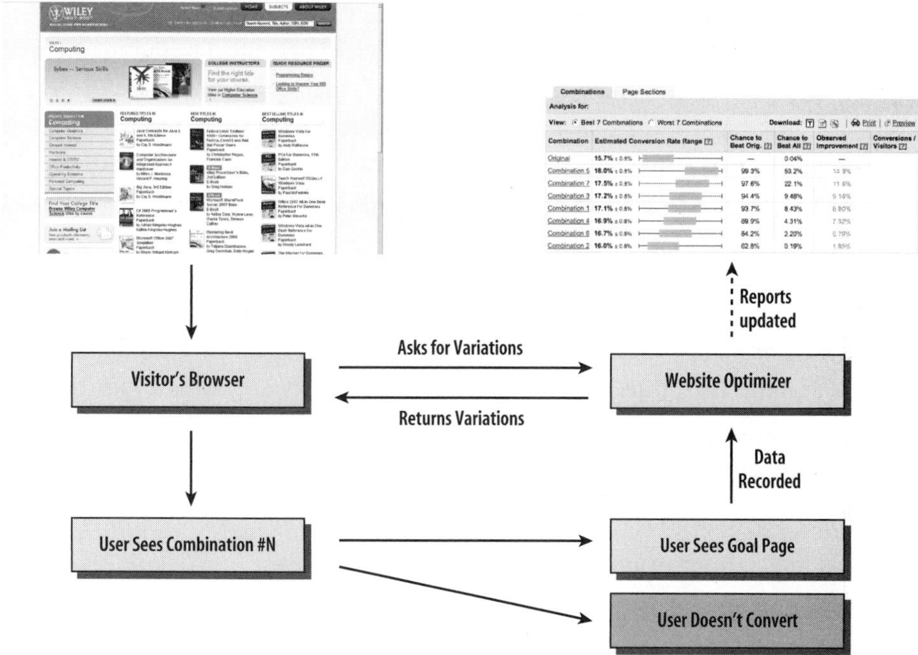

Figure 11.36 Schematic representation of how Website Optimizer works

For example, if you are testing three page sections, each with two variations, that is eight combinations to test in total (2 × 2 × 2). Each combination needs to receive approximately 100 conversions to show statistically significant test results. Assuming an average conversion rate from the test page to each goal page of 10 percent, then approximately 8,000 views of your test page are required. If that is achievable on your website within a week, then it will take approximately the same time to achieve meaningful results within Website Optimizer.

Tip: A handy calculator to help you estimate the potential duration of your experiment is available at www.google.com/analytics/siteopt/siteopt/help/calculator.html.

Once you start seeing impressions and conversions recorded in Website Optimizer, view the preliminary results by clicking "View report."

A Multivariate Case Study

This case study was produced by EpikOne (www.epikone.com)as part of their work for Calyx Flowers (www.calyxandcorolla.com) and is reproduced here with the kind permission of both parties.

As the name suggests, Calyx Flowers is a flower distribution company; it was founded in 1988 and is based in Vermont, U.S.A. Calyx Flowers had begun to invest significantly in its online marketing, particularly search engine optimization and pay-per-click. However, the company felt that the increase in visitor numbers did not match the modest increase in conversions received, i.e., flowers purchased. Furthermore, Google Analytics revealed significant exit rates for visitors who had viewed a product page but did not add to cart.

In designing the Website Optimizer experiment, EpikOne chose to test whether the product page could be more effective at producing conversions. In this example, a conversion was considered successful if a visitor added a product to the shopping cart. As shown in Figure 11.37, three sections of the product page were identified for testing:

1. Change of messaging

   Would the addition of trust factors, such as customer testimonials, help?

2. Stronger call to action

   Would larger, brighter buttons for "Buy Now" help?

3. Change of brand image

   Would a different (more emotive) product image help?

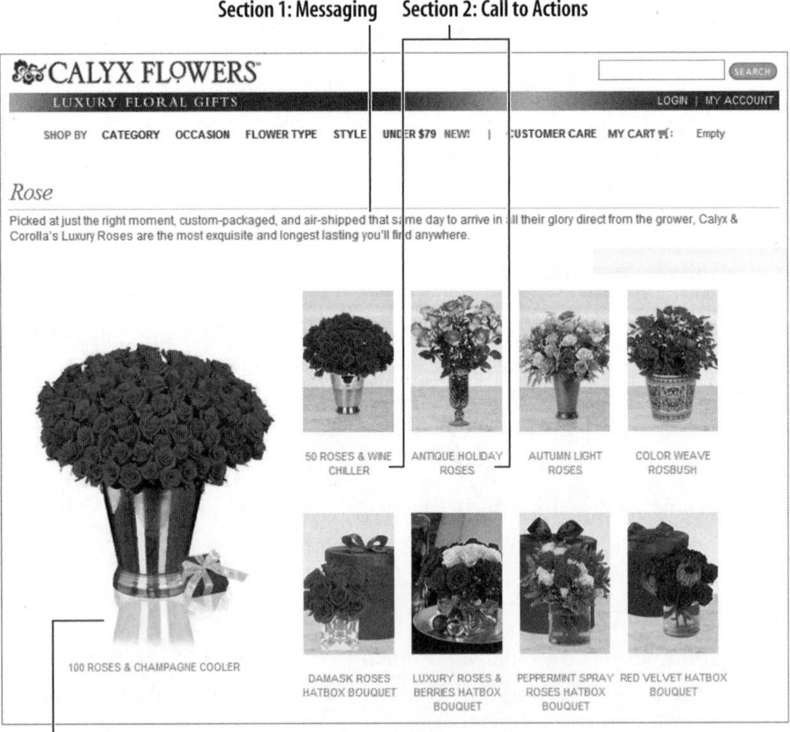

Figure 11.37 The Calyx Flowers product page, with test sections highlighted

For the experiment, each section had two combinations: the original and an alternative ($2 \times 2 \times 2 = 8$ combinations). Figure 11.38 shows the combinations with all alternatives displayed.

Figure 11.38 Alternative combinations for the Calyx Flowers product page

The experiment was launched to test which sections and which combinations would lead to better conversions. For this test, a conversion was defined as adding a product to the shopping cart. Enough conversions were gathered to complete the experiment within a week.

Results and Interpretation

When viewing results, there are two reports to consider: the Page Sections report and the Combinations report. These are shown in Figures 11.39a and b, respectively.

The Page Section report identifies which sections of the experiment have the greatest impact. This is indicated graphically with green and grey bar charts, and numerically in the adjacent table. The Chance to Beat column is a measure of the overlap of the two (grey and green bars) conversion distributions. The smaller the overlap, the greater the separation of the distributions and therefore the higher the probability of beating the original variation. In other words, was the change in the observed conversion rate real or did it just occur by chance (within error bars)? A clear separation of green and grey indicates it is real.

In Figure 11.39a, we can see that the addition of a testimonial has the greatest impact on conversion rate, closely followed by the change in product image. The enhanced call to action buttons show a negative impact (red bar)—that is, they decreased the conversion rate. However, the decrease is minimal (–0.48 percent) and the distribution overlap is large, as indicated by the chance to beat original (42.9 percent). This means there is a 57.1 percent chance that the original section could have also had the same effect. As such, the call to action section is considered to have no significant impact on conversions.

Viewing the Combinations report of Figure 11.39b, we can see that there are two superior combinations (5 and 7). Both of these contained the testimonial, with the winner also including the more emotive product image. The winner contained the original call to action links (refer to Figure 11.37).

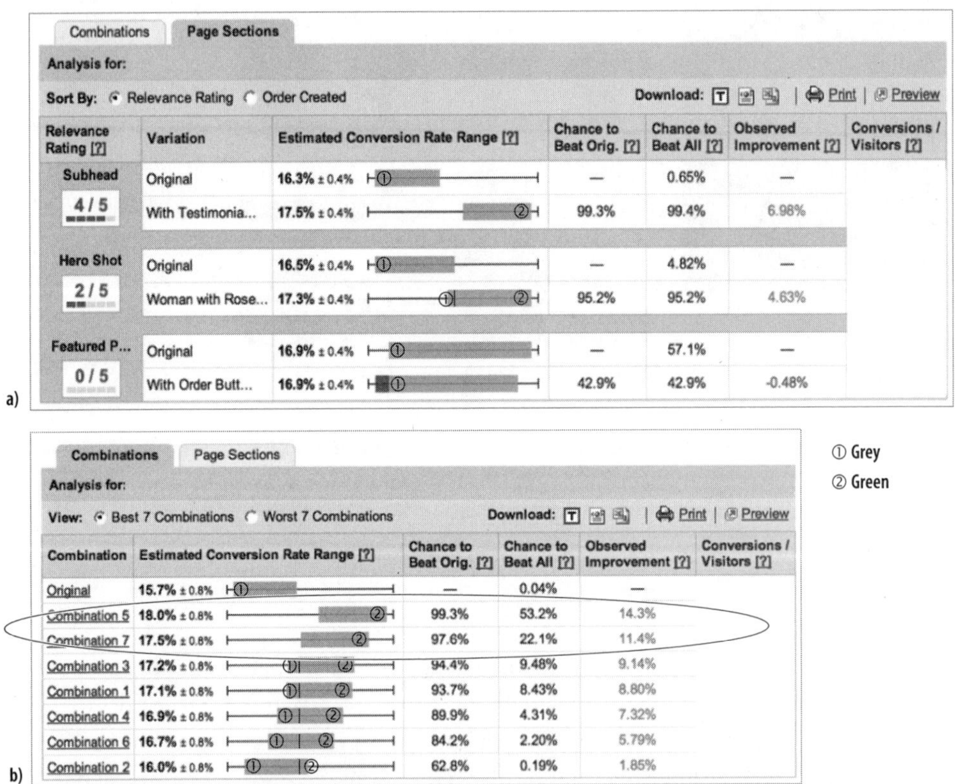

a)

Combinations | Page Sections

Analysis for:

Sort By: Relevance Rating | Order Created — Download: | Print | Preview

Relevance Rating [?]	Variation	Estimated Conversion Rate Range [?]	Chance to Beat Orig. [?]	Chance to Beat All [?]	Observed Improvement [?]	Conversions / Visitors [?]
Subhead 4 / 5	Original	16.3% ± 0.4%	—	0.65%	—	
	With Testimonia...	17.5% ± 0.4%	99.3%	99.4%	6.98%	
Hero Shot 2 / 5	Original	16.5% ± 0.4%	—	4.82%	—	
	Woman with Rose...	17.3% ± 0.4%	95.2%	95.2%	4.63%	
Featured P... 0 / 5	Original	16.9% ± 0.4%	—	57.1%	—	
	With Order Butt...	16.9% ± 0.4%	42.9%	42.9%	-0.48%	

b)

Combinations | Page Sections

Analysis for:

View: Best 7 Combinations | Worst 7 Combinations — Download: | Print | Preview

Combination	Estimated Conversion Rate Range [?]	Chance to Beat Orig. [?]	Chance to Beat All [?]	Observed Improvement [?]	Conversions / Visitors [?]
Original	15.7% ± 0.8%	—	0.04%	—	
Combination 5	18.0% ± 0.8%	99.3%	53.2%	14.3%	
Combination 7	17.5% ± 0.8%	97.6%	22.1%	11.4%	
Combination 3	17.2% ± 0.8%	94.4%	9.48%	9.14%	
Combination 1	17.1% ± 0.8%	93.7%	8.43%	8.80%	
Combination 4	16.9% ± 0.8%	89.9%	4.31%	7.32%	
Combination 6	16.7% ± 0.8%	84.2%	2.20%	5.79%	
Combination 2	16.0% ± 0.8%	62.8%	0.19%	1.85%	

Figure 11.39 (a) Page Section results; (b) Combination results

The best improvement of a 14.3 percent increase in conversions equates to a significant dollar improvement for the Calyx Flowers bottom line—of the order of millions of dollars per year. This has provided the evidence required that their online marketing efforts are working, and provide impetus to further invest in their online channel.

Summary

By describing the tasks that most analysts perform, I hoped to demystify the complexities of web analytics in this chapter. As long as you dedicate the time and resource to the tasks, you will find that this isn't rocket science.

The tasks presented here are not intended to be an exhaustive or definitive list; rather, their purpose is help you obtain useful information you can act on. Acting on your data is the single most important aspect of web analytics, yet it is this that most people stumble with.

In Chapter 11, you have learned about the following:

- How to identify poorly performing pages using a mix of methods, including a detailed funnel analysis
- How to measure the success of site search and put a dollar amount on its importance to your organization
- Optimizing your search engine marketing for both paid and non-paid search
- Monetizing a non-e-commerce website, either by assigning values to defined goals or by faking transaction calls to Google Analytics.
- Tracking offline spending by using modified landing page URLs and redirection or combining with search engine marketing
- Using Website Optimizer as a way to test a hypothesis or alternative design

Recommended Further Reading

This is not intended as an exhaustive list of reading material but more a reflection of the books and resources I have read and the blogs I have participated in over the years. If you have a relevant reading resource that I am unaware of, please email me at: brian@advanced-web-metrics.com *and I will endeavor to include it here and on the book web site itself (*www.advanced-web-metrics.com*).*

Books on Web Analytics and Related Areas

Listed in reverse published date order:

- Avinash Kaushik, *Web Analytics: An Hour a Day* (Sybex, 2007)
- Jason Burby and Shane Atchison, *Actionable Web Analytics: Using Data to Make Smart Business Decisions* (Sybex, 2007)
- David Bowen, *Spinning the Web: How to Transmit the Right Messages Online* (Bowen Craggs & Co. Limited, 2006)
- Bryan Eisenberg, Jeffrey Eisenberg, and Lisa T. Davis, *Waiting for Your Cat to Bark: Persuading Customers When They Ignore Marketing* (Thomas Nelson, 2006)
- Bryan Eisenberg, Jeffrey Eisenberg, and Lisa T. Davis, *Call to Action: Secret Formulas to Improve Online Results* (Thomas Nelson, 2006)
- Hurol Inan, *Search Analytics: A Guide to Analyzing and Optimizing Website Search Engines* (BookSurge Publishing, 2006)
- Jakob Nielsen and Hoa Loranger, *Prioritizing Web Usability* (New Riders, 2006)
- Bill Hunt and Mike Moran, *Search Engine Marketing, Inc.: Driving Search Traffic to Your Company's Web Site* (IBM Press, 2005)
- Andrew Goodman, *Winning Results with Google AdWords* (McGraw-Hill Osborne Media, 2005)
- Steve Krug, *Don't Make Me Think: A Common Sense Approach to Web Usability* (New Riders Press, 2005)
- Chris Sherman, *Google Power: Unleash the Full Potential of Google* (McGraw-Hill Osborne Media, 2005)
- Eric T. Peterson, *Web Site Measurement Hacks: Tips & Tools to Help Optimize Your Online Business* (O'Reilly Media, Inc., 2005)
- Eric T. Peterson, *Web Analytics Demystified: A Marketer's Guide to Understanding How Your Web Site Affects Your Business* (Celilo Group Media, 2004)
- Hurol Inan, *Measuring the Success of Your Website: A Customer-centric Approach to Website Management* (Longman Publishing Group, 2002)
- Jim Sterne, *Web Metrics: Proven Methods for Measuring Web Site Success* (Wiley, 2002)

Web Resources

- e-Consultancy: www.e-consultancy.com
- Interactive Advertising Bureau (IAB): www.iab.net
- Search Engine Marketing Professional Organization (SEMPO): www.sempo.org
- Site Point: www.sitepoint.com
- Web Analytics Association: www.webanalyticsassociation.org

Blog List for Web Analytics

Listed in alphabetical order (with thanks to Stephane Hamill for sharing his blog list at http://blog.immeria.net/2006/10/web-analytics-conversations.html):

Advanced Web Metrics by Brian Clifton	http://www.advanced-web-metrics.com/blog
Always be testing by Andy Edmonds	http://alwaysbetesting.com/abtest
Analytical Engine by Diamond Management & Technology Consultants	http://analyticsbyadam.blogspot.com
Analytics by Adam Berlinger	http://analyticsbyadam.blogspot.com/
AIMS Canada	http://www.blog.aimscanada.com/aims_canada/analytics
Analytical Engine—Diamond Management & Technology Consultants	http://diamondinfoanalytics.com/blog1
Analytics Insider—from the authors of Web Analytics for Dummies	http://www.analyticsinsider.com
Analytics Notes by Jacques Warren	http://www.waomarketing.com/blog
Analytics Talk by Justin Cutroni (EpikOne)	http://www.epikone.com/blog
Andy Beal's Marketing Pilgrim—Internet Marketing Blog & Consultant	http://www.marketingpilgrim.com
Applied Insights	http://www.applied-insights.co.uk
Benry	http://www.benry.net/blog
BobPage.net—Information overload	http://bobpage.net
Brad Warthan	http://bradwarthan.com
braden's web analytics, usability & online marketing blog	http://www.bradenh.ca/wordpress
Business Analytics by Khanal Bhupendra	http://analyticsbhups.blogspot.com
Canalytics	http://www.canalytics.ca/blog
Cliff Allen on Marketing	http://blog.allen.com
Coffee, Sun & Technology	http://www.coffeesuntechnology.com
Commerce360 by Craig Danuloff	http://blogs.commerce360.com
Conversion Rater	http://www.conversionrater.com
Customer Intent, Michael Stebbins at Market Motive	http://www.marketmotive.com/stebbins
Daniel Shields on Web Analytics Demystified	http://daniel.webanalyticsdemystified.com
Data Sciences Analytics by John Aitchison	http://dsanalytics.com/dsblog
Data Mining Research by Sandro Saitta	http://dataminingresearch.blogspot.com
Digital Media Analytics	http://digitalmediaanalytics.com/blog
Digital Alex by Alex Cohen	http://www.alexlcohen.com
E-consultancy—Web Measurement and Analytics—News and Blog	http://www.e-consultancy.com/news-blog
Eric T. Peterson's Analytics Weblog	http://www.webanalyticsdemystified.com/weblog

Freelance Web Analytics by Anil Batra	http://freelancewebanalytics.blogspot.com
GA Experts—Google Analytics & Urchin Software	http://www.ga-experts.co.uk/blog
Gilligan on Data by Tim Wilson	http://www.secondtree.com/data
Good Behaviour Blog by Hugh Gage	http://www.engage-digital.com
Google Analytics Blog	http://analytics.blogspot.com
Greater Returns by Aaron Gray	http://greaterreturns.blogspot.com
GrokDotCom—The Converstion Rate Marketing Blog	http://www.grokdotcom.com
immeria—an immersion into Web analytics by S.Hamel	http://shamel.blogspot.com
Influence Analytics from "Wandering" Dave Rhee	http://thewdave.com
Instant Cognition	http://blog.instantcognition.com
Juan Damia's blog	http://www.damia.com.ar
June Dershewitz on Web Analytics	http://june.typepad.com
Lies, Damned Lies...	http://www.liesdamnedlies.com
Lunametrics by Robbin Steif	http://www.lunametrics.com/blog
Marketing Productivity Blog by Jim Novo	http://blog.jimnovo.com
Mastering Web Analytics, using Google Analytics	http://www.brianjclifton.com/blog
Mymotech by Michael Helding	http://www.mymotech.com
The Dashboard Spy—Enterprise Dashboard Screenshots	http://www.enterprise-dashboard.com
The MineThatData Blog by Kevin Hillstrom	http://minethatdata.blogspot.com
Negligible quantities by Julien Coquet	http://juliencoquet.wordpress.com
Occam's Razor by Avinash Kaushik	http://www.kaushik.net/avinash
One Degree	http://www.onedegree.ca
Orthogonal Thinking by John Marshall at Market Motive	http://www.marketmotive.com/marshall
Pages vues by Benoit Arson	http://www.experiense.com/pages-vues/index.php/en
Passionate Analyst by Dylan Lewis	http://www.passionateanalyst.com
Pattern Finder	http://creese.typepad.com/pattern_finder
PlusOneAnalytics	http://www.plusoneanalytics.com/blog
Random Analytics	http://randombits.typepad.com/webanalytics
Rich Page Rambling by Rich Page	http://www.rich-page.com
SEMAngel by Gary Angel	http://semphonic.blogs.com/semangel
Share The Genie's Power :: ClickInsight Blog	http://clickinsight.blogspot.com
Signum sine tinnitu—by Guy Kawasaki	http://blog.guykawasaki.com
The Big Integration by Jacques Warren	http://www.thebigintegration.com/blog
The Blackbeak Blog.... Arr!	http://blackbeak.conversionchronicles.com/
The Dashboard Spy by Enterprise Dashboard Screenshots	http://www.enterprise-dashboard.com
The Site is Dead—Offermatica	http://www.landingpageoptimization.com
this just in	http://cutroni.com/blog
Tracking Techniques	http://trackingtechniques.com

Turn Up The Silence—iPerceptions Blog	http://blog.iperceptions.com
Unofficial Google Analytics Blog	http://www.roirevolution.com/blog
Visioactive by Ian S. Houston	http://www.visioactive.com
Web Analytics Association Blog	http://waablog.webanalyticsassociation.org
Web Analytics and Optimization Blog by Mike Sukmanowsky	http://analytics.mikesukmanowsky.com
Web Analytics: Information for the average user by Matt Lillig	http://mattlillig.blogspot.com
Web Analytique et Optimisation by Jacques Warren (in French)	http://www.waomarketing.com/blogFR/wordpress
Web Analysts.info by Lars Johansson	http://www.webanalysts.info/webanalytics
Web Analytics & Affiliate Marketing by Dennis R. Mortensen	http://visualrevenue.com/blog
Web Analysis and Online Advertising by Anil Batra	http://webanalysis.blogspot.com
WebAnalyticsBook	http://www.webanalyticsbook.com
Web Analytcs: Information for the average user	http://mattlillig.blogspot.com
Web Analytics Inside by Timo Aden (in German)	http://www.timoaden.de
Web Analytics Analyzed by Paul Strupp	http://blogs.sun.com/pstrupp
Web Analytics Applied by Paul Legutko (Semphonic)	http://legutko.typepad.com
Web Analytics Association	http://www.webanalyticsassociation.org
Web Analytics Guru by Britney	http://webanalyticsguru.blogster.com
Web Analytics in China by Florian Pihs	http://longmarch.typepad.com/web_analytics_in_china
Web Analytics Management by Phil Kemelor	http://wam.typepad.com/wam
Web Analytics Matt by Matt Hopkins	http://www.webanalyticmatt.com
Web Analytics Princess by Marianina Chaplin	http://marianina.com/blog
Web Analytics Pulse by Anil Umachigi	http://pulse-beat.blogspot.com
Web Analytics RSS Feed by Hurol Inan	http://www.hurolinan.com
Web Analytics Tool Time by Jesse Gross (Semphonic)	http://tooltime.typepad.com
Web Analytics World	http://manojjasra.blogspot.com
webanalytics at Yahoo! Groups	http://groups.yahoo.com/group/webanalytics
Web Optimization Blog by Debbie Pascoe (Maxamine)	http://weblog.maxamine.com
Web Sense by Joel Hadary (Semphonic)	http://semphonic.typepad.com/websense
Web Strategy by Jeremiah	http://www.web-strategist.com/blog
WebAnalytics.be Blog	http://webanalytics.wordpress.com
WebMetricsGuru by Marshall Sponder	http://www.webmetricsguru.com

Index

Note to the Reader: Throughout this index **boldfaced** page numbers indicate primary discussions of a topic. *Italicized* page numbers indicate illustrations.

C

D

E

F

G

H

I

J

K

L

M

N

O

P

- poorly performing pages
  - funneling, **282–289**
  - identifying, **274**
  - $Index values, **274–278**
  - Top Landing Pages reports, **279–282**, *279–280*, *282*

Q

R

S

T

U

V

W

Y

Z